Changes in Eukaryotic Gene Expression in Response to Environmental Stress

This is a volume in
CELL BIOLOGY
A series of monographs

EDITORS: D. E. BUETOW, I. L. CAMERON, G. M. PADILLA, AND A. M. ZIMMERMAN

A complete list of the books in this series appears at the end of the volume.

Changes in Eukaryotic Gene Expression in Response to Environmental Stress

Edited by

Burr G. Atkinson

Cell Science Laboratories
Department of Zoology
University of Western Ontario
London, Ontario, Canada

David B. Walden

Department of Plant Sciences
University of Western Ontario
London, Ontario, Canada

1985

ACADEMIC PRESS, INC.

(Harcourt Brace Jovanovich, Publishers)

Orlando San Diego New York London
Toronto Montreal Sydney Tokyo

ACADEMIC PRESS, INC.
Orlando, Florida 32887

United Kingdom Edition published by
ACADEMIC PRESS INC. (LONDON) LTD.
24–28 Oval Road, London NW1 7DX

Library of Congress Cataloging in Publication Data
Main entry under title:

Changes in eukaryotic gene expression in response to environmental stress.

Includes index.
1. Gene expression. 2. Heat--Physiological effect.
3. Adaptation (Physiology) I. Atkinson, B. G.
II. Walden, David B., Date.
QH450.C48 1985 575.2'2 84-12513
ISBN 0-12-066290-6 (alk. paper)

PRINTED IN THE UNITED STATES OF AMERICA

85 86 87 88 9 8 7 6 5 4 3 2 1

Contents

3 Mechanism of Translational Control in Heat-Shocked *Drosophila* Cells

DENNIS BALLINGER AND MARY LOU PARDUE

4 Coordinate and Noncoordinate Gene Expression during Heat Shock: A Model for Regulation

SUSAN LINDQUIST AND BETH DIDOMENICO

5 Intracellular Localization and Possible Functions of Heat Shock Proteins

ROBERT M. TANGUAY

B. Other Animals

6 Heat Shock Proteins in Sea Urchin Development

GIOVANNI GIUDICE

7 Heat Shock Gene Expression during Early Animal Development

JOHN J. HEIKKILA, JOHN G. O. MILLER, GILBERT A. SCHULTZ, MALGORZATA KLOC, AND LEON W. BROWDER

8 Effects of Stress on the Gene Expression of Amphibian, Avian, and Mammalian Blood Cells

BURR G. ATKINSON AND ROB L. DEAN

9 Stress Response in Avian Cells

MILTON J. SCHLESINGER

10 Stress Responses in Avian and Mammalian Cells

L. E. HIGHTOWER, P. T. GUIDON, JR., S. A. WHELAN, AND C. N. WHITE

11 Effect of Hyperthermia and LSD on Gene Expression in the Mammalian Brain and Other Organs

IAN R. BROWN

12 Thermotolerance in Mammalian Cells: A Possible Role for Heat Shock Proteins

GLORIA C. LI AND ANDREI LASZLO

II. PLANTS AND FUNGI

13 Heat Shock Genes of *Dictyostelium*

ELLIOT ROSEN, ANNEGRETHE SIVERTSEN, RICHARD A. FIRTEL, STEVEN WHEELER, AND WILLIAM F. LOOMIS

14 Plant Productivity, Photosynthesis, and Environmental Stress

DONALD R. ORT AND JOHN S. BOYER

15 Responses to Environmental Heat Stress in the Plant Embryo

JOSEPH P. MASCARENHAS AND MITCHELL ALTSCHULER

Contributors

Numbers in parentheses indicate the pages on which the authors' contributions begin.

Mitchell Altschuler (315), Department of Biological Sciences, State University of New York at Albany, Albany, New York 12222

Burr G. Atkinson (159, 349), Cell Science Laboratories, Department of Zoology, University of Western Ontario, London, Ontario, Canada N6A 5B7

A. Ayme (3), Department of Molecular Biology, University of Geneva, 1211 Geneva 4, Switzerland

Dennis Ballinger[1] (53), Department of Biology, Massachusetts Institute of Technology, Cambridge, Massachusetts 02139

Chris L. Baszczynski[2] (349), Department of Plant Sciences, University of Western Ontario, London, Ontario, Canada N6A 5B7

J. Jose Bonner (31), Program in Molecular, Cellular, and Developmental Biology, Department of Biology, Indiana University, Bloomington, Indiana 47405

John S. Boyer[3] (279), United States Department of Agriculture/Agricultural Research Service, and Departments of Plant Biology and Agronomy, University of Illinois, Urbana, Illinois 61801

Leon W. Browder (135), Department of Biology, The University of Calgary, Calgary, Alberta, Canada T2N 4N1

Ian R. Brown (211), Department of Zoology, University of Toronto, Scarborough Campus, West Hill, Ontario, Canada M1C 1A4

[1]Present address: Division of Biology, California Institute of Technology, Pasadena, California 91125.

[2]Present address: Allelix, Inc., Mississauga, Ontario, Canada L4V 1P1.

[3]Present address: Department of Soil and Crop Sciences, Texas A & M University, College Station, Texas 77843.

Eva Czarnecka[4] (327), Department of Botany, University of Georgia, Athens, Georgia 30602

Rob L. Dean (159), Cell Science Laboratories, Department of Zoology, University of Western Ontario, London, Ontario, Canada N6A 5B7

Beth DiDomenico (71), Department of Molecular Genetics and Cell Biology, University of Chicago, Chicago, Illinois 60637

Richard A. Firtel (257), Department of Biology, University of California, San Diego, La Jolla, California 92093

Giovanni Giudice (115), Department of Cellular and Developmental Biology, University of Palermo, and Institute of Developmental Biology, C.N.R., Palermo, Italy

P. T. Guidon, Jr. (197), Microbiology Section, The Biological Sciences Group, The University of Connecticut, Storrs, Connecticut 06268

John J. Heikkila (135), Department of Biology, University of Waterloo, Waterloo, Ontario, Canada N2L 3G1

L. E. Hightower (197), Microbiology Section, The Biological Sciences Group, The University of Connecticut, Storrs, Connecticut 06268

Joe L. Key (327), Department of Botany, University of Georgia, Athens, Georgia 30602

Janice Kimpel (327), Department of Botany, University of Georgia, Athens, Georgia 30602

Malgorzata Kloc[5] (135), Department of Biology, The University of Calgary, Calgary, Alberta, Canada T2N 4N1

Andrei Laszlo (227), Radiation Oncology Research Laboratory, Department of Radiation Oncology, University of California, San Francisco, California 94143

Gloria C. Li (227), Radiation Oncology Research Laboratory, Department of Radiation Oncology, University of California, San Francisco, California 94143

Chu-Yung Lin[6] (327), Department of Botany, University of Georgia, Athens, Georgia 30602

Susan Lindquist (71), Department of Molecular Genetics and Cell Biology, University of Chicago, Chicago, Illinois 60637

William F. Loomis (257), Department of Biology, University of California, San Diego, La Jolla, California 92093

Joseph P. Mascarenhas (315), Department of Biological Sciences, State University of New York at Albany, Albany, New York 12222

John G. O. Miller (135), Department of Medical Biochemistry, The University of Calgary, Calgary, Alberta, Canada T2N 4N1

M.-E. Mirault (3), Department of Molecular Biology, University of Geneva, 1211 Geneva 4, Switzerland

Ronald T. Nagao (327), Department of Botany, University of Georgia, Athens, Georgia 30602

[4]Present address: Department of Microbiology and Cell Science, University of Florida, Gainesville, Florida 32611.

[5]Present address: Department of Chemistry, Saint Mary's University, Halifax, Nova Scotia, Canada B3H 3C3.

[6]Present address: Department of Botany, National Taiwan University, Taipei, Taiwan, Republic of China.

Donald R. Ort (279), United States Department of Agriculture/Agricultural Research Service, and Department of Plant Biology, University of Illinois, Urbana, Illinois 61801

Mary Lou Pardue (53), Department of Biology, Massachusetts Institute of Technology, Cambridge, Massachusetts 02139

Elliot Rosen[7] (257), Department of Biology, University of California, San Diego, La Jolla, California 92093

Milton J. Schlesinger (183), Department of Microbiology and Immunology, Washington University School of Medicine, St. Louis, Missouri 63110

Friedrich Schöffl[8] (327), Department of Botany, University of Georgia, Athens, Georgia 30602

Gilbert A. Schultz (135), Department of Medical Biochemistry, The University of Calgary, Calgary, Alberta, Canada T2N 4N1

Annegrethe Sivertsen (257), Department of Biology, University of California, San Diego, La Jolla, California 92093

R. Southgate[9] (3), Department of Molecular Biology, University of Geneva, 1211 Geneva 4, Switzerland

Robert M. Tanguay (91), Molecular and Human Genetics Unit, CHUL Research Center, and Department of Medicine, Université Laval, Quebec, Quebec, Canada G1V 4G2

A. Tissières (3), Department of Molecular Biology, University of Geneva, 1211 Geneva 4, Switzerland

Elizabeth Vierling (327), Department of Botany, University of Georgia, Athens, Georgia 30602

David B. Walden (349), Departments of Plant Sciences and Zoology, University of Western Ontario, London, Ontario, Canada N6A 5B7

Steven Wheeler (257), Department of Biology, University of California, San Diego, La Jolla, California 92093

S. A. Whelan (197), Biochemistry/Biophysics Section, The Biological Sciences Group, The University of Connecticut, Storrs, Connecticut 06268

C. N. White (197), Microbiology Section, The Biological Sciences Group, The University of Connecticut, Storrs, Connecticut 06268

[7]Present address: Lovund Laboratory, University of Notre Dame, Notre Dame, Indiana 46556.

[8]Present address: Universität Bielefeld, Fakultät für Biologie VI (Genetik), D-4800 Bielefeld 1, Federal Republic of Germany.

[9]Present address: Department of Biochemistry and Molecular Biology, Harvard University, Cambridge, Massachusetts 02138.

Preface

The effects of environmental stresses on cells and cellular activity have been a concern of research scientists, clinicians, and agriculturists for some time. While a great number of studies in the past fifty years have exposed either the immediate lethal properties of some forms of stress or revealed a variety of long-term cellular effects of various stressors, it was recognized only recently that some forms of stress can rapidly and dramatically affect the genetic expression of cells by inducing the synthesis of a small number of proteins (the so-called stress proteins) and repressing the synthesis of most others. Research leading to this discovery was initiated by Ritossa in 1962 and, with the advent of new techniques, advanced by Tissières, Mitchell, and Tracey in 1974. Although the understanding of this cellular activity was initiated by the use of thermal stress on *Drosophila* salivary glands, such genetic response has been shown subsequently not only to be ubiquitous—in cells from bacteria to man—but also to occur under a wide variety of stress conditions.

There is little doubt that interest in the heat shock and stress responses will increase in the years to come. The study of the genes involved has proved to be convenient in both plants and animals. The stress genes serve as interesting models for the study of gene organization in *Drosophila* and pose a number of general problems of nucleotide sequence, organization, and control. In addition, the stress–response systems appear to encompass basic, relevant physiological mechanisms needed by cells for survival and/or combating stressful environmental influences.

The contributors focus on various aspects of the response of eukaryotic cells to heat stress (shock) and other stress stimuli. We have organized this volume into sections that reflect the emphasis in research utilizing *Drosophila,* a variety of animal systems, and plants. The use of similar research strategies, independent of the organism and leading to a particular understanding of the different stress responses, provides a unifying theme for this volume.

Burr G. Atkinson
David B. Walden

I

Animals

A. ***Drosophila***
Chapters 1 through 5

B. Other Animals
Chapters 6 through 12

1

Organization, Sequences, and Induction of Heat Shock Genes

R. SOUTHGATE, M.-E. MIRAULT, A. AYME, AND A. TISSIÈRES

I. INTRODUCTION

In response to heat shock or various other kinds of stress, all organisms so far examined, from bacteria, yeast, and other microorganisms to plants, insects, and higher vertebrates, such as fish, chicken, mouse, and man, react by the strong activation of a limited number of specific genes previously either silent or active only at low levels. Consequently, the proteins encoded by these genes, the so-called heat shock proteins (hsp's), are actively synthesized during stress and accumulate in such a manner as to finally represent major cellular constituents. The initial work, and to this day by far the most extensive pertaining to the hsp genes, has been done with *Drosophila melanogaster*. However, it can be expected that in the near future, detailed information on genes from bacteria, yeast, and higher organisms, including man, will become available.

Changes in Eukaryotic Gene Expression in Response to Environmental Stress

ISBN 0-12-066290-6

The first evidence that various kinds of stress lead to the activation of a small group of particular genes came from the observations of Ritossa (1962, 1964a) that, after a heat shock at 37°C or following dinitrophenol or sodium salicylate treatment, some new puffs appeared on the *Drosophila* giant salivary gland chromosomes. In addition, Ritossa (1964b) showed that these newly induced puffs were involved in RNA synthesis. In view of the speculation formulated by Beermann (1956) that puffs represented sites of active genes, it was reasonable to assume that in response to heat shock or other forms of stress a small number of specific genes was being activated. That this was in fact the case became clear when it was shown that heat shock proteins are actively synthesized following heat shock induction (Tissières *et al.*, 1974; Lewis *et al.*, 1975) as a result of the translation of specific messengers transcribed at the heat shock puff sites (McKenzie *et al.*, 1975; McKenzie and Meselson, 1977; Spradling *et al.*, 1977; Mirault *et al.*, 1978).

In *D. melanogaster* the hsp genes can be ordered into three groups, each with its own specific organization within the genome and characteristic protein gene products. The first group consists of a multigene family encoding hsp 70 and one related gene encoding hsp68. The hsp70 genes are usually organized as three copies at chromosomal site 87C1 and two copies, of a variant gene, at 87A7 in Oregon R strains. The hsp68 gene, present as a single copy at site 95D, shares strong sequence homologies with the hsp70 genes (Holmgren *et al.*, 1979). The stress activation of the genes in this group is dramatic and can be estimated to be on the order of a hundred- to perhaps a thousandfold. As has been observed in cells from many diverse origins, hsp70 represents, by far, the most abundant gene product following heat shock. Stress proteins comparable in size to hsp70 appear to have been well conserved throughout evolution. In addition, Ingolia and Craig (1982a) have shown that several genes, sharing a strong homology to the hsp70 genes, are found dispersed within the *Drosophila* genome but not at the known heat shock loci. These genes, which they call heat shock cognates, are not stress inducible.

The second group consists of the four small hsp genes encoding hsp 27, 26, 23, and 22, which are closely related to each other (Ingolia and Craig, 1981, 1982b; Southgate *et al.*, 1983). They are clustered, each as a single-copy gene on a DNA segment of 12 kb at locus 67B (Petersen *et al.*, 1979; Corces *et al.*, 1980; Craig and McCarthy, 1980; Voellmy *et al.*, 1981). It can be estimated that the stress activation of these genes is roughly 20-fold, although the actual degree of induction can be quite variable.

The third group is represented by the single hsp83 (also referred to as hsp84) gene, present at site 63BC and encoding the largest, 83-kilodalton (or 84-kilodalton), hsp (Holmgren *et al.*, 1979, 1981; Hackett and Lis, 1983), which can be observed in the absence of stress as a normal cellular protein. Its synthesis is increased only a few fold following induction, and it is located mainly within the cytoplasm (Mitchell and Lipps, 1975; Arrigo *et al.*, 1980). The heat shock protein in higher cells which appears to correspond most closely to the *Drosophila* hsp83 is about 90 kilodaltons in size. In HeLa cells there are three heat shock protein species in the molecular mass range of 90 to 110 kilodaltons (Welch and Feramisco, 1982; Welch *et al.*, 1982). Whether these three polypeptides are structurally related to each other or originate from one or several genes is not yet known.

The protein products of the first and second groups, hsp 70, 68, 27, 26, 23, and 22, are all found predominantly within the nucleus during a heat shock (Arrigo *et al.*, 1980; Velazquez *et al.*, 1980). Upon return to normal temperatures, these proteins tend to migrate back to the cytoplasm, where the small hsp's, at least, become associated into ribonucleoprotein complexes (Arrigo *et al.*, 1980; Arrigo and Ahmad-Zadeh, 1981).

The *Drosophila* hsp genes have nearly all been sequenced in their coding and flanking sequences, and some regions of homology have been noted at their 5′ ends. Are these homologies part of a common signal sequence present upstream of these genes which permits regulation of their apparent coordinate induction?

In order to study the transcriptional induction of *Drosophila* heat shock genes, monkey COS cells were transfected with hsp70 genes cloned into vectors containing a Simian virus 40 (SV40) origin of replication. These *Drosophila* hsp70 gene constructs proved to be stress inducible with heat shock, and, by testing a series of 5′ deletion mutants of the hsp70 gene, it was possible to define a sequence necessary for efficient heat induction under these experimental conditions (Mirault *et al.*, 1982; Pelham, 1982). Having observed an apparently imperfect palindromic consensus sequence in front of most of the hsp genes, Pelham and Bienz (1982) synthesized this particular sequence and, by placing it in front of the *Herpes* thymidine kinase gene in an appropriate vector, were able to render this gene heat inducible in both monkey cells and *Xenopus* oocytes. Further investigations in Cambridge and Geneva with the same kind of system are now in progress with other *Drosophila* hsp genes.

Following the initial observations of Sirotkin and Davidson (1982), several laboratories have now reported that some of the hsp genes are also expressed normally at various times during development under nonstress conditions. The mechanisms of control involved in the developmental expression of these genes can now be investigated by transformation of *Drosophila,* mediated by the P element, as described by Rubin and Spradling (1982).

II. ORGANIZATION AND SEQUENCES

A. Heat Shock Protein 70 and Related Genes

1. The hsp70 Genes

Following heat shock, mRNA from *D. melanogaster* tissue culture cells sediments into two major peaks of 20 S and 12 S, respectively (McKenzie and

Meselson, 1977; Spradling *et al.*, 1977; Mirault *et al.*, 1978; Moran *et al.*, 1978). The most abundant of the heat shock mRNA's, coding for hsp70, is found in the 20 S peak. It hybridizes *in situ* to two heat shock puff sites, 87A and 87C on chromosome 3R, indicating the existence of multiple genes encoding hsp70 (Henikoff and Meselson, 1977; Spradling *et al.*, 1977). Genomic clones bearing these hsp70 genes were isolated in several laboratories independently and have been analyzed extensively in order to decipher their organization.

The first heat shock clones to be isolated from *D. melanogaster* contained an unusual set of tandemly repeated sequences derived from locus 87C, the so called αβ repeats (Lis *et al.*, 1978). At this particular location, the transcription of these sequences is heat inducible, but they are also found dispersed at the chromocenter, where they appear not to be stress induced (Lis *et al.*, 1981). Similar sequences are also present, though not transcribed, at the chromocenter of other *Drosophila* species, such as the sibling species *D. simulans* (Livak *et al.*, 1978). It seems likely, therefore, that the αβ repeats have been brought to 87C as a result of recent evolutionary events in *D. melanogaster*. They do not appear to code for any protein product and their role at this locus remains obscure.

In addition to the αβ repeats, a variant unit called αγ has also been observed interspersed within αβ clusters at 87C. The αβ sequences are transcribed upon heat shock into RNA of three size classes, the largest of which is 3 kb long and corresponds to αβα. Interestingly, γ sequences are found at both 87C and 87A, the two sites expected to bear hsp70 genes. Lis *et al.* (1978) therefore speculated that the γ element contained sequences involved in the heat induction of both αβ and hsp70 genes and this prediction turned out to be correct.

In the meantime, a number of *D. melanogaster* clones containing hsp70 gene sequences were isolated and characterized in different laboratories (Livak *et al.*, 1978; Schedl *et al.*, 1978; Artavanis-Tsakonas *et al.*, 1979; Craig *et al.*, 1979; Moran *et al.*, 1979). These genes were derived from both loci 87A and 87C. To each of these two chromosomal sites there corresponds a characteristic gene variant. The first structural outline of the hsp70 multiple gene organization emerged from the comparison of two clones, one containing a single-gene copy from 87A, and the other, two genes in a tandem repeat from 87C (Moran *et al.*, 1979; Artavanis-Tsakonas *et al.*, 1979). Each individual hsp70 gene is organized as a basic, conserved unit Z, consisting of a 2.2-kb segment encoding hsp70 mRNA, Z_c (Z coding), and a 0.35-kb region, Z_{nc} (Z noncoding), present at the 5′ end of Z_c. No introns have been observed in these genes. Additional regions of sequence similarity, the so-called X elements, form a complex pattern of DNA homologies present upstream of the hsp70 genes at both 87A and 87C (Moran *et al.*, 1979). Several clones derived from 87C contained αβ sequences in the close vicinity of an hsp70 gene variant (Livak *et al.*, 1978; Craig *et al.*, 1979). This was the first direct evidence that these two gene types, αβ and hsp70, are closely linked physically at this locus.

The global organization of the *Drosophila* multiple hsp70 genes at both 87A and 87C was deduced by a combination of detailed genetic and molecular analyses of these two loci. The characterization of a series of overlapping deficiencies (deletions) covering one or the other locus, or both, has allowed the localization of the hsp70 genes to the two cytological bands 87A7 and 87C1 (Ish-Horowicz *et al.*, 1977, 1979a,b). Embryos lacking these two bands fail to synthesize hsp70 following heat shock, whereas this protein is made in embryos retaining either 87A7 and/or 87C1. The restriction map of the hsp70 gene sequences was examined directly in the genome of the mutant strains that lacked either the 87A7 or the 87C1 band, or both (Holmgren *et al.*, 1979; Ish-Horowicz *et al.*, 1979b; Ish-Horowicz and Pinchin, 1980). Each locus was found to contain multiple copies of its own particular hsp70 gene variant, identified by characteristic restriction sites. The overall arrangement of the hsp70 genes found at 87A7 and 87C1 of most *D. melanogaster* strains is displayed in Fig. 1.

The locus 87A7 contains two genes in opposite polarity (Mirault *et al.*, 1979; Ish-Horowicz and Pinchin, 1980), as found in clone 122 (Goldschmidt-Clermont, 1980). Polymorphic variants of this gene arrangement have been isolated, which differ by the deletion or insertion of large DNA stretches. For example, one clone contains the distal hsp70 gene of 87A7 and a copialike dispersed, middle repetitive sequence located just upstream of the heat shock gene (Artavanis-Tsakonas *et al.*, 1979); two other clone isolates lack most of the spacer DNA and the 5′ half of the proximal gene (Goldschmidt-Clermont, 1980; Udvardy *et al.*, 1982). The latter authors speculated that incorrect excision of a transposable element could have generated these apparent deletions. Another deletion mutant has a truncated proximal gene at 87A7 lacking its 3′ moiety; interestingly, embryos homozygous for this mutation and deleted of the 87C locus synthesize a new hsp70-related heat shock protein of 40 kilodaltons, in addition to the normal hsp70 encoded at 87A (Caggese *et al.*, 1979; Burke and Ish-Horowicz, 1982). This indicates that both hsp70 genes at 87A7 are activated by heat shock.

The organization of the hsp70 genes at the 87C1 locus is more complex than that at 87A7 (see Fig. 1). The hsp70 genes are found in two domains, a single proximal gene separated by about 40 kb of DNA, which contains clusters of αβ sequences, from two tandem gene copies present in the distal part of locus 87C1 (Ish-Horowicz and Pinchin, 1980). In some strains and in the Kc tissue culture cell line there are three tandem gene copies. The total number of hsp70 genes at this locus can vary between three and five copies according to genotype and fly stock (Craig *et al.*, 1979; Ish-Horowicz *et al.*, 1979b; Mirault *et al.*, 1979). Similarly, the number of αβ and αγ repeat units has also been reported to vary, from 8 to 14 and from 5 to 8, respectively (Lis *et al.*, 1978, 1981). It was thus suggested that unequal crossing-over events had generated these variations and that the tandem duplication or triplication of hsp70 genes may have arisen by the same mechanism. Such events may well have been favored by the variable

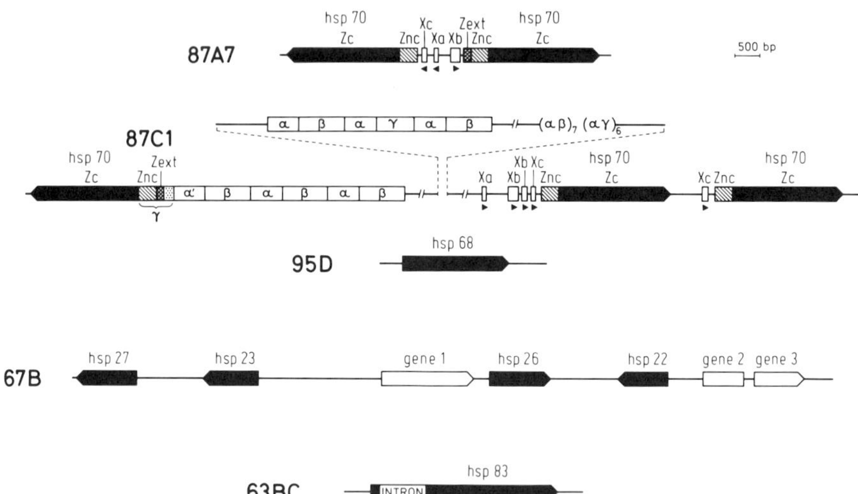

Fig. 1. The arrangement of the *Drosophila melanogaster* heat shock genes. Those regions complementary to heat shock mRNA are indicated by black arrows, the orientation of which gives the direction of transcription of each heat shock gene. The top two maps portray the organization of the hsp70 gene family at the cytological loci 87A7 and 87C1, respectively, and takes into account the sequence data from Hackett and Lis (1981), Karch *et al.* (1981), and Mason *et al.* (1982). The left-hand part of these two maps is proximal to the centromere of chromosome 3R. As described in the text, the 2.2-kb Z_c element is complementary to hsp70 mRNA. The 0.35-kb Z_{nc} regions (hatched boxes) are approximately 98% conserved at each locus and 90% homologous between the two loci (Mason *et al.*, 1982; Török *et al.*, 1982). Additional areas of homology (X_a, X_b, X_c, and Z_{ext}) are found to be both very AT rich and well conserved. X_a and X_b were first discovered by heteroduplex analysis (Moran *et al.*, 1979; Goldschmidt-Clermont, 1980). X_c and the 0.15-kb Z_{ext} (Z extension; cross-hatched boxes) were found by sequence data comparisons (Mason *et al.*, 1982). The γ element consists of the first 64 bp of Z_c, the entire Z_{nc} and Z_{ext} elements, plus a small additional region (stippled box; Lis *et al.*, 1978; Hackett and Lis, 1981; Mason *et al.*, 1982). The α, β, and γ elements are arranged as either αβ or αγ tandem arrays in the approximately 40-kb spacer which separates the single proximal hsp70 gene from two distal hsp70 gene copies at locus 87C1. The precise order of these different elements is unknown except for the αβ and αγ repeats which have been analyzed in cloned DNA's (Lis *et al.*, 1978; Holmgren *et al.*, 1979). The third map, representing the hsp68 gene at locus 95D, was taken from Holmgren *et al.* (1979, 1981). The fourth map gives the organization of the 67B locus (Corces *et al.*, 1980; Craig and McCarthy, 1980; Voellmy *et al.*, 1981; Sirotkin and Davidson, 1982). The location of the four small hsp genes is indicated by the black boxes and that of the three developmentally regulated genes by open boxes (nomenclature taken from Sirotkin and Davidson, 1982). The map of the hsp83 gene at locus 63BC comes from Holmgren *et al.* (1979, 1981). The hsp83 gene is the only *Drosophila* heat shock gene to possess an intron.

number of αβ tandem repeats. Interestingly, αβ sequences have not been detected at the 87C locus of any other *Drosophila* species investigated so far. Instead, the closely related species *D. simulans* and *D. mauritiana* and the more distantly related *D. teisseri* and *D. yakuba* all have two pairs of inverted hsp70 genes which probably represent the ancestral organization that existed at both

87A and 87C before *D. melanogaster* became separated from the other species (Leigh-Brown and Ish-Horowicz, 1981). Thus, the insertion and propagation of $\alpha\beta$ sequences, as well as the tandem duplications of the hsp70 genes, seem to be recent and related evolutionary events at the *D. melanogaster* 87C1 locus.

There is another observation that supports this interpretation. The 0.35-kb sequence Z_{nc}, which is conserved 5′ to all hsp70 genes in *D. melanogaster* (Ish-Horowicz *et al.*, 1979b; Mirault *et al.*, 1979), shows extensive homology to the part of the γ element which is found adjacent to several $\alpha\beta$ repeats (Lis *et al.*, 1981). Sequence comparison between the γ element and the two hsp70 genes at locus 87C1 revealed that 406 base pairs (bp) of the γ unit present at the 5′ end of $\alpha\beta$ repeats shared a near-perfect homology (98%) with the entire Z_{nc} and the first 64 bp of Z_c (Hackett and Lis, 1981; Mason *et al.*, 1982). These data suggest that this common upstream sequence contains *cis*-regulatory signals conferring transcriptional inducibility to both hsp70 and $\alpha\beta$ sequences. The complete element, found immediately adjacent to the proximal hsp70 gene of 87C1, is therefore thought to be the ancestral sequence which became interspersed among $\alpha\beta$ repeats by some unknown mechanism, consequently bringing them under heat shock control (Mason *et al.*, 1982; Török *et al.*, 1982). The remarkable conservation of both γ and the Z_{nc} element of hsp70 at 87C1 suggests that the dispersal of γ at this locus must be a recent event, since Z_{nc} at 87A7 has diverged significantly from both Z_{nc} and γ at 87C1 (Hackett and Lis, 1981; Karch *et al.*, 1981). The apparent ancestry of the 87C1 locus in *D. melanogaster* provides a striking example of rapid gene evolution, in that some of these genes have come under the transcriptional control of others.

Sequence comparison between an hsp70 gene from locus 87C1 (Ingolia *et al.*, 1980) and its 87A7 gene variant (Török and Karch, 1980; Karch *et al.*, 1981) indicates a 4.2% nucleotide divergence within the mRNA coding region Z_c, with complete conservation of an open, uninterrupted reading frame. Most base substitutions occur in the third codon position, and out of a total of 643 amino acids, there are only 26 substitutions and 2 insertions; this would correspond to a 2.7% divergence of the amino acid sequence (Karch *et al.*, 1981). The start of transcription, which defines the boundary between Z_c and Z_{nc}, has been localized 250 nucleotides upstream from the putative initiation codon by S1 mapping (Török and Karch, 1981). Most sequence divergence is located within the 5′-untranslated regions, as first reported by Holmgren *et al.* (1979), and in the 3′-noncoding ends of the two mRNA variants, which are totally nonhomologous (Karch *et al.*, 1981; Török *et al.*, 1982). The Z_{nc} sequences upstream of the 87A7 and 87C1 gene variants display 15% divergence. In contrast, however, at each locus both Z_{nc} and Z_c sequences are remarkably well conserved among the copies of each gene variant (Karch *et al.*, 1981; Mason *et al.*, 1982).

A structural comparison of the hsp70 genes in several species related to *D. melanogaster* led to the conclusion that these genes have apparently not evolved independently from one another (Leigh-Brown and Ish-Horowicz, 1981). These authors suggested both intralocus and, more infrequently, interlocus gene conversion events to explain the apparently concerted evolution of the hsp70 genes at these two loci. Comparison between the extragenic sequences upstream of the hsp70 genes at both 87A7 and 87C1 revealed a complex pattern of sequence homology (Moran *et al.*, 1979; Goldschmidt-Clermont, 1980; Hackett and Lis, 1981; Karch *et al.*, 1981; Mason *et al.*, 1982). The relative arrangement of the different X elements is given in Fig. 1. The origin and possible function of these sequences remain obscure, but the pattern of homology observed at the DNA level suggests that these elements could have arisen as a result of interlocus gene conversion and/or transposition events (Mason *et al.*, 1982; Török *et al.*, 1982). X elements do not appear to be essential for the heat shock induction of hsp70 genes since a mutant *Drosophila* strain possessing only one hsp70 gene copy, with no flanking X elements, is perfectly viable and responds normally to heat shock (Udvardy *et al.*, 1982). It is possible, however, that X elements are involved in more general mechanisms associated with some selective advantage, such as recombination, DNA replication, or DNA transposition. In this respect, we found that X sequences are not restricted to the 87A7 and 87C1 heat shock loci but are found at many other sites within the *D. melanogaster* genome (Lis *et al.*, 1981; M.-E. Mirault, unpublished observations).

Unexpectedly, however, hsp70 gene probes were found to hybridize *in situ* not only to the chromosomal subdivisions 87A and 87C, but also to the heat shock puff locus 95D and to site 87D, subdivisions 10, 11, and 12, which is not known to be involved in the heat shock response (Holmgren *et al.*, 1979). Both sites possess genes structurally related to hsp70, the former locus containing the gene encoding hsp68, and the latter, a heat shock cognate gene (Craig *et al.*, 1983; see Section II,A,3).

2. *The hsp68 Gene*

Two DNA clones have been described that bear hsp68 gene sequences, as shown by hybrid-arrested translation studies. One clone contains most of the gene but is missing the 5′ end (Holmgren *et al.*, 1979), whereas the other clone contains a complete gene copy (Holmgren *et al.*, 1981). Hybridization of the latter clone with whole-cell heat shock RNA protected a 2.1-kb fragment against both S1 and exonuclease VII digestion, thus indicating the absence of introns within this gene. Heteroduplexes between hsp68 and hsp70 genes revealed the existence of a strong cross-homology, but duplex melting studies indicated approximately 15% sequence divergence (Holmgren *et al.*, 1979). This homology suggests that both genes may have evolved from a common ancestral gene and that the 68- and 70-kilodalton heat shock proteins could have a similar or a shared function(s). It is perhaps no coincidence that embryos lacking the 87A7 and 87C1 loci synthesize significantly more hsp68 than do normal embryos following heat shock, possibly as a result of dosage compensation for the lack of hsp70 (Ish-Horowicz *et al.*, 1977).

3. The Heat Shock Cognate Genes

The *Drosophila* genome contains several hsp70-related genes, the transcription of which is not heat shock-inducible but regulated during development (Craig *et al.*, 1982, 1983; Ingolia and Craig, 1982a). The first of these so-called heat shock cognates, Hsc1, was isolated from the two sibling species *D. melanogaster* and *D. simulans* and was found to be located at cytological locus 70C on chromosome 3 (Ingolia and Craig, 1982a). Since then, two other heat shock cognate genes, Hsc2 and Hsc4, have also been isolated from *D. melanogaster* and found to hybridize *in situ* to the same chromosome at cytological loci 87D and 88E, respectively (Craig *et al.*, 1983). Sequence comparison of about one-third of the protein coding regions from all three cognate genes with an hsp70 gene showed approximately 76% homology at the DNA level and about 78% homology at the predicted amino acid level. Comparison of the Hsc1 gene DNA sequence from the two *Drosophila* sibling species indicated, after optimal alignment, only a few mismatches.

As shown by hybridization selection and translation experiments, RNA homologous to Hsc4 encodes a protein of approximately 70 kilodaltons with a similar, though distinguishable, electrophoretic mobility to that of hsp70. Due to the variable lengths of the Hsc 5′ leader sequences, it was possible to distinguish specific transcripts from each gene during development by means of cDNA primer extension experiments. Hsc1 and Hsc2 transcripts were undetectable in both embryonic and larval RNA preparations but were abundant in adult fly preparations. In contrast, however, Hsc4 mRNA was found to be equally abundant, at levels comparable to actin mRNA, at all three developmental stages. The relative abundance of the Hsc transcripts is not increased by heat shock.

Hsc4 contains no intron in the sequenced region encoding the first 101 amino acids, at least, but this is not the case for the other two cognates. Hsc1 is interrupted by a 1.7-kb intron inserted into the codon specifying amino acid 66, whereas Hsc2 has a 0.65-kb intron present in the codon defining amino acid 55. The sequences bordering the introns in Hsc1 and Hsc2 agree with the consensus splice site sequences in *Drosophila* and other eukaryotes (Craig *et al.*, 1983).

4. The Yeast hsp70-Related Genes

Saccharomyces cerevisiae possesses a family of genes related to the *Drosophila* hsp70 genes (Craig *et al.*, 1982; Ingolia *et al.*, 1982). Four genes were isolated from a yeast genomic library by virtue of their hybridization to a *Drosophila* hsp70 gene probe. Total genomic Southerns, using these isolated clones as probes, indicated a family of approximately 10 genes which shared different degrees of cross-homology to each other, as assessed by varying the conditions of stringency in the hybridization experiments.

The transcriptional activity of these four genes was analyzed by dot blot analysis and it was shown that two of the four genes, *YG100* and *YG102*, were inducible by heat shock. The remaining two genes, *YG101* and *YG103*, were found to be expressed only at normal temperatures and to be repressed under heat shock conditions. Seventy five percent of the protein coding region of the *YG100* DNA sequence is known and indicates that 72% of the predicted amino acid sequence is homologous with the *Drosophila* hsp70 gene. One-half of the protein coding DNA sequence of *YG101* has been determined and indicates a 64% amino acid homology with the hsp70 gene. The predicted amino acid sequences of these two yeast genes have a 65% homology.

5. The Escherichia coli hsp70 Related Gene

Following heat shock, 13 heat shock proteins are induced in *Escherichia coli* (Yamamori and Yura, 1980, 1982; Neidhardt and Van Bogelen, 1981; Neidhardt *et al.*, 1983; see also Section III,B). One of these hsp's has been identified as the protein product of the *dnaK* gene, and both hybridization studies and DNA sequencing data indicate that this gene is related to the *Drosophila* hsp70 gene. Indeed, comparison of the *dnaK* gene sequence with both *Drosophila* and yeast hsp70 heat shock genes indicates about 45–50% sequence homology at the predicted amino acid level. This homology, however, is variable, with certain domains more highly conserved than others (Craig *et al.*, 1982; Bardwell and Craig, 1984). The possible functional role of the dnaK protein in the *E. coli* heat shock response will be discussed in Section III,B.

B. Heat Shock Protein 27, 26, 23, and 22 Genes

1. Organization of the Four Small hsp Genes

In vitro translation analysis of heat shock RNA extracted from *D. melanogaster* tissue culture cells and sedimented on sucrose gradients established that the 12 S RNA peak contained those messengers encoding the four small heat shock proteins (McKenzie and Meselson, 1977; Mirault *et al.*, 1978; Moran *et al.*, 1978). This RNA fraction hybridized *in situ* exclusively to the major heat shock puff locus 67B on chromosome 3L (McKenzie and Meselson, 1977; Spradling *et al.*, 1977). The small heat shock proteins 27, 26, and 23 have also been genetically mapped to the same locus (Petersen *et al.*, 1979). This was achieved by analyzing the relative electrophoretic mobilities of variant small heat shock proteins in the progeny of genetic crosses between different *Drosophila* strains. In this way, the genes encoding these three polypeptides were mapped to the close vicinity of the heat shock puff at locus 67B, thus suggesting a close cytological linkage for these genes.

Two cDNA clones were isolated from the reverse transcripts of 12 S heat shock RNA and were found to hybridize *in situ* to locus 67B and to messengers about 1.0-kb long (Wadsworth *et al.*, 1980; Voellmy *et al.*, 1981). Both cDNA–mRNA hybrid selection or hybrid-arrested translation studies demonstrated that these cDNA's were specific to hsp26 and hsp23 transcripts, respectively. The organization of the chromosomal locus 67B was investigated by mapping various restriction sites around these two cDNA specific small hsp genes by means of total genomic Southerns. The results of these experiments indicated that these two genes were closely linked on a 12-kb DNA segment and, very probably, were unique within the haploid genome. DNA fragments, complementary to either these cDNA clones or the reverse transcripts of 12 S heat shock RNA, were isolated from genomic phage or plasmid libraries of *Drosophila* DNA and characterized (Corces *et al.*, 1980; Wadsworth *et al.*, 1980; Voellmy *et al.*, 1981).

The relationship between regions homologous to 12 S heat shock RNA and each small hsp gene was established either by Southern hybridization to characterized cDNA clones (Craig and McCarthy, 1980; Voellmy *et al.*, 1981) or by cell-free translation of heat shock RNA selected by hybridization to specific small hsp gene-bearing subclones (Corces *et al.*, 1980; Craig and McCarthy, 1980). The arrangement of these genes, as shown in Fig. 1, was confirmed by the direct observation of R loops in the electron microscope (Corces *et al.*, 1980; Voellmy *et al.*, 1981).

The direction of transcription of each gene was ascertained by R loop mapping (Voellmy *et al.*, 1981), primer extension experiments (Craig and McCarthy, 1980), or identification of the DNA strand of a specific restriction fragment (labeled at its 3′ ends) that hybridized to heat shock mRNA (Corces *et al.*, 1980). As can be seen in Fig. 1, the small hsp genes are clustered on a 12-kb DNA stretch and are not all transcribed in the same direction, which excludes a common transcriptional unit. Northern blot analysis of heat shock RNA with small hsp gene-bearing subclones showed no evidence of precursor mRNA's and sized the small hsp gene transcripts to about 1.0 kb for hsp 26, 23, and 22 and 1.25–1.3 kb for hsp 27. This indicated that each gene probably gave rise to its own transcript from its individual promoter. No introns were detected within the small hsp genes, as assessed by either electron microscopic observation of R loops or SI mapping.

Cell-free translation of heat shock mRNA, selected by hybridization to genomic subclones under conditions of reduced stringency, indicated a partial homology between the hsp27, 26, and 23 transcripts (Corces *et al.*, 1980). This homology is clearly seen in the sequences of the coding regions (see Section II,B,2). No homology was observed, however, between the small hsp genes and the genes for hsp 83, 70, and 68 (Corces *et al.*, 1980).

Close to the small heat shock gene cluster, about 10 kb downstream from hsp 27, a region of non-heat-shock-inducible transcription occurs (Craig and McCarthy, 1980). This region gives rise to an abundant RNA at 22°C for which no polypeptide product could be detected and which hybridizes to over 15 bands in total genomic Southerns, indicating that it most likely belongs to a repetitive multigene family. Another insertionlike element was found by Voellmy *et al.* (1981) about 1–2 kb to the right of the hsp22 gene in one genomic clone variant. Voellmy *et al.* also reported a region of polymorphic variation at the 3′ end of the hsp23 gene.

In addition to the four small heat shock genes, the DNA isolated from locus 67B was also found to contain at least three additional genes, which are not heat inducible (Fig. 1). By screening a *Drosophila* DNA library with developmental stage-specific cDNA probes, a genomic clone bearing these three genes, together with two small heat shock genes, was isolated and found to partially overlap the independently cloned DNA segment covering the small heat shock gene cluster (Sirotkin and Davidson, 1982). These three genes, together with the small hsp genes, are expressed abundantly during late larval and early pupal development in an apparently noncoordinate, stage-specific program (Sirotkin and Davidson, 1982; Ireland and Berger, 1982; Ireland *et al.*, 1982; Zimmerman *et al.*, 1983; Mason *et al.*, 1984). The small hsp genes, therefore, seem to be under a dual transcriptional control, inducible either by physiological stress, or by hormones, or by other developmental regulating mechanisms. Interestingly, micrococcal nuclease digestion patterns of chromatin and naked DNA from locus 67B demonstrated the existence of preferential cleavage sites in the 5′ regions of both the heat shock and the developmentally regulated genes (Keene and Elgin, 1981; Keene *et al.*, 1981). In this way the existence, location, and even probable direction of transcription of gene 1 in Fig. 1 was predicted before this gene was discovered by developmental studies (Keene *et al.*, 1981; Sirotkin and Davidson, 1982).

2. *Sequence Analysis of the Four Small hsp Genes*

Since the four clustered heat shock genes are apparently activated by common stimuli and their polypeptide products probably share both structural and physiological properties, they were analyzed and compared at the DNA and amino acid levels for putative control elements and sequence homologies (Ingolia and Craig, 1981, 1982b; Southgate *et al.*, 1983).

Each gene has a characteristic Goldberg–Hogness box situated approximately 23–25 bp 5′ to the site of initiation of transcription. The site of initiation of transcription was determined either by SI mapping (Ingolia and Craig, 1981; Southgate *et al.*, 1983) or by primer extension experiments (Ingolia and Craig, 1981), and a consensus cap site sequence, 5′ C/G-T/A-C-A-G-T/A, was derived in which the adenine in the fourth position probably follows the cap in the small hsp messages (Ingolia and Craig, 1981). The cap site in hsp70, with the exception of the fifth base, concords with this consensus sequence, as do other non-heat-shock *Drosophila* genes such as the vitellogenin (Hovemann *et al.*, 1981) and the cuticle protein genes (Snyder *et al.*, 1982).

The 5′ transcribed but untranslated sequences of the small hsp transcripts are not only unusually long (over 100 bases) but are also rich in adenine (45–50%), which is uncommon both for non-heat-shock-activated *Drosophila* genes as well as for other eukaryotic genes (see Table 1 in Ingolia and Craig, 1981; Hovemann *et al.*, 1981; Snyder *et al.*, 1982). In hsp70, the leader sequence is 244 bases long and 45% adenine rich (Ingolia *et al.*, 1980). The first 14 bases downstream from the cap site in all four small hsp genes and the hsp70 gene share a considerable degree of homology, which is possibly related to the sequence requirements of the RNA polymerase II entry site (see Fig. 8 in Ingolia and Craig, 1981). The special characteristics of these leader sequences are no doubt necessary for the translational control of the heat shock mRNA's during stress (Krüger and Benecke, 1981; Lindquist, 1980, 1981; Scott and Pardue, 1981; this volume). No homologies to the 3′-terminal portion of *Drosophila* 18 S ribosomal RNA (Jordan *et al.*, 1980) were observed.

A search of the 5′ sequences indicated no homology to the 5′-C-C-A-A-T box (Benoist *et al.*, 1980) but did show the following homologies. A sequence, 5′ A-C-T-T-T-N-A, is present 180–200 bp upstream of the cap site in all four genes and is located within a region of twofold rotational symmetry (Ingolia and Craig, 1981). A similar sequence, also part of a palindrome, occurs in a comparable position with respect to the cap site in hsp70. As was suggested by Ingolia and Craig (1981), a distance of about 200 bp corresponds, approximately, to a one-nucleosome spacing in chromatin (Kornberg, 1977), and the palindromic sequences could represent a possible site for DNA–protein interactions, implied by the transcriptional activation of these genes during stress. A 10-bp homology has been reported within these 5′-flanking regions by Southgate *et al.* (1983) between hsp22 and the *Drosophila* cuticle protein gene IV (Snyder *et al.*, 1982). A comparison of the 5′ sequence of hsp26 to those of hsp83, 70, and 68 was carried out by Holmgren *et al.* (1981) and, besides the TATA box, indicated three other notable regions of homology. One of these homologies is an imperfect dyad sequence about 25 bp upstream of the TATA box, first observed by Ingolia *et al.* (1980) in hsp70, and found in front of all four of these genes. This sequence has proved to have considerable importance for the induction of *Drosophila* heat shock genes, as will be described in Section III,A. Apart from this sequence however, the functional significance of the other reported homologies, if any, still remains to be determined.

The 3′-non-coding and -flanking regions of the small hsp's are predominantly AT rich and devoid of any obvious homology except the 5′ A-A-T-A-A-A polyadenylation signal (Southgate *et al.*, 1983). A 3′ SI mapping positioned the poly(A) addition site 14–23 bp downstream from the polyadenylation signal. The 3′-non-coding sequences of the hsp27 gene are characterized by the presence of 13 *Mbo*II restriction sites within the first 140 bp 3′ to the stop codon. This stretch of numerous direct repeats (5′ G-A-A-G-A) is very purine rich (87%). In the hsp26 gene, another stretch of direct repeats (5′ C-A-T/A) occurs just downstream from the stop codon and similar direct repeats can also be found in the hsp27 gene. Comparison of these sequences with other known 3′-non-coding regions indicates two short homologies between the hsp23 and avian α-globin genes. These 3′ sequences also contain some prominent interrupted palindromes which could possibly form hairpin loops within the mature transcripts. Interestingly, especially in light of the developmental expression of the small hsp's, the hsp26 gene has an additional TATA box and polyadenylation signal, both of which are apparently nonfunctional during heat shock, as estimated by SI mapping.

Each gene has a unique, open, uninterrupted reading frame, sufficiently large to encode a small hsp and extending from the first A-T-G downstream from the cap site. All alternative reading frames are blocked by multiple stop codons and the sequence data do not indicate evidence for introns. The molecular masses of the unmodified polypeptide chains were calculated to be 23,620, 22,997, 20,603, and 19,705 daltons after the loss of water occurring during peptide bond formation in protein synthesis is taken into account (Ingolia and Craig, 1982b; Southgate *et al.*, 1983). The two independently determined nucleotide sequences of these four coding regions indicate the same initiator and nonsense codons with a 96% homology at the amino acid level and 98% at the DNA level. Most of the 2% nucleotide differences are probably real since at least 18 restriction site differences have been detected between the two determined coding sequences (Southgate *et al.*, 1983). Although both sequenced DNA's were originally derived from the *Drosophila* Oregon R strain, these differences may represent polymorphism within the fly population (Petersen *et al.*, 1979; Buzin and Petersen, 1982).

Comparison of the coding sequences, at the amino acid level, of all four small hsp's indicates a high degree of homology over a large part of their lengths. The homologous region is 108 amino acids long, and, within this stretch, the same amino acid is present in all four small hsps at 35% of the amino acid positions and in three out of the four proteins at 71% of the positions. As predicted by hybridization studies (Corces *et al.*, 1980), hsp 27, 26, and 23 are nearly twice as homologous within the conserved domain than is hsp22 with the other small hsp's. The structure of the small hsp's has been schematically represented in Fig. 2. The central homologous block is surrounded by two heterologous regions, an amino-terminal section 58–84 amino acids in length and a short carboxyl-terminal portion 7–21 amino acids long. These variable regions account for most of the size differences between the four proteins. The homologies occurring within the 108 amino acid stretch have been depicted at both the DNA and the amino acid levels in Fig. 3. This close linkage of four partially homologous genes,

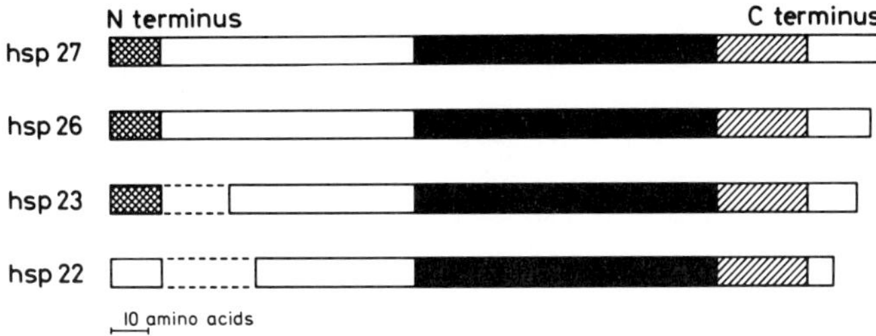

Fig. 2. A schematic diagram of the various amino acid homologies between the four small hsp genes as based on the nucleotide sequence of their coding regions (Ingolia and Craig, 1982b; Southgate *et al.*, 1983). The solid and hatched boxes represent the 108 amino acid stretch which is conserved among all four small hsp genes, whereas the solid boxes indicate the extent of the homology with mammalian α-crystallin. The cross-hatched boxes at the N termini of hsp 27, 26, and 23 indicate the 14 amino acid homology discussed in the text. The N- and C-terminal proximal regions, shown as open boxes, display little sequence homology.

within a 12-kb region, strongly suggests that this gene cluster arose, by duplication and inversion events, from a single ancestral gene.

The small hsp's were observed to be homologous to the bovine α-crystallin B_2 chain over about 40% of their lengths (Ingolia and Craig, 1982b). This homology coincides with the first 83 amino acids of the 108-amino-acid conserved stretch in the small hsp's and amino acid residues 70–152 in bovine α-crystallin (Van der Ouderaa *et al.*, 1974; Figs. 2 and 3). It has recently been shown that this conserved region is also homologous to part of the coding region of the 16-kilodalton hsp's in *Caenorhabditis elegans* (Russnak *et al.*, 1983).

Another region of weaker homology occurs among the first 14 amino acids of hsp 27, 26, and 23, but not hsp22 as is shown in Fig. 2. A computer search reported by Southgate *et al.* (1983) noted a resemblance between these 14 amino acids, part of the signal peptide from human pre-proinsulin, and residues 20–29 in bacteriorhodopsin, a bacterial transmembrane protein. Comparison of the size of the *in vivo* and the *in vitro* translation products of the messengers encoded by these genes indicates that these N-terminal regions are not cleaved during translation (Mirault *et al.*, 1978). It is possible, however, that these proteins interact with membranes by virtue of their amino-terminal regions.

An analysis of the hydrophobic/hydrophilic characteristics of the four small hsp's was reported by Southgate *et al.* (1983) using the Kyte and Doolittle (1982) computer program. Due to the amino acid homologies between each small hsp and bovine α-crystallin, the five proteins exhibit a very similar hydropathic profile in their shared regions. Outside this region, however, these proteins show no common features except the hydrophobicity of the amino-terminal regions of the small hsp's and a certain hydrophilicity of the C termini. A prominent hydrophilic peak is found in the hydropathic profile of the conserved region,

```
      1                   5                       10                      15
  27 Val Gly Lys Asp Gly Phe Gln Val Cys Met Asp Val Ser Gln Phe
  26  *   *   *   *   *   *   *   *   *   *   *   *  Ala  *   *
  23 Ile  *   *   *   *   *   *   *   *   *   *   *   *  His  *
  22  *  Asn  *   *   *  Tyr Lys Leu Thr Leu  *   *  Lys Asp Tyr
αCRY Leu Glu  *   *  Arg  *  Ser  *  Asn Leu Asn  *  Lys His  *

     GTG GGC AAA GAT GGC TTC CAG GTG TGC ATG GAT GTG TCG CAG TTC
     *** *** **G *** **A *** *** *** *** *** **C **C G*C *** ***
     A*C **A **G *** *** *** *** **C *** *** *** *** *** *** ***
     **C AA* **G *** *** *A* A*A C*C AC* C** **C **C AA* G*C *A*

                          20                      25                  30
  27 Lys Pro Asn Glu Leu Thr Val Lys Val Val Asp Asn Thr Val Val
  26  *   *  Ser  *   *  Asn  *   *   *   *   *  Asp Ser Ile Leu
  23  *   *  Ser  *   *  Val  *   *   *  Gln  *   *  Ser  *  Leu
  22 --- --- Ser  *   *  Lys  *   *   *  Leu  *  Glu Ser  *   *
αCRY Ser  *  Glu  *   *  Lys  *   *   *  Leu Gly Asp Val Ile Glu

     AAG CCC AAC GAG CTG ACC GTC AAG GTG GTG GAC AAC ACC GTG GTG
     *** *** *GT *** **C *A* **G *** *** *** *** G** T** A*C T**
     *** *** *G* **A *** GTG *** **A *** CA* *** *** T** **C C**
     --- --- *G* *** *** *AG *** *** *** C** *** G*G *G* *** **C

                       35                     40                     45
  27 --- Val Glu Gly Lys His Glu Glu Arg Glu Asp Gly His Gly Met
  26 ---  *   *   *   *   *   *   *   *  Gln  *  Asp  *   *  His
  23 ---  *   *   *  Asn  *   *   *   *   *   *  Asp  *   *  Phe
  22 Leu  *   *  Ala  *  Ser  *  Gln Gln  *  Ala Glu Gln  *  Gly
αCRY ---  *  His  *   *   *   *   *   *  Gln  *  Glu  *   *  Phe

     --- GTA GAG GGG AAG CAC GAG GAG CGC GAG GAC GGC CAT GGA ATG
     --- **C *** **C *** **T *** **A *** C** *** *A* *** **T CAC
     --- **G *** **C **C **T *** *** *** **A **T *A* *** **C T*C
     CTG **G *** *CA **A TCG *** C** *AG *** *C* *AA **A **T GGC

                          50                      55                  60
  27 Ile Gln --- Arg His Phe Val Arg Lys Tyr Thr Leu Pro Lys Gly
  26  *  Met ---  *   *   *   *   *  Arg  *  Lys Val  *  Asp  *
  23  *  Thr ---  *   *   *   *   *  Arg  *  Ala  *   *  Pro  *
  22 Tyr Ser Ser  *   *   *  Leu Gly Arg  *  Val  *   *  Asp  *
αCRY  *  Ser ---  *  Glu  *  His  *   *   *  Arg Ile  *  Ala Asp

     ATC CAG --- CGT CAC TTT GTG CGC AAG TAT ACC CTG CCC AAG GGC
     *** AT* --- **C *** *** *** *** CGC **C *AG G*T *** G*T ***
     *** *CT --- *** *** *** **C *** CGC *** G*T *** **A CCC **T
     TAT AGT TCC A*G *** **C C*C G** CGA **C GTT *** **G G*T **A

                          65                      70                  75
  27 Phe Asp Pro Asn Glu Val Val Ser Thr Val Ser Ser Asp Gly Val
  26 Tyr Lys Ala Glu Gln  *   *   *  Gln Leu  *   *   *   *   *
  23 Tyr Glu Ala Asp Lys  *  Ala  *   *  Leu  *   *   *   *   *
  22 Tyr Glu Ala Asp Lys  *  Ser  *  Ser Leu  *  Asp  *   *   *
αCRY Val  *   *  Leu Ala Ile Thr  *  Ser Leu  *   *   *   *   *

     TTT GAC CCC AAC GAG GTA GTG TCC ACT GTC TCA TCC GAC GGT GTG
     *AC A*G G*G G*G C*A **G **C **G CAG C*G **G **G **T **C ***
     *A* **G G*T G*T A** **G *CC *** **C T*G **C *** **T *** **C
     *AC **G G*G G** A** **G TCC **G T*G C*G AGC GAC *** **C **T
```

```
                    80                  85                  90
27    Leu Thr Leu Lys Ala Pro Pro Pro Pro Ser Lys Glu Gln Ala Lys
26     *   *  Val Ser Ile  *  Lys  *  Gln Ala Val  *  Asp Lys Ser
23     *   *  Ile  *  Val  *  Lys  *   *  Ala Ile  *  Asp Lys Gly
22     *   *  Ile Ser Val  *  Asn  *   *  Gly Val Gln Glu Thr Leu
αCRY   *   *  Val Asn Gly  *  Arg Lys Gln Ala  .   .   .   .   .

      CTG ACC CTC AAG GCC CCG CCG CCG CCC AGC AAG GAA CAG GCC AAG
      **C *** G** *GT ATT **C AA* *** *AG GC* GTC **G G*C AAG TCC
      *** *** A** *** *TG **C AA* **A **G GCA *TC **G G*T AAG GGC
      *** *** A** *GT *TG **C AAT **T **A G** GT* C*G G** A*A CTC

                    95                  100                 105
27    Ser Glu Arg Ile Val Gln Ile Gln Gln Thr Gly --- Pro Ala His
26    Lys  *   *   *  Ile  *   *   *   *  Val  *  ---  *   *   *
23    Asn  *   *   *   *   *   *   *   *  Val  *  ---  *   *   *
22    Lys  *   *  Glu  *  Thr  *  Glu  *   *   *  Glu  *   *  Lys

      TCG GAG CGC ATT GTC CAG ATC CAG CAA ACG GGG --- CCT GCC CAT
      AA* *** *** **C A*T **A **T *** *** GT* **A --- **C **T **C
      AAC *** *** **C **T *** *** *** **G GT* **A --- **C *** ***
      AA* *** **T GAG **G ACC *** G** **G **T **C GAG **G **A A*G

                    110
27    Leu Ser Val Lys Ala Pro
26     *  Asn  *   *   *  Asn
23     *  Asn  *   *  Glu Asn
22    Lys  *  Ala Glu Glu  *

      TTG AGC GTC AAG GCA CCG
      C*C *A* **T *** *** AAT
      C*C *AT **G *** *AG AAT
      AA* TC* *C* G** *AG **A
```

Fig. 3. Comparison of the conserved stretch among the four small hsp genes, at the amino acid and nucleotide levels, together with the homologous amino acid sequence of bovine α-crystallin B_2 chain (Van der Ouderaa *et al.*, 1974), as adapted from Ingolia and Craig (1982b) and Southgate *et al.* (1983). The top line in each block of four sequences gives either the amino acid or nucleotide sequence of hsp27. An asterisk indicates the presence of the same amino acid or nucleotide at that position as in hsp27. A minimal number of gaps had to be introduced in order to obtain an optimal alignment of the sequence data (dashes).

which probably represents a major surface structural domain common to all five proteins.

The homology between these *Drosophila* proteins and mammalian α-crystallin possibly represents an evolutionary, highly conserved protein domain necessary for some shared physiological function of these proteins. The 20–30 S protein–RNA complexes, apparently containing all four small hsp's (A. P. Arrigo, personal communication), have been observed in the cytoplasm of *Drosophila* tissue culture cells following heat shock (Arrigo and Ahmad-Zadeh, 1981) and perhaps exist in the nucleus during heat shock (Levinger and Varshavsky, 1981). Mammalian α-crystallin can form large aggregates of up to 800,000 daltons in the vertebrate eye, and, by analogy, the region of sequence homology with this protein may be involved in the aggregation of the small hsp's, necessary for a yet undetermined function.

C. Heat Shock Protein 83 Gene

Most of the information concerning the organization of the *Drosophila* hsp83 gene came from the study of three cloned DNA segments. The first clone to be isolated, pPW244, contained a randomly sheared fragment of Oregon R DNA, bearing most of the hsp83 gene but lacking its 3′ end (Holmgren *et al.*, 1979). It hybridized *in situ* to chromosomal locus 63BC and was homologous to the heat shock messenger encoding hsp83, as demonstrated by hybrid-arrested translation. These results showed that this plasmid contained most of the hsp83 gene, but, since the region homologous to the hsp83 messenger extended to one end of the DNA insert, it was necessary to determine how much of the hsp83 gene was missing from this plasmid. A 2.3-kb fragment from plasmid pPW244 and a 2.6-kb fragment of embryonic DNA homologous to this plasmid, were found to be protected against S1 digestion after hybridization to poly(A)$^+$ heat shock RNA. This indicated that the hsp83 messenger was complementary to a 2.6-kb genomic DNA sequence of which only 2.3 kb were cloned in pPW244. The direction of transcription, established by S1 mapping, showed that the missing 0.3 kb of DNA were located at the 3′ end of the hsp83 gene. Restriction analysis of embryonic DNA, as probed by Southern hybridization using labeled plasmid DNA, indicated that the hsp83 gene was probably unique within the haploid genome.

Two additional clones, 301.1 and λ6, each bearing the entire hsp83 gene, were subsequently isolated. Clone 301.1 is an 8.2-kb *Eco*RI fragment of Oregon R DNA inserted into pBR322, and λ6 is from a Charon 4 library of *D. melanogaster* Canton S embryonic DNA (Holmgren *et al.*, 1981).

The presence of an intron in this gene was first suggested by the finding in heat shock nuclear RNA of hsp 83 transcripts larger than the expected 3-kb polyadenylated mRNA (Holmgren *et al.*, 1979). The observation that hybridization with whole-cell heat shock RNA protected a 2.6-kb fragment, from clone 301.1, against endonuclease S1 and a 3.7-kb segment against exonuclease VII clearly demonstrated the existence of this intron. This indicated that the transcription unit of hsp83 is 3.7 kb long and consists of two exons interrupted by a 0.9-kb intron. The first exon is 0.15 kb long, whereas the second is 2.6 kb (Holmgren *et al.*, 1981). The length of the intron in the clone isolated from the Canton S strain is 0.2 kb smaller than that found in the Oregon R strain DNA. Restriction analysis of embryonic DNA shows only the presence of the 0.9-kb intron in Oregon R DNA (Fig. 1) whereas both the 0.9-kb and 0.7-kb introns are found in the Canton S strain DNA. It should be noted that hsp83 is the only *Drosophila* heat shock protein gene to possess an intron.

As mentioned in Section I, the translation product of the hsp83 gene is observed in all cells, even in the absence of stress, at a fairly high level, and the activation by heat shock, approximately 5- to 6-fold, is much lower than that for other hsp genes (Lindquist, 1980; O'Connor and Lis, 1981). Hackett and Lis (1983) have recently reported a partial sequence analysis of the hsp83 gene.

III. INDUCTION

A. Induction of *Drosophila* Heat Shock Genes

When it became clear that an essential feature of the heat shock response consisted of the vigorous activation of a small number of specific genes (Tissières *et al.*, 1974; McKenzie and Meselson, 1977; Spradling *et al.*, 1977), it appeared natural that their study could result in the understanding of the mechanism of their induction and coordinate regulation. Is there a signal at the DNA level which is involved directly with the induction mechanism and is this hypothetical signal common to some extent, at least, to all the *Drosophila* stress-induced genes?

In order to investigate these questions, *Drosophila* hsp70 genes were introduced into mouse cells by cotransformation with the *Herpes* thymidine kinase gene (Corces *et al.*, 1981), into monkey COS cells by transfection with vectors derived from SV40 (Mirault *et al.*, 1982; Pelham, 1982; Pelham and Bienz, 1982), or into *Xenopus* oocytes by injection (Pelham and Bienz, 1982; Voellmy and Rungger, 1982). In each of these three instances, it was found that the *Drosophila* hsp70 genes were only actively transcribed when the recipient cells were heat shocked. In some experiments arsenite poisoning was used to replace heat shock as an inducer of transcription. It could be concluded that in these different cell types, as well as in *Drosophila,* the mechanism of transcriptional induction by stress of the hsp70 gene was remarkably well conserved. Moreover, the heat shock inducibility of this gene in foreign cells provides a convenient means of locating the putative regulatory signals at the DNA level.

To define the extent of the DNA region necessary for control of heat shock transcription, Corces *et al.* (1981) constructed chimeric genes containing a 1.3-kb segment of *Drosophila* DNA from upstream of an hsp70 gene that was fused in the leader sequences to a *Herpes* thymidine kinase or human growth hormone gene. This 1.3-kb segment of DNA, extending to 198 bp downstream of the hsp70 cap site, was found to be sufficient to confer heat inducibility to both chimeric genes after introduction into mouse cells by transformation. Analysis of a series of *Bal*31 5′ deletions upstream of a native hsp70 gene suggested that no more than 51 bp 5′ to the cap site were necessary for heat shock transcriptional control, although some of the induction results obtained in different transformants were not unequivocal due to variable integration sites in the genome.

Using similar *Bal*31 deletion mutants in monkey COS cells transfected by SV40 derived vectors, it was found that 5′ deletions to positions −68 or −66 from the cap site did not disturb the stress induction of the hsp70 gene (Mirault *et al.*, 1982; Pelham, 1982). However, 5′ deletions to positions −53 or −44 were found to drastically reduce hsp70 transcription upon heat shock. Similar results were found by Bienz and Pelham (1982) in frog oocytes. Thus, the data from mouse, monkey, and frog cells indicate that an essential region for transcriptional control of the *Drosophila* hsp70 gene lies between positions −44 and −66. A DNA sequence comparison upstream of the different *D. melanogaster* heat shock genes have revealed interesting common features (Ingolia *et al.*, 1980; Holmgren *et al.*, 1981; Pelham, 1982; Southgate *et al.*, 1983). These promoter regions appear to contain many different types of repeats, some with dyad symmetry (Holmgren *et al.*, 1981; Mirault *et al.*, 1982), which have also been noted upstream of other eukaryotic genes (Mellon *et al.*, 1981; Brinster *et al.*, 1982). A palindromic consensus sequence of the type 5′ C-T-N-G-A-A-N-N-T-T-C-N-A-G has been observed upstream of the hsp genes as shown in Fig. 4 (Pelham, 1982). Recently, Pelham and Bienz (1982) have replaced the "upstream element" of the *Herpes* thymidine kinase promoter (McKnight *et al.*, 1981) with short synthetic oligonucleotides bearing 8 to 10 bases matching this consensus sequence, and have shown that the *tk* gene consequently became heat inducible, in both COS cells and *Xenopus* oocytes. An approximate correlation was observed between the degree of base-pair match to the consensus sequence and the extent of heat inducibility of the chimeric genes. This observation was confirmed by the finding that, of the four small heat shock genes tested in COS cells, the two which are most efficiently activated by heat shock, hsp22 and hsp26, are those which have the best matches to the consensus sequence (Pelham and Lewis, 1983; A. Ayme and R. Southgate, unpublished observations).

A reasonable hypothesis is that a specific DNA binding protein interacts with the consensus sequence and is involved in the activation of the heat shock promoters in stressed cells. Such protein(s), acting positively, either stoichiometrically or catalytically, in the regulation of transcription, would appear to have been conserved in widely divergent species, since *Drosophila* heat shock genes remain heat inducible in other eukaryotic cells. Such a contention has recently received direct support. Parker and Topol (1984b) have succeeded in isolating a fracton from *Drosophila melanogaster* nuclear extracts enriched for a protein factor which they call HSTF for "heat shock transcription factor". This factor is required for the active transcription of an hsp70 gene at heat shock temperatures, in addition to RNA polymerase II and another transcription factor from *Drosophila,* the A factor (Parker and Topol, 1984a). Footprint analysis of the HSTF on the hsp70 gene reveals that it binds specifically to a 55-base pair region upstream from the T-A-T-A box which includes the consensus sequence shown to be essential for the heat induction of hsp70 (Pelham and Bienz, 1982). Ex-

```
                                                   NUMBER OF MATCHES TO
"PELHAM BOX":CT GAA  TTC AG                        CONSENSUS SEQUENCE
hsp83        ** **_  *** **
CCTCTAGAAGTTTCTAGAGACTTCCAGTTCGGTCGGGTTTTTCTATAAA              9/10
hsp70        ** ***  *** _*
CGAGAGAGCGCGCCTCGAATGTTCGCGAAAAGAGCGCCGGAGTATAAA               9/10
hsp68        ** ***  *** __
CTCGCAGGGAAATCTCGAATTTTCCCCTCCCGGCGACAGAGTATAAA                8/10
hsp27        _* __*  __* **
GTTCCGTCCCTGGTTGCCATGCACTAGTGTGTGTGAGCCCAGCGTCAGTATAAAA        5/10
hsp26        *_ **_  *** **
TTTCTGTCACTTTCCGGACTCTTCTAGAAAAGCTCCAGCGGGTATAAAA              8/10
hsp23        *_ ***  *** _*
GATATTTTCAGCCCGAGAAGTTTCGTGTCCCTTCTCGATGTTTGTGCCCCCTAG         8/10
CACACAGACACGACGCGCACACACACAGCGCCGACGGGCGCACGCACACTTCGA
CAGCAAGCGGTTGTATAAA
hsp22        *_ *_*  *** **
ATTCGAGAGAGTGCCGGTATTTTCTAGATTATATGGATTTCCTCTCTGTCAAGAGTATAAA 8/10
```

Fig. 4. Comparison of the promoter regions of the *Drosophila* hsp genes as adapted from Pelham and Lewis (1983). An asterisk indicates the same nucleotide as found in the ''Pelham box'' (Pelham, 1982). The number of base-pair matches to this consensus sequence in each gene is given in the column on the right of the figure. Heat shock protein 23 differs from the other *Drosophila* hsp genes in that its ''Pelham box'' apparently occurs much further upstream.

onuclease chromatin digestion studies on the hsp70 gene also suggest the presence of proteins bound to this same region under heat shock conditions (Wu, 1984).

B. Induction of *Escherichia coli* Heat Shock Genes

Escherichia coli normally grows between 30 and 37°C but can survive temperatures as high as 50°C for limited periods of time. Upon transfer to high temperatures, such as 42°C, *E. coli* responds by a very rapid change in its protein synthesis pattern. One of these modifications is the transient induction of a specific set of 13 heat shock proteins. Kinetic experiments have shown this induction to be coordinate and to occur at the level of transcription, the extent of the hsp induction being proportional to the degree of heat shock over a certain "critical point" (Yamamori and Yura, 1980, 1982; Neidhardt and Van Bogelen, 1981). Genetic evidence indicates that the coordinate induction of these 13 proteins, upon a shift to a high temperature, is controlled by a single chromosomal gene called either *hin* (Yamamori and Yura, 1980, 1982) or *htpR* (Neidhardt and Van Bogelen, 1981). These 13 heat shock proteins and the *hin* or *htpR* gene constitute a regulon, that is, a group of genes/proteins sharing a common regulator element. A mutation in the *hin* or *htpR* gene abolishes the coordinate response of all 13 proteins to heat shock so that, when kept at 42°C, mutant cells rapidly die (Neidhardt and Van Bogelen, 1981; Yamamori and Yura, 1982). Mutants in the *hin* or *htpR* gene provide, therefore, an opportunity to study not only the mechanism of induction but also the protection against thermal killing in bacteria. The *hin* or *htpR* gene has recently been cloned (Neidhardt *et al.*, 1983). A simple hypothesis is that the product of this *hin* or *htpR* gene is a protein required for the heat shock induction (transcriptional activation) of several *E. coli* hsp genes. Some of these heat shock proteins would appear to be necessary for cell growth at high temperatures. Several of these heat shock proteins have been identified and their genes mapped (Neidhardt and Van Bogelen, 1981).

One of the major *E. coli* hsp's is the dnaK protein, which is homologous to the *Drosophila* hsp70 gene, as described in Section II,A,5. Mutations in the *dnaK* gene are known to affect both DNA and RNA synthesis, and it has recently been demonstrated that the *dnaK* gene product also modulates the *E. coli* heat shock response (Tilly *et al.*, 1983). The latter is not only normally induced very rapidly but is also switched off swiftly after about 15 min incubation at high temperatures (Yamamori *et al.*, 1978; Yamamori and Yura, 1982; Neidhardt and Van Bogelen, 1981). In the case of the mutation *dnaK* 756, however, this response does not terminate and synthesis of the *E. coli* heat shock proteins continues at a high rate for at least 2 hr (Tilly *et al.*, 1983). At the same time, the normal, non-heat-shock proteins are clearly underproduced. The *dnaK* gene mutants, which overproduce the dnaK protein at all temperatures, manifest a greatly reduced the heat shock response, implying that the high intracellular level of dnaK protein negatively modulates the heat shock response in *E. coli*. It, therefore, seems that the *dnaK* gene product is involved directly in shutting off the *E. coli* heat shock response. This might be the consequence of the dnaK protein interfering with the stimulatory effect of the *hin* or *htpR* gene product. Although other explanations can be proposed, experiments reported by Tilly *et al.* (1983) tend to support this model. *In vitro* experiments have shown that the purified dnaK protein possesses both an ATPase and an autophosphorylating activity and participates in DNA replication (Zylicz *et al.*, 1983). Autophosphorylation of the dnaK protein has also been observed *in vivo*.

The remarkable homologies of the heat shock response in very divergent species and, in particular, the sequence similarities to the *Drosophila* hsp70 gene in yeast and *E. coli* suggest that at least part of the mechanism of heat shock response might be similar in *E. coli* and eukaryotes. It should therefore be envisaged that a gene related to the *E. coli hin* or *htpR* gene exists in eukaryotes. If this is so, it can be expected that similar mutants will eventually be isolated from eukaryotic organisms, such as yeast or perhaps even *Drosophila,* allowing a detailed study of the heat shock response in these species. Additionally, in light of the apparent modulator function of the *E. coli* dnaK protein in this organism's heat shock response, it is important to ascertain whether the structural similarities between the dnaK protein and both yeast and *Drosophila* hsp70 are also reflected at the functional level. This is particularly interesting since the *Drosophila* hsp70 has been implicated in an autoregulatory mechanism both at the transcriptional and posttranscriptional levels (DiDomenico *et al.*, 1982).

Note Added in Proof

The *htpR* gene has recently been sequenced (Landick *et al.*, 1984). The predicted primary structure of the corresponding polypeptide was found to resemble part of the sequence of the *E. coli* σ-subunit of RNA polymerase and to show similarity to protein–DNA contact points conserved in known DNA binding proteins. This suggests that the *htpR* protein activates heat shock genes by interacting both with DNA and with RNA polymerase.

Acknowledgments

We thank B. Brun for her excellent assistance in typing the manuscript, O. Jenni and Y. Epprecht for preparing the figures, and M. Goldschmidt-Clermont, L. Hall, and C. Southgate for critical reading of the manuscript. This work was supported by the Swiss National Science Foundation grant no. 3.079.81 to A. Tissières and by the E. and L. Schmidheiny Foundation.

REFERENCES

Arrigo, A. P., and Ahmad-Zadeh, C. (1981). Immunofluorescence localization of a small heat shock protein (hsp23) in salivary gland cells of *Drosophila melanogaster*. *Mol. Gen. Genet.* **184,** 73–79.

Arrigo, A. P., Fakan, S., and Tissières, A. (1980). Localization of the heat shock induced proteins in *Drosophila melanogaster* tissue culture cells. *Dev. Biol.* **78,** 86–103.

Artavanis-Tsakonas, S., Schedl, P., Mirault, M.-E., Moran, L., and Lis, J. (1979). Genes for the 70,000 dalton heat shock protein in two cloned *Drosophila melanogaster* DNA segments. *Cell* **17,** 9–18.

Bardwell, T. C. A., and Craig, E. A. (1984). Major heat shock gene of *Drosophila* and the *Escherichia coli* heat inducible *dna*K gene are homologous. *Proc. Natl. Acad. Sci. U.S.A.* **81,** 848–852.

Beerman, W. (1956). Nuclear differentiation and functional morphology of chromosomes. *Cold Spring Harbor Symp. Quant. Biol.* **21,** 217–232.

Benoist, C., O'Hare, K., Breathnach, R., and Chambon, P. (1980). The ovalbumin gene-sequence of putative control regions. *Nucleic Acids Res.* **8,** 127–142.

Bienz, M., and Pelham, H. R. B. (1982). Expression of a *Drosophila* heat shock protein in *Xenopus* oocytes: conserved and divergent regulatory signals. *EMBO J.* **1,** 1583–1588.

Brinster, R. L., Chen, H. Y., Warren, R., Sarthy, A., and Palmiter, R. D. (1982). Regulation of metallothionein—thymidine kinase fusion plasmids injected into mouse eggs. *Nature (London)* **296,** 39–42.

Burke, J. F., and Ish-Horowicz, D. (1982). Expression of *Drosophila* heat shock genes is regulated in Rat-1 cells. *Nucleic Acids Res.* **10,** 3821–3830.

Buzin, C. H., and Petersen, N. S. (1982). A comparison of the multiple *Drosophila* heat shock proteins in cell lines and larval salivary glands by two dimensional-gel electrophoresis. *J. Mol. Biol.* **158,** 181–201.

Caggese, C., Caizzi, R., Morea, M., Scalenghe, F., and Ritossa, F. (1979). Mutation generating a fragment of the major heat shock-inducible polypeptide in *Drosophila melanogaster. Proc. Natl. Acad. Sci. U.S.A.* **76,** 2385–2389.

Corces, V., Holmgren, R., Freund, R., Morimoto, R., and Meselson, M. (1980). Four heat shock proteins of *Drosophila melanogaster* coded within a 12 kilobase region in chromosome subdivision 67B. *Proc. Natl. Acad. Sci. U.S.A.* **77,** 5390–5393.

Corces, V., Pellicer, A., Axel, R., and Meselson, M. (1981). Integration, transcription and control of a *Drosophila* heat shock gene in mouse cells. *Proc. Natl. Acad. Sci. U.S.A.* **78,** 7038–7042.

Craig, E. A., and McCarthy, B. J. (1980). Four *Drosophila* heat shock genes at 67B: characterization of recombinant plasmids. *Nucleic Acids Res.* **8,** 4441–4457.

Craig, E. A., McCarthy, B. J., and Wadsworth, S. C. (1979). Sequence organization of two recombinant plasmids containing genes for the major heat shock induced protein of *Drosophila melanogaster. Cell* **16,** 575–588.

Craig, E. A., Ingolia, T. D., Slater, M., Manseau, L. J., and Bardwell, J. (1982). *Drosophila,* yeast and *E. coli* genes related to the *Drosophila* heat shock genes. *In* "Heat Shock, from Bacteria to Man" (M. J. Schlesinger, M. Ashburner, and A. Tissières, eds.), pp. 11–18. Cold Spring Harbor Lab., Cold Spring Harbor, New York.

Craig, E. A., Ingolia, T. D., and Manseau, L. J. (1983). Expression of *Drosophila* heat shock cognate genes during heat shock and development. *Dev. Biol.* **99,** 418–426.

DiDomenico, B. J., Bugaisky, G. E., and Lindquist, S. (1982). The heat shock response is self-regulated at both the transcriptional and posstranscriptional levels. *Cell* **31,** 593–603.

Goldschmidt-Clermont, M. (1980). Two genes for the major heat-shock protein of *Drosophila melanogaster* arranged as an inverted repeat. *Nucleic Acids Res.* **8,** 235–252.

Hackett, R. W., and Lis, J. T. (1981). DNA sequence analysis reveals extensive homologies of regions preceding hsp70 and αβ heat shock genes in *Drosophila melanogaster. Proc. Natl. Acad. Sci. U.S.A.* **78,** 6196–6200.

Hackett, R. W., and Lis, J. T. (1983). Localization of the hsp83 transcript within a 3292 nucleotide sequence from the 63B heat shock locus of *D. melanogaster. Nucleic Acids Res.* **11,** 7011–7030.

Henikoff, S., and Meselson, M. (1977). Transcription at two heat shock loci in *Drosophila. Cell* **12,** 441–451.

Holmgren, R., Livak, K., Morimoto, R., Freund, R., and Meselson, M. (1979). Studies of cloned sequences from four *Drosophila* heat shock loci. Cell. **18,** 1359–1370.

Holmgren, R., Corces, V., Morimoto, R., Blackman, R., and Meselson, M. (1981). Sequence homologies in the 5′ regions of four *Drosophila* heat-shock genes. *Proc. Natl. Acad. Sci. U.S.A.* **78,** 3775–3778.

Hovemann, B., Galler, R., Welldorf, U., Küpper, H., and Bautz, E. F. K. (1981). Vitellogenin in *Drosophila melanogaster:* Sequence of the yolk protein I gene and its flanking regions. *Nucleic Acids Res.* **9,** 4721–4734.

Ingolia, T. D., and Craig, E. A. (1981). Primary sequence of the 5′ flanking regions of the

Drosophila heat shock genes in chromosome subdivision 67B. *Nucleic Acids. Res.* **9,** 1627–1642.

Ingolia, T. D., and Craig, E. A. (1982a). *Drosophila* gene related to the major heat shock-induced gene is transcribed at normal temperatures and not induced by heat shock. *Proc. Natl. Acad. Sci. U.S.A.* **79,** 525–529.

Ingolia, T. D., and Craig, E. A. (1982b). Four small *Drosophila* heat shock proteins are related to each other and to mammalian α-crystallin. *Proc. Natl. Acad. Sci. U.S.A.* **79,** 2360–2364.

Ingolia, T. D., Craig, E. A., and McCarthy, B. J. (1980). Sequence of three copies of the gene for the major *Drosophila* heat shock induced protein and their flanking regions. *Cell* **21,** 669–679.

Ingolia, T. D., Slater, M. R., and Craig, E. A. (1982). *Saccharomyces cerevisiae* contains a complex multigene family related to the major heat shock-inducible gene of *Drosophila. Mol. Cell Biol.* **2,** 1388–1398.

Ireland, R. C., and Berger, E. (1982). Synthesis of low molecular weight heat shock peptides stimulated by ecdysterone in a cultured *Drosophila* cell line. *Proc. Natl. Acad. Sci. U.S.A.* **79,** 855–859.

Ireland, R. C., Berger, E., Sirotkin, K., Yund, M. A., Osterbur, D., and Fristrom, J. (1982). Ecdysterone induces the transcription of four heat shock genes in *Drosophila* S3 cells and imaginal discs. *Dev. Biol.* **93,** 498–507.

Ish-Horowicz, D., and Pinchin, S. M. (1980). Genomic organization of the 87A7 and 87C1 heat-induced loci of *Drosophila melanogaster. J. Mol. Biol.* **142,** 231–245.

Ish-Horowicz, D., Holden, J. J., and Gehring, W. J. (1977). Deletions of two heat-activated loci in *Drosophila melanogaster* and their effects on heat-induced protein synthesis. *Cell* **12,** 643–652.

Ish-Horowicz, D., Pinchin, S. M., Gausz, J., Gyurkovics, H., Bencze, G., Goldschmidt-Clermont, M., and Holden, J. J. (1979a). Deletion mapping of two *Drosophila melanogaster* loci that code for the 70,000 dalton heat induced protein. *Cell* **17,** 565–571.

Ish-Horowicz, D., Pinchin, S. M., Schedl, P., Artavanis-Tsakonas, S., and Mirault, M.-E. (1979b). Genetic and molecular analysis of the 87A7 and 87C1 heat-inducible loci of *Drosophila melanogaster. Cell* **18,** 1351–1358.

Jordan, B. J., Latil-Damotte, M., and Jourdan, R. (1980). Sequence of the 3′ terminal portion of *Drosophila melanogaster* 18 S rRNA and of the adjoining spacer. *FEBS Lett.* **117,** 227–231.

Karch, F., Török, I., and Tissières, A. (1981). Extensive regions of homology in front of the two hsp70 heat shock variant genes in *Drosophila melanogaster. J. Mol. Biol.* **148,** 219–230.

Keene, M. A., and Elgin, S. C. R. (1981). Micrococcal nuclease as a probe of DNA sequence organization and chromatin structure. *Cell* **27,** 57–64.

Keene, M. A., Corces, V., Lowenhaupt, K., and Elgin, S. C. R. (1981). DNase I hypersensitive sites in *Drosophila* chromatin occur at the 5′ ends of regions of transcription. *Proc. Natl. Acad. Sci. U.S.A.* **78,** 143–146.

Kornberg, R. (1977). Structure of chromatin. *Annu. Rev. Biochem.* **46,** 931–954.

Krüger, C., and Benecke, B.-J. (1981). *In vitro* translation of *Drosophila* heat shock and non heat shock mRNAs in heterologous and homologous cell-free systems. *Cell* **23,** 595–603.

Kyte, J., and Doolittle, R. F. (1982). A simple method for displaying the hydropathic character of a protein. J. Mol. Biol. **157,** 105–132.

Landick, R., Vaughn, V., Lau, E. T., Van Bogelen, R. A., Erickson, J. W., and Neidhardt, F. C. (1984). Nucleotide sequence of the heat shock regulatory gene of *E. coli* suggests its protein product may be a transcription factor. *Cell,* **38,** 175–182.

Leigh-Brown, A. J., and Ish-Horowicz, D. (1981). Evolution of the 87A and 87C heat shock loci in *Drosophila. Nature* (*London*) **290,** 677–682.

Levinger, L., and Varshavsky, A. (1981). Heat shock proteins of *Drosophila* are associated with nuclease resistant, high salt resistant nuclear structures. *J. Cell Biol.* **90,** 793–796.

Lewis, M. J., Helmsing, P., and Ashburner, M. (1975). Parallel changes in putting activity and patterns of protein synthesis in salivary glands of *Drosophila*. *Proc. Natl. Acad. Sci. U.S.A.* **72,** 3604–3608.

Lindquist, S. L. (1980). Varying patterns of protein synthesis in *Drosophila* during heat shock: implications for regulation. *Dev. Biol.* **77,** 463–479.

Lindquist, S. L. (1981). Regulation of protein synthesis during heat shock. *Nature (London)* **293,** 311–314.

Lis, J. T., Prestidge, L., and Hogness, D. S. (1978). A novel arrangement of tandemly repeated genes at a major heat shock site in *Drosophila melanogaster*. *Cell* **14,** 901–919.

Lis, J. T., Neckameyer, W., Mirault, M.-E., Artavanis-Tsakonas, S., Lall, P., Martin, G., and Schedl, P. (1981). DNA sequences flanking the starts of the hsp70 and αβ heat shock genes are homologous. *Dev. Biol.* **83,** 291–300.

Livak, K. T., Freund, R., Schweber, M., Wensink, P. C., and Meselson, M. (1978). Sequence organization and transcription at two heat shock loci in *Drosophila*. *Proc. Natl. Acad. Sci. U.S.A.* **75,** 5613–5617.

McKenzie, S. L., and Meselson, M. (1977). Translation *in vitro* of *Drosophila* heat-shock messages. *J. Mol. Biol.* **117,** 279–283.

McKenzie, S. L., Henikoff, S., and Meselson, M. (1975). Localization of RNA from heat-induced polysomes at puff sites in *Drosophila melanogaster*. *Proc. Natl. Acad. Sci. U.S.A.* **72,** 1117–1121.

McKnight, S. L., Gavis, E. R., Kingsbury, R., and Axel, R. (1981). Analysis of transcriptional regulatory signals of the HSV thymidine kinase gene: Identification of an upstream control region. *Cell* **25,** 385–398.

Mason, P. J., Török, I., Kiss, I., Karch, F., and Udvardy, A. (1982). Evolutionary implications of a complex pattern of DNA sequence homology extending far upstream of the hsp 70 genes at loci 87A7 and 87C1 in *Drosophila melanogaster*. *J. Mol. Biol.* **156,** 21–36.

Mason, P. J., Hall, L. M. C., and Gausz, J. (1984). The expression of heat shock genes during normal development in *Drosophila melanogaster* (heat shock/abundant transcripts/developmental regulation). *Mol. Gen. Genet.* **194,** 73–78.

Mellon, P., Parker, V., Gluzman, Y., and Maniatis, T. (1981). Identification of DNA sequences required for transcription of the human α 1-globin gene in a new SV40 host-vector system. *Cell* **27,** 279–288.

Mirault, M.-E., Goldschmidt-Clermont, M., Moran, L., Arrigo, P. A., and Tissières, A. (1978). The effect of heat shock on gene expression in *Drosophila melanogaster*. *Cold Spring Harbor Symp. Quant. Biol.* **42,** 819–827.

Mirault, M.-E., Goldschmidt-Clermont, M., Artavanis-Tsakonas, S., and Schedl, P. (1979). Organization of the multiple genes for the 70,000 dalton heat shock protein in *Drosophila melanogaster*. *Proc. Natl. Acad. Sci. U.S.A.* **76,** 5254–5258.

Mirault, M.-E., Southgate, R., and Delwart, E. (1982). Regulation of heat shock genes: a DNA sequence upstream of *Drosophila* hsp70 genes is essential for their induction in monkey cells. *EMBO J.* **1,** 1279–1285.

Mitchell, H. K., and Lipps, L. S. (1975). Rapidly labeled proteins on the salivary gland chromosomes of *Drosophila melanogaster*. *Biochem. Genet.* **13,** 585–602.

Moran, L., Mirault, M.-E., Arrigo, P. A., Goldschmidt-Clermont, M., and Tissières, A. (1978). Heat shock of *Drosophila melanogaster* induces the synthesis of new messenger RNAs and proteins. *Philos. Trans. R. Soc. London Ser. B* **283,** 391–406.

Moran, L., Mirault, M.-E., Tissières, A., Lis, J., Schedl, P., Artavanis-Tsakonas, S., and Gehring, W. J. (1979). Physical map of two *Drosophila melanogaster* DNA segments containing sequences coding for the 70,000 dalton heat shock protein. *Cell* **17,** 1–8.

Neidhardt, F. C., and Van Bogelen, R. A. (1981). Positive regulatory gene for temperature-controlled proteins in *Escherichia coli*. *Biochem. Biophys. Res. Commun.* **100,** 894–900.

Neidhardt, F. C., Van Bogelen, R. A., and Lau, E. T. (1983). Molecular cloning and expression of a gene that controls the high temperature regulon of *Escherichia coli*. *J. Bacteriol.* **153,** 597–603.

O'Connor, D., and Lis, J. T. (1981). Two closely linked transcription units within 63B heat shock puff locus of *D. melanogaster* display strikingly different regulation. *Nucleic Acids Res.* **9,** 5075–5092.

Parker, C. S., and Topol, J. (1984a). A *Drosophila* RNA polymerase II transcription factor contains a promoter-region-specific DNA binding activity. *Cell* **36,** 357–369.

Parker, C. S., and Topol, J. (1984b). A *Drosophila* polymerase II transcription factor binds to the regulatory site of an hsp70 gene. *Cell* **37,** 273–283.

Pelham, H. R. B. (1982). A regulatory upstream promoter element in the *Drosophila* hsp70 heat-shock gene. *Cell* **30,** 517–528.

Pelham, H. R. B., and Bienz, M. (1982). A synthetic heat-shock promoter element confers heat inducibility on the *Herpes* simplex virus thymidine kinase gene. *EMBO J.*, **1,** 1473–1477.

Pelham, H. R. B., and Lewis, M. (1983). Assay of natural and synthetic heat shock promoters in monkey COS cells: requirements for regulation. *In* "Gene Expression: UCLA Symposia on Molecular and Cellular Biology" (D. Hamer and M. Rosenberg eds.), Vol. 8, pp. 75–86. Liss, New York.

Petersen, N., Moller, G., and Mitchell, H. (1979). Genetic mapping of the coding regions for three heat-shock proteins in *Drosophila melanogaster*. *Genetics* **92,** 891–902.

Ritossa, F. M. (1962). A new puffing pattern induced by heat shock and DNP in *Drosophila*. *Experientia* **18,** 571–575.

Ritossa, F. M. (1964a). Experimental activation of specific loci in polytene chromosomes of *Drosophila*. *Exp. Cell Res.* **35,** 601–607.

Ritossa, F. M. (1964b). Behaviour of RNA and DNA synthesis at the puff level in salivary gland chromosomes of *Drosophila*. *Exp. Cell. Res.* **36,** 515–523.

Rubin, G. M., and Spradling, A. C. (1982). Genetic transformation of *Drosophila* with transposable element vectors. *Science* **218,** 348–353.

Russnak, R. H., Jones, D., and Candido, E. P. M. (1983). Cloning and analysis of cDNA sequences coding for two 16 kilodalton heat shock proteins (hsps) in *Caenorhabditis elegans:* Homology with the small hsps in *Drosophila*. *Nucleic Acids Res.* **11,** 3187–3205.

Schedl, P., Artavanis-Tsakonas, S., Steward, R., Gehring, W. J., Mirault, M.-E., Goldschmidt-Clermont, M., Moran, L., and Tissières, A. (1978). Two hybrid plasmids with *Drosophila melanogaster* DNA sequences complementary to mRNA coding for the major heat shock protein. *Cell* **14,** 921–929.

Scott, M., and Pardue, M. L. (1981). Translational control in lysates of *Drosophila melanogaster* cells. *Proc. Natl. Acad. Sci. U.S.A.* **78,** 3353–3357.

Sirotkin, K., and Davidson, N. (1982). Developmentally regulated transcription from *Drosophila melanogaster* chromosomal site 67B. *Dev. Biol.* **89,** 196–210.

Snyder, M., Hunkapiller, M., Yuen, D., Silvert, D., Fristrom, J., and Davidson, N. (1982). Cuticle protein genes of *Drosophila:* Structure, organization and evolution of four clustered genes. *Cell* **29,** 1027–1040.

Southgate, R., Ayme, A., and Voellmy, R. (1983). Nucleotide sequence analysis of the *Drosophila* small heat shock gene cluster at locus 67B. *J. Mol. Biol.* **165,** 35–57.

Spradling, A., Pardue, M. L., and Penman, S. (1977). Messenger RNA in heat-shocked *Drosophila* cells. *J. Mol. Biol.* **109,** 559–587.

Tilly, K., McKittrick, N., Zylicz, M., and Georgopoulos, C. (1983). The *dnaK* protein modulates the heat shock response of *Escherichia coli*. *Cell* **34,** 641–646.

Tissières, A., Mitchell, H. K., and Tracy, U. M. (1974). Protein synthesis in salivary glands of *Drosophila melanogaster:* Relation to chromosome puffs. *J. Mol. Biol.* **84,** 389–398.

Török, I., and Karch, F. (1980). Nucleotide sequences of heat shock activated genes in *Drosophila melanogaster*. I. Sequences in the regions of the 5′ and 3′ ends of the hsp70 gene in the hybrid plasmid 56H8. *Nucleic Acids Res.* **8,** 3105–3123.

Török, I., Mason, P. J., Karch, F., Kiss, I., and Udvardy, A. (1982). Extensive regions of homology associated with heat-induced genes at loci 87A7 and 87C1 in *Drosophila melanogaster*. *In* "Heat Shock, from Bacteria to Man" (M. J. Schlesinger, M. Ashburner, and A. Tissières, eds.), pp. 19–25. Cold Spring Harbor Lab., Cold Spring Harbor, New York.

Udvardy, A., Sümegi, J., Csordas-Toth, E., Gausz, J., Gyurkovics, H., Schedl, P., and Ish-Horowicz, D. (1982). Genomic organization and functional analysis of a deletion variant of the 87A7 heat shock locus of *Drosophila melanogaster*. *J. Mol. Biol.* **155,** 267–280.

Van der Ouderaa, F. J., de Jong, W. W., Hilderink, A., and Bloemendal, H. (1974). The amino acid sequence of the α B_2 chain of bovine α-crystallin. *Eur. J. Biochem.* **49,** 157–168.

Velazquez, J. M., DiDomenico, B. J., and Lindquist, S. (1980). Intracellular localization of heat shock proteins in *Drosophila*. *Cell* **20,** 679–689.

Voellmy, R., and Rungger, D. (1982). Transcription of a *Drosophila* heat shock gene is heat-induced in *Xenopus* oocytes. *Proc. Natl. Acad. Sci. U.S.A.* **79,** 1776–1780.

Voellmy, R., Goldschmidt-Clermont, M., Southgate, R., Tissières, A., Levis, R., and Gehring, W. J. (1981). A DNA segment isolated from chromosomal site 67B in *Drosophila melanogaster* contains four closely linked heat shock genes. *Cell* **23,** 261–270.

Wadsworth, S., Craig, E. A., and McCarthy, B. J. (1980). Genes for three *Drosophila* heat-shock induced proteins at a single locus. *Proc. Natl. Acad. Sci. U.S.A.* **77,** 2134–2137.

Welch, W. J., and Feramisco, J. R. (1982). Purification of the major mammalian heat shock proteins. *J. Biol. Chem.* **257,** 14,949–14,959.

Welch, W. J., Garrels, J. I., and Feramisco, J. R. (1982). The mammalian stress proteins. *In* "Heat Shock, from Bacteria to Man" (M. J. Schlesinger, M. Ashburner, and A. Tissières, eds.), pp. 257–266. Cold Spring Harbor Lab., Cold Spring Harbor, New York.

Wu, C. (1984). Two protein-binding sites in chromatin implicated in the activation of heat shock genes. *Nature* (*London*) **309,** 229–234.

Yamamori, T., and Yura, T. (1980). Temperature-induced synthesis of specific proteins in *Escherichia coli:* evidence for transcriptional control. *J. Bacteriol.* **142,** 843–851.

Yamamori, T., and Yura, T. (1982). Genetic control of heat shock protein synthesis and its bearing on growth and thermal resistance in *Escherichia coli* K12. *Proc. Natl. Acad. Sci. U.S.A.* **79,** 860–864.

Yamamori, T., Ito, K., Nakamura, Y., and Yura, T. (1978). Transient regulation of protein synthesis in *Escherichia coli* upon shift up of growth temperature. *J. Bacteriol.* **134,** 1133–1140.

Zimmerman, J. L., Petri, W., and Meselson, M. (1983). Accumulation of a specific subset of *D. melanogaster* heat shock mRNAs in normal development without heat shock. *Cell* **32,** 1161–1170.

Zylicz, M., LeBowitz, J. H., McMacken, R., and Georgopoulos, C. (1983). The *dnaK* protein of *Escherichia coli* possesses an ATPase and autophosphorylating activity and is essential in an *in vitro* DNA replication system. *Proc. Natl. Acad. Sci. U.S.A.* **80,** 6431–6435.

2

Mechanism of Transcriptional Control during Heat Shock

J. JOSE BONNER

I. INTRODUCTION

When stressed by heat shock, cells of most organisms respond with the induction of specific protein synthesis. A few of the induced proteins are synthesized at detectable levels in unstressed cells, while most are not. What are the controls that operate to effect this induction, and what is its function? The function of the heat shock-induced proteins (hsp's) may be summarized briefly. It appears to be one of thermotolerance, or protection from lethality at high temperatures. This is suggested by the good correlation between treatments which induce the hsp's and treatments which confer thermotolerance, correlations which have now been made in a variety of biological systems (see, for example, Schlesinger *et al.*, 1982). Although the role which the hsp's play in thermotolerance is very poorly understood, the ubiquity of the phenomenon and the similarity of the hsp's in different organisms argue that the heat shock response has been of fundamental importance throughout much of evolution and that the biochemical mechanisms of the response may be nearly alike in unrelated organisms.

Changes in Eukaryotic Gene Expression
in Response to Environmental Stress

ISBN 0-12-066290-6

Studies of the mechanism of induction have centered primarily on the control of translation and the control of transcription. At least in *Drosophila,* both of these processes show a similar trend: in response to heat shock, the synthesis of normal RNA's and proteins decreases while production of the hsp's is maximized. The translational control is most dramatic since it seems to work via a translational block to which the hsp mRNA's are immune. This is discussed by Ballinger and Pardue in Chapter 3 of this volume. [One should note, however, that this type of translational control does not seem to be a feature of the heat shock response in yeast, an organism with relatively shorter mRNA lifetimes than those of *Drosophila* (Lindquist, 1981).]

It is the control of transcription during heat shock which is the subject of this chapter. Transcriptional controls have been observed in nearly all cell types which have been examined. Only two significant deviations from this generalization have been seen. One is the preblastula embryo, in which the induction of transcription of the heat shock genes appears to be incompatible with the rapid cleavage divisions (e.g., Dura, 1981; Roccheri *et al.*, 1981). The second is the *Xenopus* oocyte, which, although capable of increasing the transcription of heat shock-inducible genes, appears to be fully loaded with hsp mRNA. Using this store of preexisting hsp mRNA, even enucleated oocytes can activate the translational control mechanism to effect the synthesis of hsp's (Bienz and Gurdon, 1982). On the whole, however, the induction of specific gene transcription is an aspect of the response to heat shock which nearly all cells undergo. An understanding of the mechanism should be important, not only from the standpoint of the molecular biology of regulatory phenomena, but also with regard to the evolutionary variations of the mechanism in organisms with diverse metabolic needs.

II. PHENOMENOLOGY OF TRANSCRIPTIONAL CONTROL

That transcriptional activation of the genes encoding the hsp's of *Drosophila melanogaster* does, in fact, occur has been reviewed previously (Ashburner and Bonner, 1979) and will not be discussed in detail here. In essence, the argument rests on the very low abundance of hsp mRNA's in nonstressed cells compared to cells which have been heat shocked (e.g., Findly and Pederson, 1981), and the demonstration, by Spradling *et al.* (1975), McKenzie *et al.* (1975), and Bonner and Pardue (1976), that RNA complementary to the heat shock genes is efficiently synthesized only in shocked cells. While these experiments measured primarily RNA accumulation and not transcription per se, similar results have been obtained using short pulse-labeled mRNA's (Lengyel and Pardue, 1975) and RNA synthesized *in vitro* by elongation transcription in isolated nuclei

(Bonner and Kerby, 1982). The latter results confirm that regulation occurs at the level of transcription. Transcriptional activation is furthermore consistent with the failure to observe hsp synthesis in heat-shocked cells that have been poisoned with inhibitors of RNA synthesis, such as actinomycin D (Linquist, 1980). Since the publication of the several experiments cited above, numerous reports, utilizing various hybridization schemes to cloned DNA probes, have provided data consistent with transcriptional activation in organisms besides *Drosophila.* Although the majority of the data does not rule out nontranscriptional controls that are formally possible, there has been no evidence in any system for synthesis followed by rapid degradation of hsp mRNA's in nonstressed cells or for mechanisms similar to attenuation, as seen in prokaryotes.

In polytene cells of *Drosophila* (for example, the salivary gland), the unusual chromosome structure offers a number of cytological approaches for visualizing transcriptional activation due to heat shock. Because of their pertinence to the discussions which follow, these will be reviewed in some detail. It should be recalled that the induction of transcription of the heat shock genes was first discovered as a cytological phenomenon: heat shock induced the formation of puffs on the polytene chromosomes. The exact significance of puffing, however, remains unclear. It has long been suspected that puffing represents the ''opening up'' of the chromosome to make the DNA available to RNA polymerase. Indeed, several authors report the induction of puffs in salivary glands which have been poisoned, probably incompletely, with inhibitors of RNA synthesis (e.g., Berendes, 1968). Alternatively, it has been suggested that puffs may be the result of RNA storage near the chromosome, since full length transcripts can be isolated from microdissected puffs (Bisseling *et al.*, 1976). It is, furthermore, clear that many chromosome bands undergo transcription without puffing. For example, the histone genes which reside in region 39DE in *D. melanogaster* (Pardue *et al.*, 1977) transcribe efficiently in salivary glands yet fail to puff.

Although puff formation is a phenomenon which is poorly understood, there is, nonetheless, reasonably convincing evidence that puffing and transcription occur simultaneously at the major heat shock-inducible genes. For example, the puff sites at 87A and 87C, the locations of the genes encoding hsp70 (hsp of 70,000 daltons), exhibit very low levels of RNA synthesis in the nonpuffed condition while, after heat shock, the induced puffs are very active. Autoradiography, such as shown in Fig. 1, provides a measurement of chromosomal activity, though this is somewhat removed from a measurement of transcription itself due to the possibility of a variable efficiency of RNA transport away from the chromosome. The closest approximation to an autoradiographic demonstration of transcription utilizes isolated nuclei or nuclei in detergent-permeabilized salivary glands. Here, RNA processing efficiencies and transcription rates are low, allowing relatively short pulse times; again, puffs are considerably more active than the nonpuffed chromosomes (Bonner and Kerby, 1982).

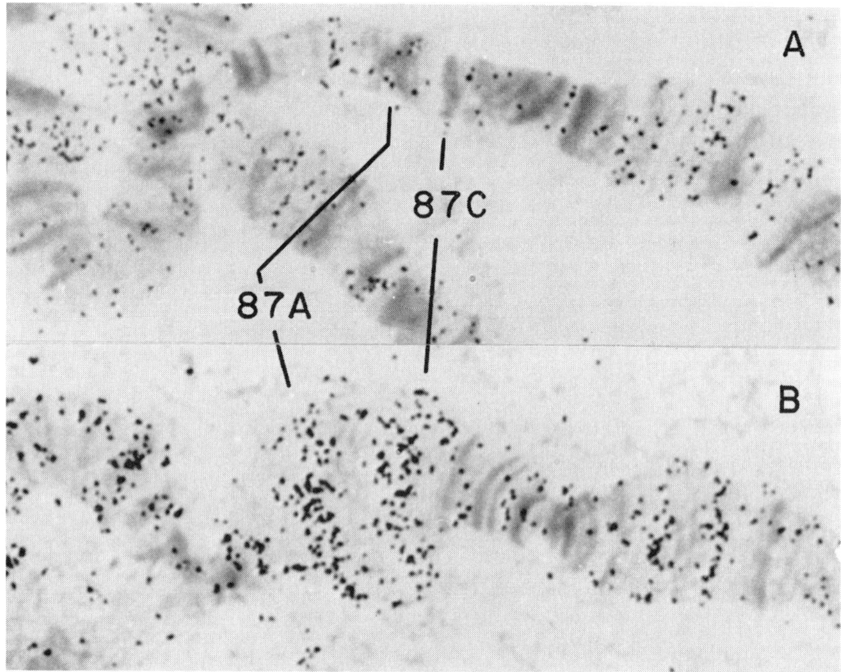

Fig. 1. Autoradiographic analysis of heat shock transcription. [^{3}H]Uridine incorporation by (A) nonshocked salivary glands and (B) salivary glands heat shocked at 37°C. Glands were incubated in 80% Grace's insect tissue culture medium, with 40 μCi/ml [^{3}H]uridine. Labeling was for 20 min at 22°C (A) or for the last 20 min of a 30-min 37°C incubation (B). Autoradiographic exposure (Kodak NTB-2 Film) was for 11 days (A) and 9 days (B).

Simultaneous with the changes in transcription pattern in *Drosophila* salivary glands is the redistribution of RNA polymerase II, the polymerase responsible for the transcription of mRNA's, and the heat shock mRNA's in particular (Bonner and Kerby, 1982). Immunochemical localization of RNA polymerase II on polytene chromosomes, shown in Fig. 2, indicates that this polymerase migrates to, and accumulates at, the puff sites. Indeed, this polymerase is lost from previously active puff sites after heat shock. The redistribution of RNA polymerase and the increased incorporation of RNA precursors both agree with activation at the level of transcription.

With regard to the "normal" genes whose transcription was in progress prior to the onset of heat shock, there have been somewhat conflicting reports. Tissières *et al.* (1974) provided autoradiographic evidence for the drastic curtailment of transcription of these normal mRNA's, consistent with RNA polymerase relocalization. In Fig. 1, however, such curtailment is much less pronounced. Furthermore, Mayrand and Pederson (1983) have observed that normal RNA's may be found in the hnRNA fraction of heat-shocked cells, thus arguing that RNA processing, rather than transcription, is primarily defective in heat-shocked cells. Interestingly, the majority of the mRNA's encoding hsp's lack intervening sequences and would thus be immune to a block in RNA splicing. It is most likely that these several observations can be reconciled by a simple model in which normal transcription occurs fairly efficiently during relatively mild heat shocks, whereas it is curtailed dramatically when the shock becomes "severe." Because the severity of the stress varies not only with temperature but with growth conditions and local cellular environment (e.g., batch of tissue culture medium; Lindquist, 1980), it is very likely that the 37°C-shock of Tissières *et al.* (1974) was "more severe" than the 37°C-shock used in preparing the autoradiograms of Fig. 1.

Any model of the molecular mechanism of transcriptional control must take into account not only the activation of the genes encoding the hsp's but also the diminution of transcription of "normal" genes. These transcriptional phenomena may be expressed more simply in terms of RNA polymerase II activity;

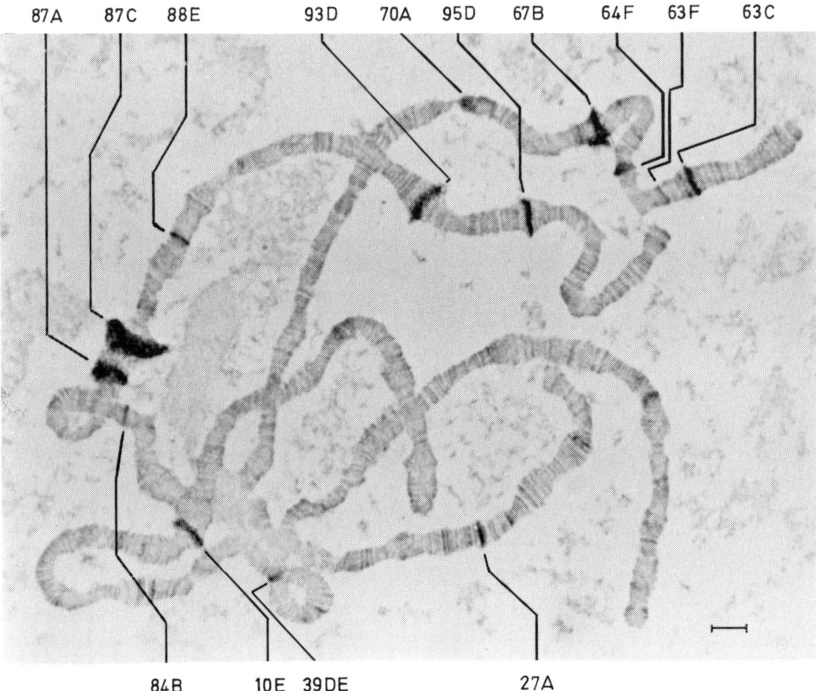

Fig. 2. RNA polymerase II localization after heat shock. Chromosomes from heat-shocked salivary glands were stained immunochemically using antibody to *Drosophila* RNA polymerase II and horseradish peroxidase. (From Bonner and Kerby, 1982.)

under normal conditions, this polymerase recognizes the promoters of a wide variety of genes and transcribes them efficiently, whereas, after the onset of stress, only a small number of genes is selected. Whether it is likely that "normal" genes are actively repressed remains to be seen. Perhaps they are not actively repressed; it is conceivable that the selectivity of RNA polymerase, after stress, is based solely on competition for a limited amount of polymerase, with the promoters of the active heat shock genes being the more effective competitors.

Thus, the more important goal in this regulatory system is to determine the mechanism of activation of heat shock gene transcription. Only when this is understood will we be able to assess the significance of additional phenomena, such as migration of RNA polymerase from previously active genes to the heat shock genes. With regard to the activation of the heat shock genes, several different models are equally likely on purely theoretical grounds. The genes for the hsp's could, for example, be under a state of repression, in which case heat shock would result in their derepression. Alternatively, they could be under positive control, with an inducer required for their activation. A third model, perhaps more attractive becuase of its ability to account for the redistribution of RNA polymerase described above, would be one of direct modification of RNA polymerase. A precedent for the last model exists in the modification of *Bacillus subtilis* RNA polymerase by replacement of its σ subunit by bacteriophage SP01 and during sporulation (Losick and Pero, 1981). The data presented in Section III are most consistent with the second alternative.

III. IDENTIFICATION OF REGULATORY MECHANISM

A number of different approaches have been used to attempt to elucidate the molecular mechanism of transcriptional regulation by heat shock. They may be roughly divided into three categories: examination of the DNA sequences responsible for gene activation, isolation of the protein factors involved, and analysis of chromatin structure. The third category is conceptually the most complex. Although paradigms may exist in other systems, most notably the control of yeast mating-type chromatin structure (Nasmyth, 1982), the mechanisms are not well understood. Interest in chromatin structure as a regulatory mechanism is based on the observation that active and inactive genes appear to adopt different conformations. The most clear-cut probe of these conformational differences has been DNase I, which digests the DNA of active-conformation genes much more efficiently than the DNA of inactive-conformation genes (Weintraub and Groudine, 1976). Superimposed on the overall sensitivity to digestion by DNase I is the extreme sensitivity of localized regions in the vicinity of the 5′ ends of the transcription units, first shown by Wu (1980). Because these

DNase I "hypersensitive sites" are especially accessible to DNase I, it is argued that they should be especially accessible to regulatory molecules *in vivo*. Both of these structural characteristics seem to be required for gene activity (see, for example, Groudine and Weintraub, 1981), but they are not necessarily sufficient without additional factors.

It would appear that the structural parameters of the heat shock genes change little in response to activation by stress. Like the chicken globin genes analyzed by Weintraub and Groudine (1976), which retain the active configuration after the repression of transcription during terminal differentiation, the heat shock-inducible genes appear to be in the active, DNase I-sensitive conformation in nonstressed cells (Wu *et al.*, 1979a). Similarly, the "hypersensitive sites" are present both before and after shock (Wu, 1980), unlike the situation with respect to the repressed versus active yeast mating-type cassettes (Nasmyth, 1982). There does appear to be some effect of heat shock on the overall structure—an increase in DNase I sensitivity and a lessening of the sharpness of the bands seen on electrophoresis of micrococcal nuclease-digested chromatin (Wu *et al.*, 1979b). The latter observation may be a reflection of the effect in the assay of the high density of RNA polymerase molecules on the very actively transcribed genes.

In an *in vitro* assay, however, Craine and Kornberg (1981) have detected a difference in chromatin structure in the vicinity of the hsp70 genes. They utilized as an assay the incubation of isolated nuclei in extracts from heat-shocked cells, followed by incubation of the repurified nuclei with bacterial RNA polymerase. RNA synthesized in this second incubation was analyzed by hybridization to cloned hsp70 DNA. They identified in the extracts from heat-shocked cells a protein component which rendered the DNA near the hsp70 gene relatively more actively transcribed by the bacterial polymerase compared to untreated nuclei. Because the bacterial polymerase does not recognize eukaryotic transcription signals, and because the experiments were performed using intact nuclei, the interpretation of the results is that the effect of the protein factor was to alter the conformation of the chromatin, thus making the DNA more accessible to the polymerase. Although Craine and Kornberg argue that such a conformational change would be likely to occur on transcriptional activation, the direct involvement of the factor in transcription remains to be shown.

In the final analysis, the basis of selection of the genes encoding the hsp's for efficient transcription during heat shock must reside in the DNA in or near these genes. There must be specific recognition sequences which target these genes for activation, whether the activation be achieved through chromatin conformational changes or through direct interaction with a regulatory molecule. Corces *et al.* (1981, 1982) and Pelham and colleagues (Pelham, 1982; Pelham and Beinz, 1982) have sought to identify this sequence for the hsp70 genes. By constructing modified genes which link the hsp70 promoter to the transcription units of other proteins, these groups have produced heat shock-inducible, easily identifiable genes. Two major results have come from these analyses. First, the *Drosophila* promoter sequence functions efficiently in mouse, monkey, or frog cells. After transformation or microinjection, transcription of the modified DNA's was readily induced by heat shock at the temperature to which the "host" cells normally respond. This remarkable observation argues that the factors involved in the regulation of heat shock transcription are well conserved among these different species.

The second major finding results from the analysis of various gene constructs in which greater and greater amounts of promoter sequence were deleted. It would appear that the critical DNA sequences, which render transcription inducible by stress, reside within the 60–70 nucleotides nearest the start site of transcription. Indeed, Pelham and Bienz (1982) have successfully synthesized an artificial "heat shock promoter" of similar size and sequence. DNA more distal than this appears to be dispensable for regulation by heat shock. Interestingly, the DNase I "hypersensitive site" identified by Wu (1980) is in, or very near, this same region (some 20–180 bases nearest the start site). This latter observation argues that the "hypersensitive site" represents a chromatin configuration which makes the critical regulatory DNA sequence accessible to regulatory molecules.

Given that there are DNA sequences which are required for activation of transcription, it may be presumed that there must be proteins which recognize those sequences. Wu (1984) has obtained evidence that this is likely to be so. Using a very elegant technique for mapping regions within DNase I "hypersensitive sites" of chromatin which are protected from exonuclease III digestion, he observed that nuclei from heat-shocked cells (and only heat-shocked cells) exhibit protection of the specific regulatory DNA sequence identified by Pelham (1982). This observation indicates the presence of a protein(s) binding to the regulatory sequence and suggests that the transcription of heat shock-inducible genes is under the positive control of this DNA binding protein.

Jack and colleagues (Jack *et al.* 1981; Jack and Gehring, 1982) were the first to search for such a regulatory protein directly, taking advantage of the fact that protein, but not double-stranded DNA, would bind to nitrocellulose filters. Using this assay, they isolated a protein from the nuclei of heat-shocked cells which bound specifically to sequences near the hsp70 gene. The same factor bound the DNA found near several other heat shock-inducible genes but not DNA of histone or alcohol dehydrogenase genes. This protein is a reasonable candidate for a regulatory factor. However, the binding sites vary in distance from the start sites of transcription, even for separate isolates of hsp70 genes (Jack and Gehring, 1982), and seem to be distal to the region identified by Pelham (1982) and Corces *et al.* (1982). Thus, its role in transcriptional activation remains to be proved.

The protein factors of Craine and Kornberg (1981) and Jack and colleagues (Jack *et al.* 1981; Jack and Gehring, 1982) were isolated using assays different from the activation of transcription. The activities which these proteins have been shown to possess are tantalizing. However, their true significance will not be known until they can be examined in an accurate *in vito* transcription system that can faithfully reproduce the events of *in vitro* transcriptional regulation.

IV. TRANSCRIPTIONAL INDUCTION *IN VITRO*

In our laboratory, we have sought to devise an *in vitro* transcription system and to apply it to this problem. In doing so, we have taken advantage of the cytological peculiarities of *Drosophila* salivary glands—the polytene chromosomes, in which induction of transcription of heat shock genes can be seen morphologically as puffs. It was learned relatively early (Compton and Bonner, 1978; Bonner, 1981) that salivary gland nuclei could be prepared under conditions in which they retained their activity. Such nuclei, when exposed to extracts from heat-shocked cells, responded with the induction of the normal set of heat shock puffs. Puffs were not induced when nuclei were exposed to nonstressed extracts. This system should provide a satisfactory assay for components of the transcriptional regulation mechanism and should be useful to assess the functional significance of protein factors identified by other indirect means.

The potential disadvantage of the *in vitro* puffing system for these kinds of analyses, however, is our limited understanding of the nature of the puff, reviewed earlier. It is, thus, of utmost importance to determine whether puff induction in isolated polytene nuclei represents induction of transcription or a chromatin conformational change which may be uncoupled from transcription. By a number of criteria, the events which accompany puff induction in isolated nuclei are indistinguishable from those occurring *in vivo*. For example, the morphological limits of the puffs are identical with those of *in vivo*-induced puffs, the puffs are quite active in RNA synthesis (shown autoradiographically by Comptom and McCarthy, 1978), and the migration of RNA polymerase II occurs as it does *in vivo*. The last phenomenon, demonstrated in Fig. 3, provides a critical test of RNA polymerase activity. Furthermore, none of these changes—including puff expansion itself—occurs under conditions in which transcription is abolished by low concentrations of α-amanitin. These cytologically assayable phenomena indicate that puffing *in vitro* is dependent on RNA polymerase II and argue that puffing is a consequence of, rather than a prerequisite for, RNA synthesis.

To be certain that puff induction *in vitro* is an accurate measure of transcriptional activation, however, it is necessary to demonstrate that the RNA tran-

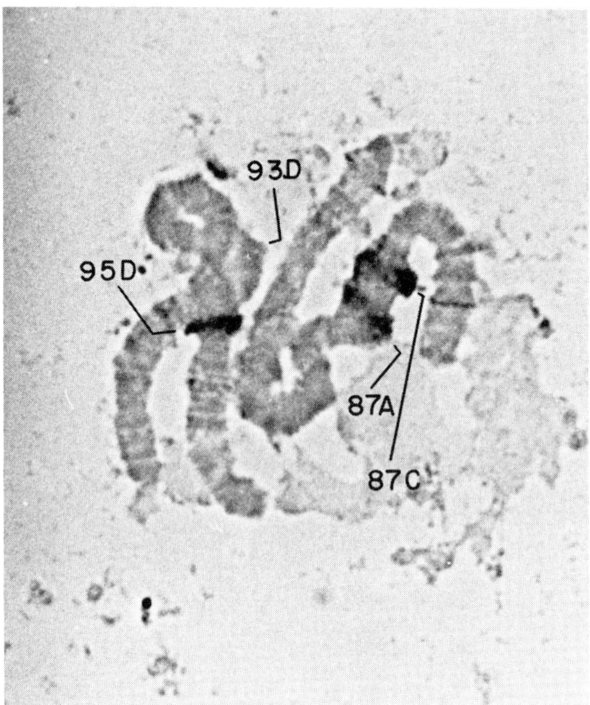

Fig. 3. *In vitro* induction of heat shock puffs results in RNA polymerase relocalization. Chromosome 3R from an *in vitro* incubation has been stained immunochemically for RNA polymerase II. The newly induced puffs are stained intensely. (From Bonner, 1981.)

scripts of the heat shock genes are, in fact, initiated in the isolated nuclei. This necessity is made more pressing by the observation that in many nuclear transcription systems, RNA polymerase II has shown little or no ability to reinitiate transcription once the elongation of *in vivo*-initiated transcripts has been completed. We have assayed initiation through the use of [γ-S]ATP, an analog of adenosine 5′-triphosphate in which one of the OH groups on the terminal phosphate has been replaced with an SH group. On incorporation of [γ-S]ATP (or [γ-S]GTP into RNA, the SH group will be lost if the analog is used during elongation, but it will be retained in the 5′-terminal position. Thus, RNA molecules initiated with [γ-S]ATP or [γ-S]GTP will be isolable by mercury–Sepharose chromatography. Table I provides the results of several experiments employing this nucleotide analog to investigate initiation by isolated polytene nuclei.

On average, about 1% of the RNA synthesized in these isolated nuclei appears to have been covalently linked to an SH group and thus retained on the mercury–Sepharose column. Given that salivary gland extracts are relatively rich in ribonuclease activity and that this activity is not completely removed during the nuclear isolation procedure (J. J. Bonner, unpublished observations), it is likely that the actual initiation rate is somewhat higher. It is furthermore possible that a significant fraction of the RNA's initiated *in vitro* were also capped *in vitro,* thereby eliminating the 5′-terminal SH group.

The two control experiments listed in Table I indicate that the vast majority of RNA isolated by virtue of its SH label does, in fact, represent initiated RNA's rather than spuriously labeled RNA molecules. We measured "thiophosphate transfer," or kinase activity, in two ways. First, we included labeled RNA in a nuclear transcription reaction containing nonradioactive nucleotides. From this reaction, 0.06% of the labeled RNA was retained on the column. This would indicate that only some 5% of the column-bound material in the normal reactions can be accounted for by transfer of the thiolated phosphate group to preexisting RNA molecules. In a second control experiment, we allowed nuclei to incorporate [^{32}P]UTP in the first half of the reaction in the absence of the thiolated analogs. During the second half of the reaction [^{3}H]CTP and thiolated ATP and GTP were added, along with an excess of unlabeled UTP. In this experiment, true initiation would result in 5′-terminal SH groups only on the RNA's synthe-

TABLE I

Initiation of Transcription in Isolated Nuclei[a]

Experiment	Unbound (cpm)	Bound (cpm)	% Bound
1	77,500	680	0.87
2	86,370	990	1.1
3	390,000	9000	2.3
4 (control)	264,000	160	0.06
5 (control)			
^{3}H	13,100	90	0.68
^{32}P	8,800	10	0.11

[a] Nuclei were incubated, under heat shock puff induction conditions, with the nucleotide analogs [γ-S]ATP and [γ-S]GTP replacing ATP and GTP and with [^{3}H]UTP and [^{3}H]CTP. RNA was isolated after 60 min and passed over mercury–Sepharose. After extensive washing, bound RNA was eluted with dithiothreitol. In experiment 4, labeled RNA was added at the onset of the incubation, and UTP and CTP were unlabeled. In experiment 5, [^{32}P]UTP was added at the beginning of incubation, with other nucleotides at half their normal levels. After 30 min [γ-S]ATP, [γ-S]GTP, and excess unlabeled UTP were added. In a parallel experiment, both ^{32}P and ^{3}H nucleotides were added at 30 min to estimate the amount of ^{32}P incorporated during the second half of the reaction. In the last reaction, the ratio of ^{3}H to ^{32}P was 3:15; a similar ratio in experiment 5 would indicate that 4158 cpm of ^{32}P incorporation occurred after addition of γ-S nucleotides. Of this, 0.68% = 28 cpm, enough to account for all of the ^{32}P retained on the column.

sized during the second half of the reaction. In this case, the percentage of ^3H retained on the column would be significantly greater than the percentage of ^{32}P. The amount of ^{32}P which would be retained by virtue of its incorporation during the second half of the reaction can be calculated from a parallel experiment in which [^{32}P]UTP was added during the second half of the reaction along with [^3H]CTP. This calculation indicates that all of the ^{32}P retained on the column can be accounted for by RNA synthesized in the second half of the reaction, in the presence of the thiolated nucleotides, and not by transfer of the thiophosphate to previously synthesized RNA. It is thus reasonable to conclude that the SH-labeled RNA, isolated by mercury–Sepharose chromatography, represents *in vitro*-initiated RNA.

In vitro-initiated RNA was prepared with internal ^3H label from CTP and UTP, isolated by mercury–Sepharose chromatography, and hybridized to poly-

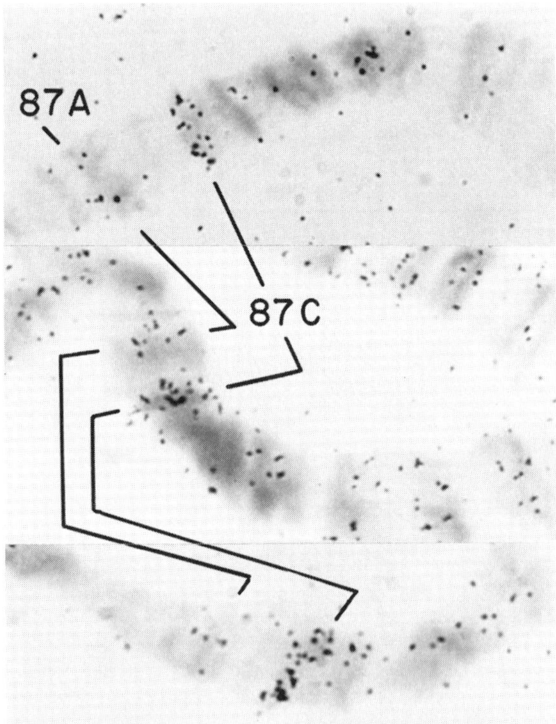

Fig. 4. *In vitro*-initiated RNA hybridized to polytene chromosomes *in situ*. RNA initiated with [γ-S]ATP or [γ-S]GTP (Table I, experiment 3) was isolated on mercury–Sepharose and hybridized *in situ* (procedure of Bonner and Kerby, 1982). Hybridization to 87A and 87C is indicated. Exposure time, 120 weeks.

tene chromosomes to determine whether the initiated RNA's are derived from the *in vitro*-induced heat shock puffs. As can be seen in Fig. 4, hybridization is clearly detectable at 87A and 87C, the locations of the genes encoding hsp70.

By all the criteria which we have measured, both cytological and biochemical as described earlier, the isolated nuclei perform in a manner which is very similar to the nuclei of whole cells. We have been unable to induce puffs without transcription, either by salt treatments (see, for example, Lezzi and Robert, 1972) or in the presence of RNA polymerase inhibitors. It therefore seems unlikely that *in vitro* puff induction is merely a morphological phenomenon independent of transcription; rather, it is an indication of the induction of transcription in a fully competent system.

V. NATURE OF INDUCER

Using as an assay the ability to induce puffs in isolated nuclei, we have sought to identify the components of the transcriptional regulation mechanism. We have partially purified, from *Drosophila* tissue culture cells, a heat-stable protein which mediates the induction of heat shock puffs when incubated with polytene nuclei under transcription conditions. In the absence of this factor, puff induction fails to occur. The ability of the *in vitro* puffing system to reproduce the events of transcriptional control seen *in vivo* suggests that this factor is a good candidate for a transcriptional regulator. Although not completely purified, we can estimate the size of the protein as 12,000–20,000 daltons, as this is the size range of the major proteins in the most highly purified preparation (Fig. 5). This small size is consistent with the heat stability, as the likelihood of regaining activity on renaturation of a polypeptide after heat denaturation decreases with increasing size and complexity of the molecule.

We have taken advantage of the heat stability of the puff-inducing factor to analyze the parameters affecting its activation in heat-shocked cells. Figure 6B demonstrates that the puff-inducing activity appears rapidly after the onset of the stress and remains at a more or less constant level for at least 100 min. Interestingly, the untreated extracts from cells shocked for 100 min failed to induce puffs while the same extracts were quite active in puff induction after heat treatment. This suggests that the puff-inducing activity in cells heat shocked for such long periods is masked by a heat-labile component. This observation is consistent with the kinetics of puffing during heat shock *in vivo;* the puffs are induced rapidly, remain active for some 50 min, and then regress (Ashburner and Bonner, 1979). Regression *in vivo* is blocked by inhibitors of protein synthesis (M. Ashburner, personal communication), suggesting that it may be a result of the activity of one of the heat shock-induced proteins. A number of reports have

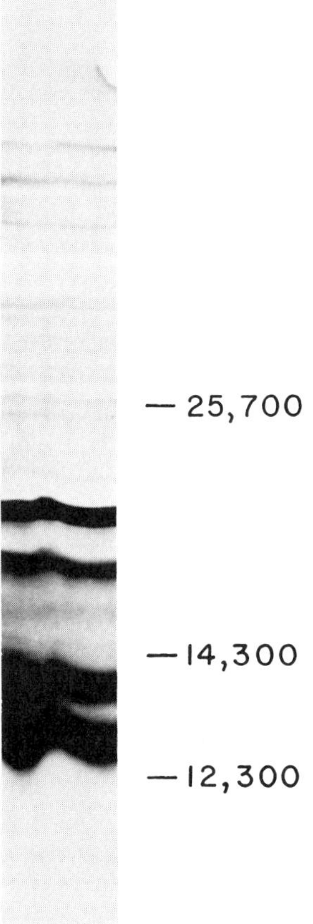

Fig. 5. Partially purified puff inducer preparation on 13.3% acrylamide–SDS gel, silver-stained. Puff-inducing fractions following thermal denaturation, hydroxylapatite chromatography, and DNA–agarose chromatography were isolated.

appeared suggesting that various hsp's play a role in feedback regulation of the transcriptional response (Bonner, 1982; DiDomenico *et al.*, 1982), but a detailed mechanism has by no means been determined.

The rapidity of appearance of the puff-inducing activity, as shown in Fig. 6B, is consistent with the rapid induction of the heat shock genes *in vivo*. This fact, and the observation that the puffs can be induced in salivary glands even after prolonged incubation in the presence of inhibitors of protein synthesis, argues that the puff-inducing protein preexists in cells which have not been heat shocked, although in an inactive form. If this were the case, it should be possible to activate the puff inducer *in vitro*, starting with extracts from normal, non-shocked cells. In doing so, it has proved inappropriate to incubate cellular extracts at elevated temperatures in an effort to mimic heat shock; the extracts coagulate. A satisfactory stress, which can be applied without thermal shock, however, is to be found in the biochemical response to oxygenation after periods of prior anoxia (van Breugal, 1966). This treatment results in the induction of the heat shock puffs in salivary glands. By analogy with the induction of puffs *in*

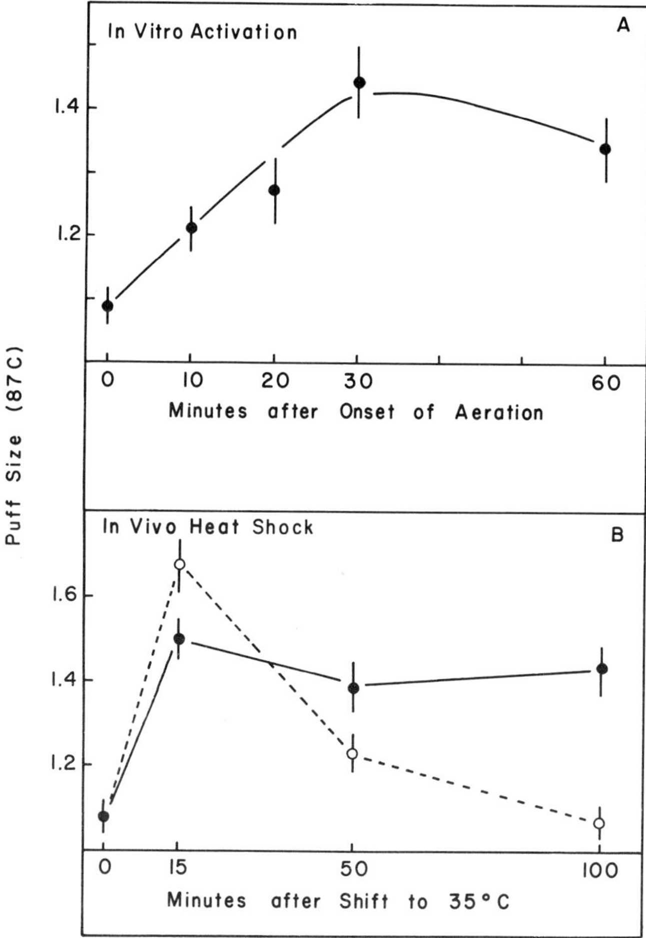

Fig. 6. Kinetics of activation and of production of the puff inducer. (A) A cytoplasmic extract from Kc cells was aerated for the times indicated, boiled 2 min, centrifuged to remove precipitated proteins, and tested for the ability to induce puffs in polytene nuclei, according to Bonner (1981). (B) Kc cells were heat shocked at 35°C for the times indicated and cytoplasmic extracts were prepared and tested for puff-inducing ability either directly (○) or after boiling as in panel A (●). The puff at 87C was measured relative to the reference band, 87F. (From Bonner, 1982.)

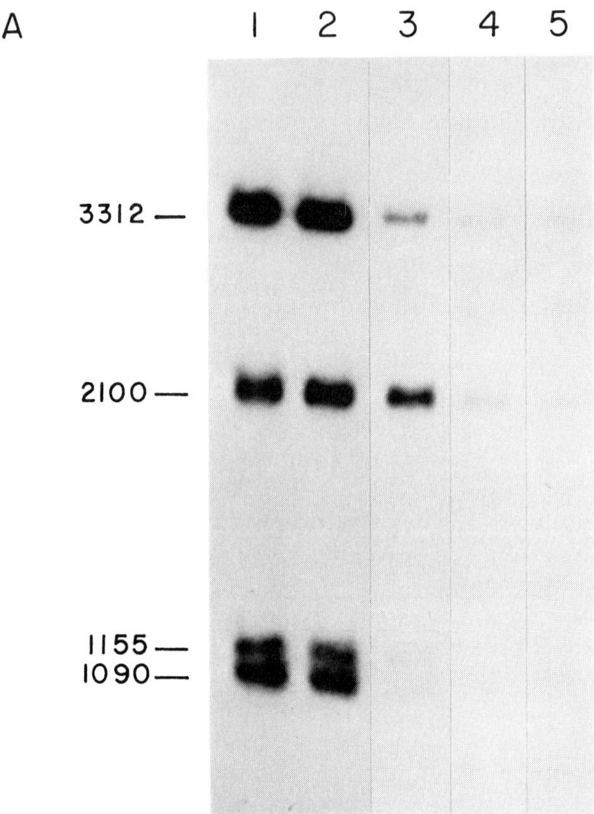

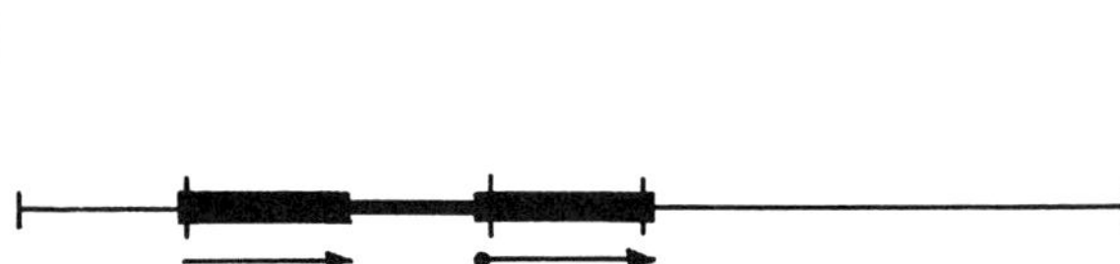

Fig. 7. Puff-inducing proteins bind DNA in a sequence-specific manner. (A) The cloned hsp70 gene B8 of Craig *et al.* (1979), shown in (B), was restricted with *Pst*I and labeled with T4 polymerase and [α-^{32}P]dATP. Labeled DNA was incubated with puff-inducing protein (Fig. 5) and passed over nitrocellulose, and the bound material eluted and run on a 1.2% agarose gel. Lanes: 1, DNA starting material; 2, DNA bound by undiluted protein sample; 3, DNA bound by 1:135 dilution; 4, DNA bound by 1:1215 dilution; 5, no protein control. The 2100-base fragment exhibits much more significant binding than do the other fragments. (B) The hsp70 clone used here is a *Bam*HI subclone from a tandem repeat of hsp70 genes. Thin lines, pBR322; boxes, hsp70 gene; arrows, direction of transcription of this "circularly permuted" gene; knob on right-hand arrow = 5′ end; cross hatches, *Pst*I sites.

vivo on recovery from anoxia, cytoplasmic extracts from normal cells—although unable to induce puffs in isolated polytene nuclei without further treatment—may be rendered capable of puff induction by aeration (Fig. 6A). The biochemistry of this phenomenon is poorly understood; it may reflect the relatively anoxic state of tissue culture cells grown in large volumes, or it may reflect abnormal biochemical events occurring in the cytoplasmic extracts. It is nonetheless clear that the prerequisite steps occur under these conditions to convert inactive puff inducer to its active state.

Once activated, what is the mechanism whereby the puff-inducing protein interacts with the chromosome to effect the induction of transcription? The results of Jack and colleagues (Jack *et al.*, 1981; Jack and Gehring, 1982) and Pelham (1982) would suggest that it might function through direct interaction with DNA in the vicinity of heat shock-inducible genes. This possiblity was investigated by the nitrocellulose filter-binding assay (following the procedures of Jack *et al.*, 1981). ^{32}P-labeled DNA fragments from a recombinant clone containing the hsp70 gene (clone B8 of Craig *et al.*, 1979) were allowed to interact with the most highly purified preparation of puff inducer available (the preparation shown in Fig. 5) and fractionated by filtration through nitrocellulose. Protein-bound DNA was eluted and analyzed by gel electrophoresis as shown in Fig. 7. At limiting dilutions of protein, one DNA fragment binds considerably more efficiently than the others—indicative of a relatively high affinity binding site for at least one of the proteins in the preparation. The fragment thus selected contains DNA sequences flanking the hsp70 gene and includes both the promoter region identified by Pelham (1982) and the binding site of the protein identified by Jack and Gehring (1982).

VI. MECHANISM OF TRANSCRIPTIONAL CONTROL

The experiments described in Section V are sufficient to propose a simple model of the basic events of the transcriptional control mechanism, shown schematically in Fig. 8. It is clear from the work of Pelham (1982) and Pelham and Bienz (1982) that the promoter of at least one of the heat shock genes is composed of a short stretch of DNA which resides a rather short distance from the start site of transcription. DNA more distal than this is unnecessary for inducibility by heat shock. This DNA sequence must be recognized by a regulatory protein whose activity is manifested only after the cell has been stressed. A good candidate for such a protein is the one present in the partially purified puff-inducing preparation described earlier. The important characteristics of the puff-inducing activity are that it appears rapidly on stress and it mediates the induction of puffs in isolated nuclei. The puff inducing activity correlates with the activity

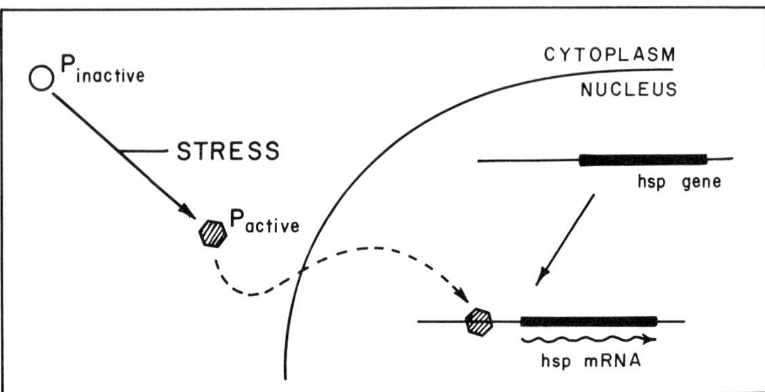

Fig. 8. A basic model of the events leading to induction of transcription of hsp genes. The cytoplasmic, inactive puff inducer is activated by stress, whereupon it binds the DNA of the hsp genes, activating transcription.

seen *in vivo* in that it is present in nonstressed cells but inactive. It is converted into its active state through a biochemical process which, *in vivo,* does not depend on protein synthesis and which may be reproduced *in vitro.*

There are several arguments, however, which suggest that the transcriptional induction mechanism is not as simple as suggested in Fig. 8. A very critical, yet undetermined, detail of the control mechanism outlined here is the relationship of the binding site of the puff inducing factor to the DNA sequence identified as a heat shock-specific promoter. At the time of this writing, it is not possible to rule out binding to a site more distal than the promoter element. This possibility is made likely by two considerations. First, there is remarkable similarity between the factor described here and the protein identified by Jack *et al.* (1981). Both are relatively low molecular weight, heat stable proteins. It is furthermore possible that the DNA binding and puff inducing activities described here reside in different proteins in the preparation. A second and more thought provoking consideration is presented by the very recent report of Parker and Topol (1984). They describe the isolation of a transcription factor which binds specifically and tightly to the heat shock promoter element and is required for transcription of isolated heat shock-gene DNA *in vitro.* The purification and stability characteristics of this factor are dramatically different from those of the puff inducing factor described above. The two factors may prove to be different, yet each appears to be specifically required for the induction of transcription of heat shock genes, one in whole nuclei and the other on purified DNA. It is conceivable that the nuclei used for the *in vitro* puffing studies contain an inactive form of the transcription factor identified by Parker and Topol (1984), and that this factor is rendered active by interaction with the heat stable factor described here. Such a model would not be unreasonable although it certainly requires a more complex view of the transcriptional mechanism than has heretofore been envisioned.

REFERENCES

Ashburner, M., and Bonner, J. J. (1979). The induction of gene activity in *Drosophila* by heat shock. *Cell* **17,** 241–254.

Berendes, H. D. (1968). Factors involved in the expression of gene activity. *Chromosoma* **24,** 418–437.

Bienz, M., and Gurdon, J. B. (1982). The heat shock response in *Xenopus* oocytes is controlled at the translational level. *Cell* **29,** 811–819.

Bisseling, T., Berendes, H. D., and Lubsen, N. H. (1976). RNA synthesis in puff 2-48BC after experimental induction in *Drosophila hydei*. *Cell* **8,** 299–304.

Bonner, J. J. (1981). Induction of *Drosophila* heat shock puffs in isolated polytene nuclei. *Dev. Biol.* **86,** 409–418.

Bonner, J. J. (1982). Regulation of the *Drosophila* heat shock response. *In* "Heat Shock, from Bacteria to Man" (M. J. Schlesinger, M. Ashburner, and A. Tissières, eds.), pp. 147–153. Cold Spring Harbor Lab., Cold Spring Harbor, New York.

Bonner, J. J., and Kerby, R. L. (1982). RNA polymerase II transcribes all of the heat shock induced genes of *Drosophila melanogaster*. *Chromosoma* **85,** 93–108.

Bonner, J. J., and Pardue, M. L. (1976). The effect of heat shock on RNA synthesis in *Drosophila* tissues. *Cell* **8,** 43–50.

Compton, J. L., and Bonner, J. J. (1978). An *in vitro* assay for the specific induction and regression of puffs in isolated polytene nuclei of *Drosophila melanogaster*. *Cold Spring Harbor Symp. Quant. Biol.* **42,** 835–838.

Compton, J. L., and McCarthy, B. J. (1978). Induction of the *Drosophila* heat shock response in isolated polytene nuclei. *Cell* **14,** 191–201.

Corces, V., Pellicer, A., Axel, R., and Meselson, M. (1981). Integration, transcription, and control of a *Drosophila* heat shock gene in mouse cells. *Proc. Natl. Acad. Sci. U.S.A.* **78,** 7038–7042.

Corces, V., Pellicer, A., Axel, R., Mei, S.-Y., and Meselson, M. (1982). Approximate location of sequences controlling transcription of a *Drosophila* heat shock gene. *In* "Heat Shock, from Bacteria to Man" (M. J. Schlesinger, M. Ashburner, and A. Tissières, eds.), pp. 27–34. Cold Spring Harbor Lab., Cold Spring Harbor, New York.

Craig, E. A., McCarthy, B. J., and Wadsworth, S. C. (1979). Sequence organization of two recombinant plasmids containing genes for the major heat shock-induced protein of *D. melanogaster*. *Cell* **16,** 575–588.

Craine, B. L., and Kornberg, T. (1981). Activation of the major *Drosophila* heat shock genes *in vitro*. *Cell* **25,** 671–682.

DiDomenico, B. J., Bugaisky, G. E., and Lindquist, S. (1982). The heat shock response is self-regulated at both the transcriptional and post-transcriptional levels. *Cell* **31,** 593–603.

Dura, J. M. (1981). Stage dependent synthesis of heat shock induced proteins in early embryos of *Drosophila melanogaster*. *Mol. Gen. Genet.* **184,** 381–385.

Findly, R. C., and Pederson, T. (1981). Regulated transcription of the genes for actin and heat shock proteins in cultured *Drosophila* cells. *J. Cell Biol.* **88,** 323–328.

Groudine, M., and Weintraub, H. (1981). Activation of globin genes during chicken development. *Cell* **24,** 393–402.

Jack, R. S., and Gehring, W. J. (1982). A *Drosophila* DNA-binding protein showing specificity for sequences close to heat shock genes. *In* "Heat Shock, from Bacteria to Man" (M. J. Schlesinger, M. Ashburner, and A. Tissières, eds.), pp. 155–160. Cold Spring Harbor Lab., Cold Spring Harbor, New York.

Jack, R. S., Gehring, W. J., and Brack, C. (1981). Protein component from *Drosophila* larval nuclei showing sequence specificity for a short region near a major heat shock protein gene. *Cell* **24,** 321–332.

Lengyel, J. A., and Pardue, M. L. (1975). Analysis of hnRNA made during heat shock in *Drosophila melanogaster* cultured cells. *J. Cell Biol.* **67,** 240a.

Lezzi, M., and Robert, M. (1972). Chromosomes isolated from unfixed salivary glands of *Chironomous. In* "Results and Problems in Cell Differentiation," Vol. 4: Developmental Studies on Giant Chromosomes (W. Beermann, ed.), pp. 35–57. Springer-Verlag, Berlin and New York.

Lindquist, S. (1980). Varying patterns of protein synthesis in *Drosophila* during heat shock: implications for regulation. *Dev. Biol.* **77,** 463–479.

Lindquist, S. (1981). Regulation of protein synthesis during heat shock. *Nature* (*London*) **293,** 311–314.

Losick, R., and Pero, J. (1981). Cascades of sigma factors. *Cell* **25,** 582–584.

McKenzie, S., Henikoff, S., and Meselson, M. (1975). Localization of RNA from heat-induced polysomes at puff sites in *D. melanogaster. Proc. Natl. Acad. Sci. U.S.A.* **72,** 1117–1121.

Mayrand, S., and Pederson, T. (1983). Heat shock alters nuclear ribonucleoprotein assembly in *Drosophila* cells. *Mol. Cell Biol.* **3,** 161–171.

Nasmyth, K. A. (1982). The regulation of yeast mating-type chromatin structure by SIR: an action at a distance affecting both transcription and transposition. *Cell* **30,** 567–578.

Pardue, M. L., Kedes, L. H., Weinberg, E. S., and Birnstiel, M. L. (1977). Localization of sequences coding for histone messenger RNA in the chromosomes of *Drosophila melanogaster. Chromosoma* **63,** 135–152.

Parker, C. S., and Topol, J. (1984). A *Drosophila* RNA polymerase II transcription factor binds to the regulatory site of an hsp70 gene. *Cell* **37,** 273–283.

Pelham, H. R. B. (1982). A regulatory upstream promoter element in the *Drosophila* hsp70 heat shock gene. *Cell* **30,** 517–528.

Pelham, H. R. B., and Bienz, M. (1982). A synthetic heat shock promoter element confers heat-inducibility on the Herpes simplex virus thymidine kinase gene. *EMBO J.* **1,** 1473–1477.

Roccheri, M. C., Di Bernardo, M. G., and Giudice, G. (1981). Synthesis of heat shock protein in developing sea urchins. *Dev. Biol.* **83,** 173–177.

Schlesinger, M. J., Ashburner, M., and Tissières, A., eds. (1982). "Heat Shock, from Bacteria to Man." Cold Spring Harbor Lab., Cold Spring Harbor, New York.

Spradling, A., Penman, S., and Pardue, M. L. (1975). Analysis of *Drosophila* mRNA by *in situ* hybridization: sequences transcribed in normal and heat shocked culture cells. *Cell* **4,** 395–404.

Tissières, A., Mitchell, H. K., and Tracy, U. M. (1974). Protein synthesis in salivary glands of *D. melanogaster*. Relation to chromosome puffs. *J. Mol. Biol.* **84,** 389–398.

van Breugal, F. M. A. (1966). Puff induction in larval salivary gland chromosomes of *D. hydei. Genetica* (*The Hague*) **37,** 17–28.

Weintraub, H., and Groudine, M. (1976). Chromosomal subunits in active genes have an altered conformation. *Science* **193,** 848–855.

Wu, C. (1980). The 5′ ends of *Drosophila* heat shock genes in chromatin are hypersensitive to DNase I. *Nature* (*London*) **286,** 854–860.

Wu, C. (1984). Two protein-binding sites in chromatin implicated in the activation of heat shock genes. *Nature* (*London*) **309,** 229–235.

Wu, C., Bingham, P. M., Livak, K. J., Holmgren, R., and Elgin, S. C. R. (1979a). The chromatin structure of specific genes: I. Evidence for higher order domains of defined DNA sequences. *Cell* **16,** 797–806.

Wu, C., Wong, Y.-C., and Elgin, S. C. R. (1979b). The chromatin structure of specific genes: II. Disruption of chromatin structure during gene activity. *Cell* **16,** 807–814.

3

Mechanism of Translational Control in Heat-Shocked *Drosophila* Cells

DENNIS BALLINGER AND MARY LOU PARDUE

I. INTRODUCTION

The very limited set of polypeptides that is synthesized during heat shock (Tissières *et al.*, 1974; Lewis, *et al.*, 1975) is the result of changes at several levels in the metabolism of the stressed cell (reviewed in Ashburner and Bonner, 1979; Schlesinger *et al.*, 1982). During heat shock there is a large decrease in the number of species of RNA molecules that are transcribed, processed, and transported to the cytoplasm (Spradling *et al.*, 1975, 1977). In addition, there is a corresponding decrease in the number of cytoplasmic RNA species that are selected for translation (McKenzie *et al.*, 1975; McKenzie, 1976). This translational selection is not seen under all conditions that induce the small set of heat shock proteins (hsp's). For example, at 33°C *Drosophila* cells produce both the hsp's and the proteins made at control temperatures (25° proteins), while at 36°C only the hsp's are efficiently translated (Fig. 1). However, under more extreme conditions of stress, translational discrimination of specific mRNA's has been

Changes in Eukaryotic Gene Expression
in Response to Environmental Stress

ISBN 0-12-066290-6

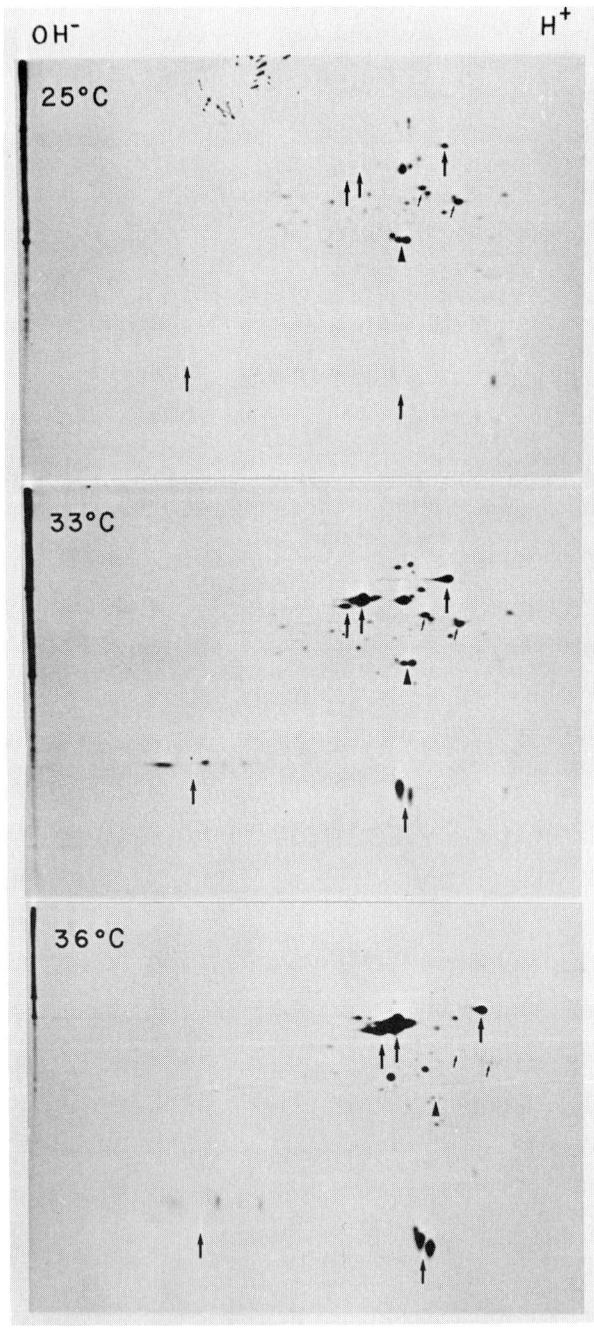
OH⁻
H⁺
25°C
33°C
36°C

observed in *Drosophila* (McKenzie, 1976), yeast (Lindquist, 1981), and mammalian cells (Thomas *et al.*, 1982). Thus, such control seems likely to be a general feature of the stress response in eukaryotic cells. The undertranslated mRNA's (25° mRNA's) are sequestered but still translatable in *Drosophila* and mammalian cells (McKenzie, 1976; Mirault *et al.*, 1978; Storti *et al.*, 1980; Peterson and Mitchell, 1981; Thomas *et al.*, 1982). In contrast, the preexisting mRNA's are degraded during heat shock in yeast (Lindquist, 1981). This review will focus primarily on *Drosophila* cells.

The evolutionary conservation of the heat shock response in bacteria, animals, and plants strongly suggests that this response has one or more functions important to cell survival. We do not yet understand how this response aids cell survival, but what is known about its biology indicates that it may be a homeostatic mechanism for coping with environmental stress. There are at least two ways that translational control might be useful as a component of a response for coping with unfavorable environmental conditions. First, translational discrimination allows the cell to concentrate its resources on production of hsp's very rapidly. The normal lifetime of much of the mRNA in animal cells is many hours. Unless sequestered or specifically degraded, the 25° mRNA's present in the cytoplasm at the time of heat shock could direct a significant part of the cell's protein synthesis for several hours. In addition, the sequestering of preexisting mRNA's allows cells in a developing organism to preserve the population of mRNA species that exists before heat shock. For example, many cells in embryos have populations of maternally encoded mRNA's that were produced during the complex process of oogenesis. It is unlikely that an embryonic cell would be able to continue development if a heat shock were to destroy these maternal mRNA's.

Although the heat shock response appears to protect against major perturbations in development, it can cause minor alterations in the developmental pro-

---

Fig. 1. Two-dimensional separation of proteins labeled with [^{35}S]methionine in *Drosophila* cells incubated at 25, 33, and 36°C. Schneider 2-L cells were grown as described in Ballinger and Pardue (1983). Equal aliquots of cells were incubated for 45 min at the indicated temperature. Lyophilized [^{35}S]methionine was added to a final concentration of 200 μCi/ml and the cells were incubated for an additional 30 min. The cells were processed for electrophoresis essentially as described by Garrels (1979). Equal aliquots of each culture (representing about 2×10^6 cells) were separated by isoelectric focusing in tube gels (O'Farrell, 1975), followed by separation on polyacrylamide slab gels containing SDS (described in Ballinger and Pardue, 1983). The slab gels were treated with Enhance (New England Nuclear) and dried. Autoradiographic exposure was carried out for 36 hr at −70°C. The arrows indicate the same position in each gel (some positions were confirmed by the mobility of partially purified proteins coelectrophoresed in the same gels as the labeled proteins). The large arrowhead indicates the position of two actin electrophoretic variants. The small arrows indicate the mobility of α- and the more acidic β-tubulin. The medium arrows indicate the positions of the major hsp's, in descending M_r: hsp83 (synthesized at all three temperatures); hsp70; hsp68; hsp27 and hsp26; and hsp23 and hsp22.

gram. Heat shocks at certain critical periods produce phenocopies, morphological abnormalities which resemble the phenotypes of known mutations (Mitchell and Lipps, 1978). It appears that these phenocopies are the results of differences in rates of recovery of precisely timed developmental processes (Peterson and Mitchell, 1982). The abnormalities are generally not severe, and flies carrying them are viable; however, phenocopies give an indication of the potential magnitude of developmental perturbations which might be produced by complete destruction of the cell's mRNA if replacement were not exact.

II. STUDIES OF TRANSLATIONAL CONTROL IN INTACT *DROSOPHILA* CELLS

Analysis of polypeptides synthesized by cells incubated at various temperatures (i.e., under varying degrees of heat stress) shows a sharp cutoff point for the synthesis of the 25° proteins. The exact temperature of this cutoff point may vary with cell line, growth conditions, and the rate at which the incubation temperature is increased. However, the cutoff temperature is very reproducible when these variables are held constant. With Schneider 2-L cells, under the conditions we use (see Ballinger and Pardue, 1983), the cutoff point is 34°C. Hsp's are synthesized at temperatures below 34°C, but synthesis of 25° proteins is also vigorous (e.g., 33°C in Fig. 1). Above 34°C abundant synthesis of hsp's continues, but the synthesis of 25° proteins is curtailed (e.g., 36°C in Fig. 1). At 36–37°C the 25° proteins are produced at 2–10% of the 25°C level (Lindquist, 1980a; D. Ballinger, unpublished observations).

At 36°C the preferential synthesis of the hsp's is due to translational selection of heat shock mRNA's from a pool which includes preexisting 25° mRNA's. The presence of the 25° mRNA's in heat-shocked cells was first inferred by McKenzie (1976), who showed that the 25° proteins are produced by heat-shocked cells which have been allowed to recover at 25°C in the presence of actinomycin D. Control experiments indicated that actinomycin D had inhibited transcription of new RNA during the recovery. Thus, the heat-shocked cells must have preserved the preexisting 25° mRNA's during heat shock, although the mRNA's were not translated.

In contrast to heat-shocked cells, which preferentially select heat shock mRNA's, control cells show no discrimination in favor of 25° mRNA's. After recovery at 25°C, cells actively translate both 25° mRNA and any remaining heat shock mRNA (McKenzie, 1976; Storti *et al.*, 1980). The lack of mRNA discrimination seen in control (25°) cells is also seen in cells after low levels of heat shock. For example, cells growing at 33°C translate both 25° and heat shock mRNA's efficiently (Fig. 1).

The production of hsp's or heat shock mRNA's does not seem to be necessary for induction of heat shock translational control. Cells that have been preincubated with actinomycin D and then heat shocked do not produce hsp's since most new mRNA transcription is blocked. Such cells stop translating the 25° mRNA present in the cytoplasm. However, if the cells are producing the small endogenous RNA virus, HPS-1, there is no decrease in synthesis of the viral polypeptides. The viral polypeptides are normally made in heat-shocked cells. Thus, the viral mRNA's appear to be translated as the heat shock mRNA's. The synthesis of viral polypeptides in actinomycin-treated heat-shocked cells shows that the cells are capable of translation and message discrimination even though they have no heat shock mRNA and can make no hsp's (Scott *et al.*, 1980).

Although they seem to play no role in the induction of translational control, the hsp's have been implicated in potentiating the recovery of 25° protein synthesis when cells are returned to control temperatures. When cultured cells or pupae are incubated at temperatures at which they do not synthesize heat shock proteins (39–41°C) and then are returned to 25°C, they do not recover 25° protein synthesis with the kinetics seen when they are allowed to produce functional hsp's (Lindquist, 1980a; Peterson and Mitchell, 1981). Peterson and Mitchell (1981) demonstrated that the production of functional hsp's had a direct effect on the protein synthetic machinery and no effect on the stability of 25° mRNA's. Further evidence for the role of hsp's in stabilizing the translational machinery comes from analog-substitution experiments (DiDomenico *et al.*, 1982). If cells are heat shocked in the presence of canavanine (an arginine analog which should interrupt the function of the hsp's produced) and then returned to 25°C, the cells never recover the ability to synthesize 25° proteins. The effect of canavanine is partially reversed by the addition of arginine to the cells during the recovery phase at 25°C. After a lag during which the cells produce non-analog-substituted hsp's, the cells recover the snythesis of 25° proteins (DiDomenico *et al.*, 1982). Under all of these regimes, functional hsp's are produced by the cells before 25° proteins are synthesized. A gradual shift in temperature allows for the synthesis of hsp's at temperatures above 39°C, whereas cells shifted directly to 39°C do not synthesize hsp's (Lindquist, 1980a; DiDomenico *et al.*, 1982). Cells subjected to gradual increases in temperature also recover 25° protein synthesis much more rapidly than shocked cells, again implicating the hsp's in the recovery of 25° protein synthesis.

There are at least two possible explanations for these data. The heat shock may alter the translational preference of the cell in a way that can only be reversed by the synthesis or accumulation of functional hsp's (DiDomenico *et al.*, 1982). Alternatively, increased temperature may damage the translational apparatus in such a way that it can no longer translate 25° mRNA's, and hsp's may be able to reverse the damage or substitute for the damaged component (Peterson and Mitchell, 1981).

One additional aspect of the mechanism of translational discrimination has come from *in vivo* studies. Lindquist (1981) has studied protein synthesis in cells which had been subjected to an initial 25-min incubation at 37°C and then returned to 25°C in the presence of actinomycin D, which prevented further RNA synthesis. When such cells were subjected to a series of subsequent heat shocks at 37°C followed by periods of recovery at 25°C, they synthesized only hsp's at 37°C but synthesized both hsp's and 25° proteins during the periods of recovery. These data argue against the possibility that only new mRNA's entering the cytoplasm during heat shock are translated, perhaps by virtue of their association with a regulatory molecule that becomes irreversibly dissociated during recovery at 25°C.

III. STUDIES USING CELL-FREE TRANSLATION SYSTEMS

The initial studies suggested that heat shock mRNA's might simply be translated more efficiently than 25° mRNA's and so be able to successfully compete with them for limiting components of the translational machinery. However, attempts to show competition between *Drosophila* heat shock and 25° mRNA's in cell-free translation systems made from rabbit reticulocytes or from wheat germ have been unsuccessful (Storti *et al.*, 1980; Lindquist, 1981; Krüger and Benecke, 1981). These two *in vitro* translation systems translate both classes of *Drosophila* mRNA's indiscriminately. In this respect, these *in vitro* systems resemble nonstressed cells.

It is possible to reproduce heat shock-translational selection in a cell-free system from cultured *Drosophila* cells (Storti *et al.*, 1980; Scott and Pardue, 1981; Krüger and Beneke, 1981). When lysates made from *Drosophila* cells which have been heat shocked at 36°C for 1 hr are used to translate mixtures of heat shock and 25° mRNA's, only the heat shock RNA's are translated. Similar lysates made from cells growing at 25°C will translate both 25° and heat shock mRNA's (Fig. 2). Such experiments show that heat shock induces a stable change (or changes) in the translational machinery of the cell. Although the change may be induced by high temperature, it is not immediately reversed by a decrease in temperature. Lysates made from both heat-shocked and 25° cells are optimally active at 28°C, and all of our *in vitro* translations are done at that temperature. Translational control in the lysate from heat-shocked cells is as complete at 28°C as it was in the intact cell at 36°C.

It would be very interesting to know whether translational control could be induced in these cell-free cytoplasmic systems. Unfortunately, we have not been able to answer this question. In our experiments to date, incubation at 36°C destroys the translational ability of both control and heat shock lysates. There are several possible explanations for this failure. For example, increased activity of enzymes released from cell compartments during preparation of the lysate may destroy essential translational components. In any case, we have not yet been able to use the *in vitro* system to study induction of translational control. It would also be interesting to know whether lysates from heat-shocked cells could recover the ability to synthesize 25° proteins after return to control temperatures. In intact cells, recovery of normal protein synthesis is much slower than induction of the heat shock response. Our lysates are maximally active only for about 30 min. This is less than the time needed for recovery by normal cells, making it impossible to use this lysate to study recovery *in vitro*.

The *in vitro* translation studies strongly suggest that the cell's translational machinery is selecting mRNA's on the basis of RNA sequence or conformation. The early experiments on intact cells had left open the possibility that each mRNA exported from the nucleus during heat shock carried a special protein tag to ensure its translation. However, the RNA that is added to the lysates for *in vitro* translation has been treated with protease and with phenol/chloroform or precipitated with guanidinium hydrochloride (Storti *et al.*, 1980; Scott and Pardue, 1981; Scott, 1980). Translational discrimination of this apparently protein-free RNA occurs during *in vitro* translation.

We do not yet know what feature of RNA sequence, conformation or modification, is being recognized by the translational machinery in the heat-shocked cells. Heat shock mRNA's appear to be normally polyadenylated (Spradling *et al.*, 1977) and have normal cap structures (Levis, 1977). Whatever the feature that is recognized, it is not unique to *Drosophila* heat shock mRNA's. As already noted, transcripts of the *Drosophila* virus HPS-1 are treated like heat shock mRNA's. We have also studied mRNA's of vesicular stomatitus virus (VSV), grown in Chinese hamster ovary cells (Scott and Pardue, 1981). Lysates from heat-shocked and control *Drosophila* cells translate mRNA's for the VSV polypeptides N and NS equally well, but the M polypeptide is undertranslated in lysates from heat-shocked cells. Thus, the N and NS mRNA's are treated like heat shock mRNA's. The M mRNA is recognized as a 25° mRNA.

Ultimately, it should be possible to use the *Drosophila* lysates to determine which component of the translational machinery is changed by heat shock. When lysates from heat-shocked and control cells are mixed prior to the addition of mRNA, the proteins synthesized appear to be an average of the proteins each lysate makes individually (Scott and Pardue, 1981). These data give no evidence of dominant, exchangeable inhibitors or activators of 25° mRNA translation in either lysate. In other experiments, micrococcal nuclease-treated lysates were fractionated by centrifugation to yield a crude ribosomal pellet and a supernatant fraction. These mRNA-free fractions were then used to supplement either the homologous or the heterologous lysate (Scott and Pardue, 1981). Neither of the supernatant fractions had an effect on the pattern of translation when it was added to the other lysate. However, the crude ribosomal pellet from 25° cells, when added to a heat shock lysate, produced a dramatic restoration of 25° protein synthesis. Crude ribosomes from heat shock lysates had no effect on the pattern of protein synthesis in either control or heat shock lysates (Fig. 2).

There is evidence that the effect of the crude ribosomal pellet in the restoration of 25° protein synthesis is not due to contamination by polysomes. First, the ribosomal pellet does not direct synthesis of polypeptides in cell-free systems made from rabbit reticulocytes. Such reticulocyte systems produce polypeptides very efficiently from bona fide *Drosophila* polysomes. Second, the ribosomal pellet very effectively restores synthesis of the VSV-M polypeptide in lysates of heat-shocked cells. The restoration of translation by the control (25°) ribosomal pellet strengthens our belief that the M mRNA is recognized with 25° mRNA's. The restoration of VSV polypeptide synthesis cannot be caused by contaminating polysomes since there was no VSV in the cells from which the crude ribosomal pellet was made.

The ability of 25° crude ribosomes to restore 25° protein synthesis in heat-shocked lysates indicates that these lysates lack a function necessary for translation of 25° mRNA's. The change in the crude ribosomal pellet that affects message discrimination has not yet been characterized. After a high-salt wash, the crude ribosomes from a control (25°) lysate become less effective in rescuing translation of 25° mRNA's in a heat shock lysate. Surprisingly, high-salt-washed ribosomes from heat shock lysates are able to restore the translation of 25° protein synthesis to approximately the same level as the salt-washed 25° ribosomes. Perhaps a ribosome component needed for translation of normal messages but not for translation of heat shock messages is inactivated in response to heat shock, and the inactivating agent is removed or reversed by high-salt treatment. Alternatively, the salt added with the salt-washed ribosomes may have a direct effect on translational discrimination *in vitro*. The effect of high salt on normal ribosomes appears to be paradoxical but may simply reflect artifactual damage to the ribosomes during the high-salt wash.

Since there appears to be no requirement for new RNA or protein synthesis in the induction of heat shock translational control (see Section II), it seems likely that the induction causes modification of existing components. One such modification is known to accompany heat shock *in vivo*. In *Drosophila* cells grown at 25°C and S6-like ribosomal protein is phosphorylated in about 40% of the ribosomes. Within 10 min of a shift to 36°C this protein is quantitatively dephosphorylated (Glover, 1982). It is possible that the presence of phosphorylated S6 is required for translation of 25° mRNA's but not for efficient translation of heat shock mRNA's. The state of ribosomal phosphorylation in cell-free lysates is not known, nor is it known what effect a high-salt wash would have on ribosomal phosphorylation or on the enzymes which regulate the level of this phosphorylation.

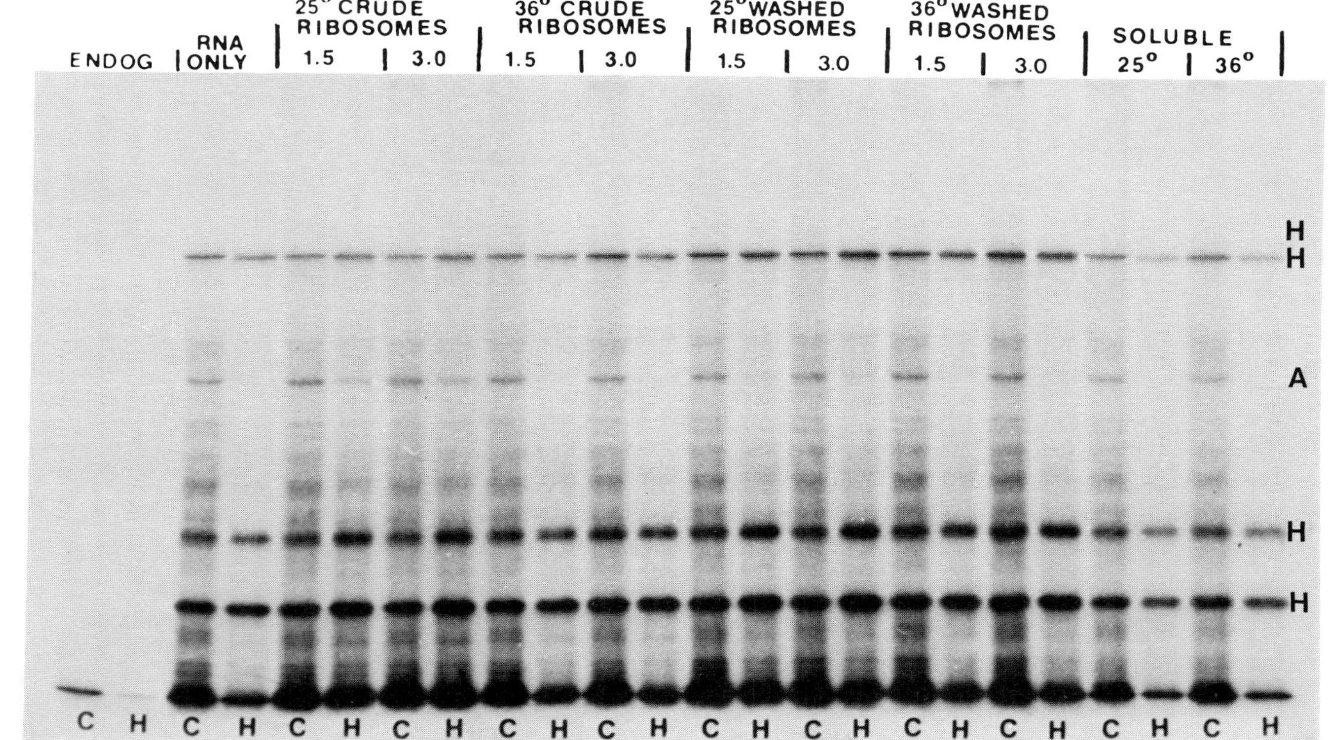

Fig. 2. Proteins synthesized *in vitro* by lysates of *Drosophila* cells incubated at 25°C (C) and incubated for 1 hr at 36°C (H): effects of supplements on translational control. The same 1:1 mixture of poly(A)$^+$ RNA from *Drosophila* embryos kept at 22°C or kept for 1 hr at 36°C was translated in all reactions. Reactions were supplemented with 1.5, 3.0, and 5.0 μl of lysate fractions prepared as described in Scott and Pardue (1981). Two microliters of the crude ribosomes would equal the amount already present in each reaction. The same amount of *in vitro* reaction mixture was loaded onto each lane of the 10% (w/v) polyacrylamide gel and autoradiographic exposure was for three days. ENDOG, synthesis without added RNA; RNA only, RNA without added supplements; H, heat shock lysate; C, control lysate. The migration of major heat shock proteins is indicated with H and actin with A. (From Scott and Pardue, 1981.)

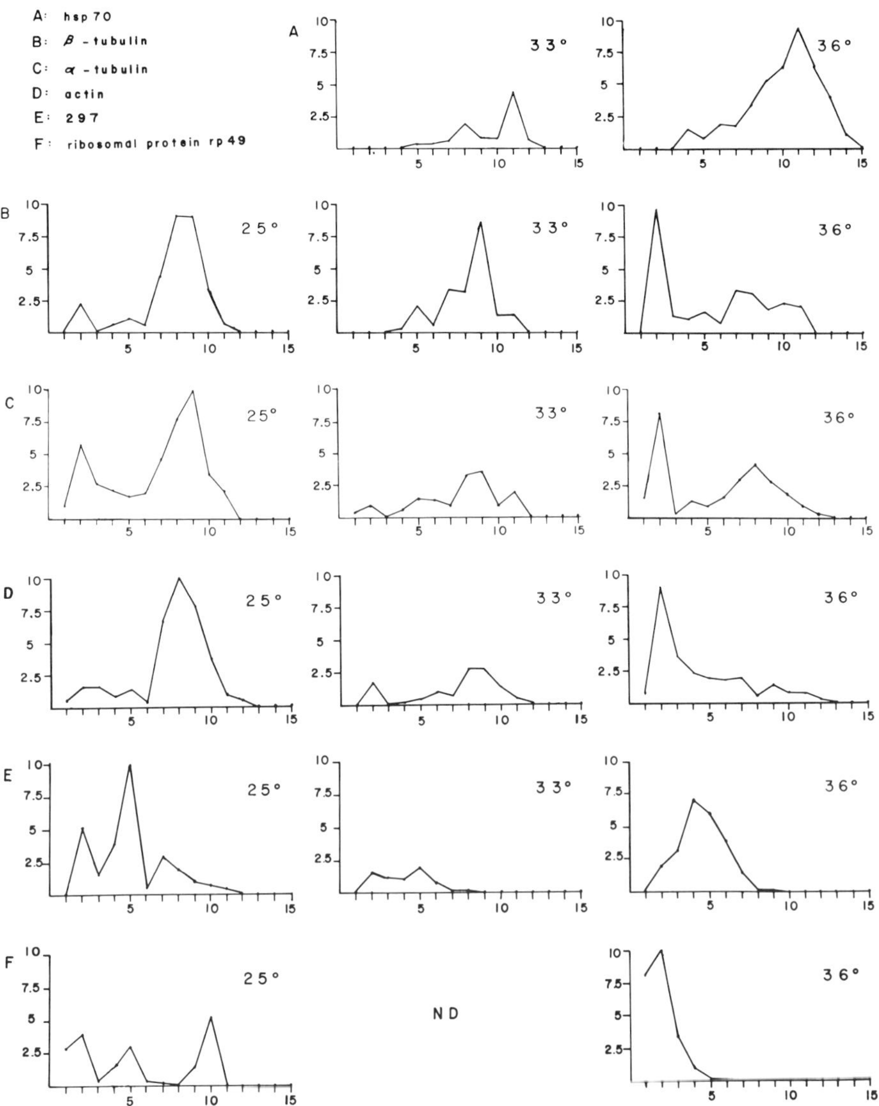

Fig. 3. The hybridization of specific cloned DNA's to RNA extracted from fractions of polysomal sucrose gradients. Cytoplasmic extracts of equal numbers of cells incubated at 25°C, or incubated for 1 hr at 33 or 36°C were analyzed on 0.5–1.5 *M* sucrose in a buffer containing 0.5 *M* KCl (Ballinger and Pardue, 1983). The optical density profiles of the 25 and 36°C samples are the same as those shown in Fig. 4 (0 and 60 min, respectively). The top of all gradients is to the left. Each graph shows the plot of the fraction number (abscissa) against the relative amount of the particular RNA (ordinate). RNA extracted from gradient fractions was denatured with glyoxal, separated on agarose gels, transferred to nitrocellulose, and hybridized to cloned DNA nick-trans-

IV. STUDIES DEFINING THE STEPS AT WHICH PROTEIN SYNTHESIS IS ALTERED IN HEAT-SHOCKED CELLS

There have been many studies characterizing protein synthesis during and after the heat shock response in *Drosophila* cells and tissues (see section II). However, few of these have focused directly on the mechanism of the translational repression of 25° mRNA's in heat-shocked cells. We have recently shown that the reduced levels of 25° protein synthesis during heat shock is due to a 15- to 30-fold reduction in the rates of both initiation and elongation of 25° mRNA's in *Drosophila* cells incubated at 36°C (Pardue *et al.*, 1981; Ballinger and Pardue, 1982, 1983). In contrast, the heat shock mRNA's are efficiently translated in these cells (Lindquist, 1980b).

The step of protein synthesis that is affected by a translational block can be assessed by observing the distribution of translationally inhibited messages between polysomes and RNP's. For instance, a block of initiation will completely release affected mRNA's from polysomes, whereas a reduction in the rate of initiation without a change in the rate of elongation will cause the average number of ribosomes associated with an affected message to decrease. A slowing of the rate of elongation without a change in the rate of initiation will increase the number of ribosomes on the message.

We have monitored the distribution between RNP's and polysomes of mRNA's in heat-shocked and control cells in several ways. We have prepared cytoplasmic extracts of cells incubated at 25°C or incubated for 1 hr at 33 or 36°C by detergent lysis. After removal of nuclei and mitochondria by centrifugation, the cytoplasmic extracts are adjusted to 0.5 *M* KCl and analyzed by velocity sedimentation through 0.5–1.5 *M* sucrose gradients or by equilibrium sedimentation in metrizamide density gradients. Total RNA is extracted from fractions of these gradients for further analysis. The distribution of mRNA's is assayed indirectly by analyzing the *in vitro* translation products encoded by gradient fraction RNA's or directly by hybridization of cloned DNA probes to nitrocellulose filter-bound RNA from the gradient fractions. Both of these methods have been used to detect only abundant mRNA's (we have not used any cloned DNA encoding rare messages). The hybridization data for six cloned DNA probes are shown in Fig. 3.

There are three populations of messages which can be identified by their polysomal distribution in control and heat-shocked cells. Class I mRNA's in-

---

lated in the presence of ^{32}P-labeled nucleotides. The density of hybridization to particular RNA's was measured from autoradiograms of two serial dilutions of each RNA and averaged to determine the relative amount of RNA in each fraction, in arbitrary units. (Reproduced with permission from Ballinger and Pardue, 1983. Copyright 1983 M.I.T. Press.)

clude all of the major heat shock mRNA's and are present and associated with polysomes only in cells incubated at elevated temperatures (e.g., hsp70, mRNA, Fig. 3A). One exception is hsp83 mRNA, which is found and translated at all temperatures studied (Fig. 1).

Class II mRNA's are found in both RNP's and polysomes of cells incubated at all three temperatures. This class includes most of the abundant mRNA's observed by translation of sucrose gradient fraction RNA's (Pardue *et al.*, 1981; Ballinger and Pardue, 1982, 1983), and contains the messages for α- and β-tubulin and for actin and the mRNA from the intermediate repeat family 297 (Fig. 3B–E). The proteins encoded by Class II mRNA's are not efficiently produced in cells grown at 36°C *in vivo* (see Fig. 1). Lindquist (1980a) has estimated that 25° proteins are produced at 2–10% of the 25°C levels in cells grown at 37°C. In spite of the reduced synthesis of these proteins, their mRNA's are found in substantial quantities in the polysomes of heat-shocked cells. Furthermore, the average number of ribosomes per message is the same at all three temperatures, indicating that the ratio of initiation to elongation is unchanged (Ballinger and Pardue, 1983; Fig. 3B–E).

An estimate of the potential output of protein from an RNA is given by the total number of ribosomes associated with the message and the output of protein per ribosome. For the Class II mRNA's mentioned earlier, there is a 2- to 3-fold reduction in the total number of ribosomes per message in cells grown at 36°C relative to cells grown at 25°C (Ballinger and Pardue, 1983). The output of protein per ribosome should increase by a factor of 6 with this increase in temperature (discussed in Ballinger and Pardue, 1983). Given the combined effect of the reduced number of ribosomes and the increased output of protein per ribosome, one would predict that the Class II mRNA's should produce 2- to 3-fold more protein in cells incubated at 36°C than in cells incubated at 25°C. There is, however, a 10- to 50-fold reduction in the synthesis of these proteins in cells incubated at 36–37°C relative to cells grown at 25°C (Lindquist, 1980a; D. Ballinger, unpublished observations). Thus, we observe at least a 20- to 30-fold reduction in the synthesis of these proteins relative to the expected synthesis at 36°C calculated in this way. The most likely explanation of these results is that the rates of both elongation and initiation are reduced 20- to 30-fold on Class II mRNA's in heat-shocked cells, relative to the efficiently translated heat shock messages in these cells.

Although class II mRNA's might be associated with proteins in particles that migrate fortuitously in the polysomal region of sucrose gradients, this possibility is very unlikely. The RNA's have the density characteristics of polysomes, as demonstrated by their sedimentation in metrizamide gradients (Ballinger and Pardue, 1983). Furthermore, 25° mRNA's can be directly demonstrated to be associated with ribosomes in heat-shocked cells. If cytoplasmic extracts of cells incubated at 36°C are added directly into the reticulocyte lysate translation system in the presence of a potent inhibitor of initiation (10^{-7} *M* edeine), then the proteins synthesized *in vitro* will consist entirely of runoff products from ribosomes which have initiated *in vivo*. In such an experiment substantial quantities of 25° proteins appear among the *in vitro* runoff products produced by extracts of heat-shocked cells (D. Ballinger, unpublished observations). This directly demonstrates the association of ribosomes which have correctly initiated and are synthesizing 25° proteins in heat-shocked cells.

The final class of mRNA's (Class III) is exemplified by the mRNA for a ribosomal protein (rp 49; Fig. 3F). Class III mRNA's are primarily associated with RNP's in cells incubated at 36°C, and their synthesis thus appears to be blocked specifically at the level of initiation in heat-shocked cells (Ballinger and Pardue, 1983). Ribosomal proteins in other organisms are known to feedback inhibit their own translation in cases where individual ribosomal proteins are overproduced (Dean *et al.*, 1981a,b). Therefore, the block of initiation on the ribosomal protein mRNA's in Class III may be a consequence of reduced processing of the rRNA precursor in heat-shocked cells (Lengyel and Pardue, 1975).

These studies suggest that the mechanism of translational inhibition of 25° mRNA's in heat-shocked cells involves at least two levels of control. One control is an altered rate of protein synthesis initiation, whereas the other is an altered rate of polypeptide elongation. For Class II messages, the reduction in the rate of initiation is associated with an equivalent reduction in the rate of elongation. For Class III messages, the reduced rate of initiation dominates any reduced rate of elongation. These data predict that heat-shocked cells must contain at least two classes of ribosome–mRNA complex, and perhaps two classes of ribosome, per se. One class consists of ribosomes from the polysomes which efficiently produce heat shock proteins, and the other consists of ribosomes which produce protein 20- to 30-times less efficiently and are associated with Class II 25° mRNA's in heat-shocked cells. Although lysates from heat-shocked *Drosophila* cells discriminate between heat shock and 25° mRNA's (see Section III), we do not yet know whether they faithfully reproduce all of these aspects of translational control.

Several laboratories have previously studied the mRNA's associated with polysomes in heat-shocked cells. McKenzie and Meselson (1977) failed to detect substantial quantities of 25° proteins among the *in vitro* translation products encoded by preparations of poly(A) containing mRNA from the polysomes of heat-shocked cells. However, the fact that most of the 25° mRNA's are unable to bind to oligo(dT)cellulose after 60 min of heat shock (Storti *et al.*, 1980) would preclude their detection in such an assay. The mRNA's which do not bind oligo(dT)cellulose can hybridize [^{3}H]oligo(U), due either to short poly(A) tails or to internal A-rich regions (Storti *et al.*, 1980). These data may account for the results of Biessman *et al.* (1978), who found that 50% of the oligo(dT)-primed cDNA copied from heat-shocked polysomal RNA was complementary to 25°

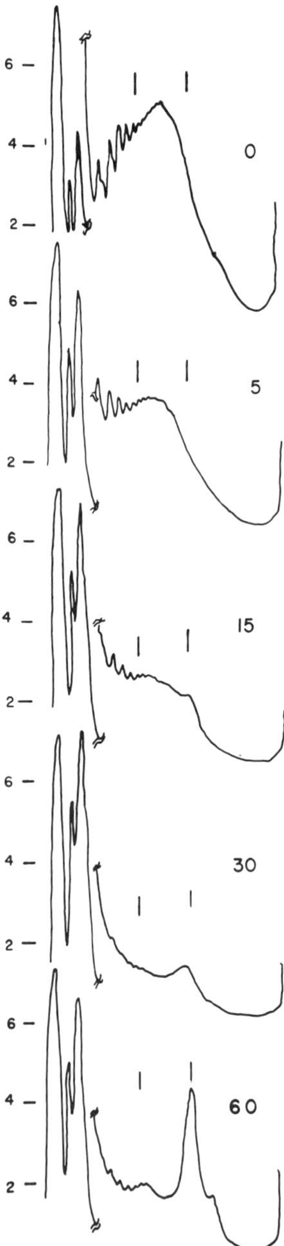

Fig. 4. Optical density profiles of sucrose gradient-separated cytoplasmic extracts of cells incubated at 25°C (zero) and incubated at 36°C for 5, 15, 30, and 60 min. Cells were grown and processed exactly as described in Ballinger and Pardue (1983). For heat shock, the cells were shifted to a water bath at 36°C, the culture temperature increased at an initial rate of at least 6.5°C/min and

mRNA. Finally, preexisting mRNA has been found by *in vitro* translation to be associated with the polysomes of heat-shocked *Drosophila* cells (Sondermeijer and Lubsen, 1978; Krüger and Benecke, 1981).

The association of 25° mRNA's with polysomes is difficult to reconcile with the observation that polysomes disappear completely within 10 min after a 37°C heat shock and then reform, with similar kinetics to the activation of hsp synthesis (McKenzie *et al.*, 1975). We have attempted to resolve this discrepancy by observing the polysomes of heat-shocked cells shortly after a shift to 36°C (cells were grown and processed exactly as described in Ballinger and Pardue, 1983). Under our conditions, the polysomes do not disappear completely; rather there appears to be a gradual shift in the polysomal profile from a unimodal distribution, characteristic of cells incubated at 25°C, to the bimodal distribution, characteristic of cells incubated at 36°C for 1 hr (Fig. 4). This gradual shift in the polysome profile is consistent with the idea that the cells must change state from one permissive for 25° protein synthesis to one which inefficiently translates these proteins. A similar gradual shift is observed in the spectrum of proteins labeled during short pulses after varying lengths of incubation at elevated temperatures. For instance, Lindquist (1980a) estimates that after a 40-min incubation at 37°C, 35% of the [^{3}H]leucine is incorporated into 25° proteins; this proportion drops to 10% after 60 min at 37°C. A gradual reduction in 25° protein synthesis is also seen during the first hour of heat shock at 36°C by Storti *et al.* (1980). It appears that polysome collapse, like other aspects of the heat shock response (Lindquist, 1980a), is sensitive to minor alterations in cell type and culture conditions. The cell line, growth conditions, and heat shock temperature that we use differ from those used in experiments in which polysome collapse was observed (McKenzie *et al.*, 1975). In this volume, Lindquist and Di-Domenico (see Chapter 4) observe the proteins pulse labeled at short times after shifts to 35° and 37°C. The time course of the decay of 25° protein synthesis is very dependent on heat shock temperature, even when all other variables are held constant. It is interesting, however, that 1 hr after temperature increase both the proteins made and the distribution of ribosomes in polysomes are similar under all these conditions. Thus, the observed differences may reflect complexities in the approach to the heat shock steady state.

V. SUMMARY

One aspect of the control of gene expression in heat-shocked *Drosophila* cells is preferential translation of heat shock mRNA's from a pool of cytoplasmic

---

reached 36°C within 4 min. Detergent extracts from approximately 5×10^7 cells were analyzed on 0.5–1.5 *M* sucrose gradients in buffer containing 0.5 *M* KCl. The slashes mark equivalent positions in each gradient, corresponding to the two major peaks of polysomes in cells incubated for 1 hr at 36°C. The tops of all gradients are to the left. The margin scales give relative optical density; there is a 4-fold decrease in full-scale absorbance at the breaks in the traces.

mRNA's that includes the preexisting mRNA's. This translational control makes possible the rapid diversion of cellular protein synthesis to the synthesis of hsp's. Preservation of preexisting mRNA's facilitates resumption of normal protein synthesis when the stress conditions are removed. A combination of *in vivo* and *in vitro* studies indicates that heat shock induces a change in the cellular translational machinery such that heat-shocked cells lack a function required for the translation of 25° mRNA's. This change is not dependent on the induced heat shock proteins. It involves either ribosomes or something closely associated with ribosomes and causes message selection on the basis of some feature of RNA sequence or conformation. The decreased rate of synthesis of many abundant 25° mRNA's in heat-shocked cells reflects significant reductions in both the rate of initiation of protein synthesis and the rate of peptide elongation. This suggests that heat-shocked cells lack at least two functions necessary for the translation of 25° mRNA's. One function is required for the efficient initiation of protein synthesis, and the other is required for rapid polypeptide elongation on a specific class of 25° mRNA's.

Acknowledgments

This work was supported by a grant from the National Institutes of Health. D. G. B. was a graduate fellow of the Whitaker Health Sciences Fund during these studies.

REFERENCES

Ashburner, M., and Bonner, J. J. (1979). The induction of gene activity in *Drosophila* by heat shock. *Cell* **17,** 247–254.

Ballinger, D. G., and Pardue, M. L. (1982). The subcellular localization of messenger RNAs in heat shocked *Drosophila* cells. *In* "Heat Shock, from Bacteria to Man" (M. J. Schlesinger, M. Ashburner, and A. Tissières, eds.), pp. 183–190. Cold Spring Harbor Lab., Cold Spring Harbor, New York.

Ballinger, D. G., and Pardue, M. L. (1983). The control of protein synthesis during heat shock in *Drosophila* cells involves altered polypeptide elongation rates. *Cell* **33,** 103–114.

Biessman, H., Levy, W. B., and McCarthy, B. J. (1978). *In vitro* transcription of heat shock-specific RNA from chromatin of *Drosophila melanogaster* cells. *Proc. Natl. Acad. Sci. U.S.A.* **75,** 759–763.

Dean, D., Yates, J. L., and Nomura, M. (1981a). *Escherichia coli* ribosomal protein S8 feedback regulates part of the *spc* operon. *Nature (London)* **289,** 89–91.

Dean, D., Yates, J. L., and Nomura, M. (1981b). Identification of ribosomal protein S7 as a repressor of translation within the *str* operon of *Escherichia coli*. *Cell* **24,** 413–419.

DiDomenico, B. J., Bugaisky, G. E., and Lindquist, S. (1982). The heat shock response is self-regulated at both the transcriptional and post-transcriptional levels. *Cell* **31,** 593–603.

Garrels, J. I. (1979). Two-dimensional gel electrophoresis and computer analysis of proteins synthesized by clonal cell lines. *J. Biol. Chem.* **254,** 7961–7977.

Glover, C. V. C. (1982). Heat shock induces rapid dephosphorylation of a ribosomal protein in *Drosophila*. *Proc. Natl. Acad. Sci. U.S.A.* **79,** 1781–1782.

Krüger, C., and Benecke, B.-J. (1981). *In vitro* translation of *Drosophila* heat shock and non-heat shock mRNAs in heterologous and homologous cell-free systems. *Cell* **23,** 595–603.

Lengyel, J. A., and Pardue, M. L. (1975). Analysis of hnRNA made during heat shock in *Drosophila melanogaster* cultured cells. *J. Cell Biol.* **67,** 240a.

Levis, R. W. (1977). The nuclear RNA of cultured *Drosophila* cells. Ph.D. thesis, M. I. T., Cambridge, Massachusetts.

Lewis, M., Helmsing, P. J., and Ashburner, M. (1975). Parallel changes in puffing activity and patterns of protein synthesis in salivary glands of *Drosophila. Proc. Natl. Acad. Sci. U.S.A.* **72,** 3604–3608.

Lindquist, S. (1980a). Varying patterns of protein synthesis in *Drosophila* during heat shock: implications for regulation. *Dev. Biol.* **77,** 463–479.

Lindquist, S. (1980b). Translational efficiency of heat-induced messages in *Drosophila melanogaster* cells. *J. Mol. Biol.* **137,** 151–158.

Lindquist, S. (1981) Regulation of protein synthesis during heat shock. *Nature* (*London*) **293,** 311–314.

McKenzie, S. L. (1976). Protein and RNA synthesis induced by heat treatment in *Drosophila melaongaster* tissue culture cells. Ph.D. thesis, Harvard University, Cambridge, Massachusetts.

McKenzie, S. L., and Meselson, M. (1977). Translation *in vitro* of *Drosophila* heat shock messages. *J. Mol. Biol.* **117,** 279–283.

McKenzie, S. L., Henikoff, S., and Meselson, M. (1975). Localization of RNA from heat-induced polysomes at puff sites in *Drosophila melanogaster. Proc. Natl. Acad. Sci. U.S.A.* **72,** 1117–1121.

Mirault, M.-E., Goldschmidt-Clermont, M., Moran, L., Arrigo, A. P., and Tissières, A. (1978). The effect of heat shock on gene expression in *Drosophila melanogaster. Cold Spring Harbor Symp. Quant. Biol.* **42,** 819–827.

Mitchell, H. K., and Lipps, L. S. (1978). Heat shock and phenocopy induction in *Drosophila. Cell* **15,** 907–918.

O'Farrell, P. H. (1975). High resolution two-dimensional electrophoresis of proteins. *J. Biol. Chem.* **250,** 4007–4021.

Pardue, M. L., Ballinger, D. G., and Scott, M. P. (1981). Expression of the heat shock genes in *Drosophila melanogaster. In* "Developmental Biology Using Purified Genes" (D. D. Brown, ed.), pp. 415–427. Academic Press, New York.

Peterson, N. S., and Mitchell, H. K. (1981). Recovery of protein synthesis after heat shock: prior heat treatment affects the ability of cells to translate mRNA. *Proc. Natl. Acad. Sci. U.S.A.* **78,** 1708–1711.

Peterson, N. S., and Mitchell, H. K. (1982). Effects of heat shock on gene expression during development: induction and prevention of the multihair phenocopy in *Drosophila. In* "Heat Shock, from Bacteria to Man" (M. J. Schlesinger, M. Ashburner, and A. Tissières, eds.), pp 345–352. Cold Spring Harbor Lab., Cold Spring Harbor, New York.

Schlesinger, M. J., Ashburner, M., and Tissières, A., eds. (1982). "Heat Shock, from Bacteria to Man." Cold Spring Harbor Lab., Cold Spring Harbor, New York.

Scott, M. P. (1980). Translational control of protein synthesis in *Drosophila*. Ph.D. thesis, M.I.T. Cambridge, Massachusetts.

Scott, M. P., and Pardue, M. L. (1981). Translational control in lysates of *Drosophila melanogaster* cells. *Proc. Natl. Acad. Sci. U.S.A.* **78,** 3353–3357.

Scott, M. P., Fostel, J. M., and Pardue, M. L. (1980). A new type of virus from cultured *Drosophila* cells: characterization and use in studies of the heat-shock response. *Cell* **22,** 929–941.

Sondermeijer, P. J. A., and Lubsen, N. H. (1978). Heat-shock peptides in *Drosophila hydei* and their synthesis *in vitro. Eur. J. Biochem.* **88,** 331–339.

Spradling, A., Penman, S., and Pardue, M. L. (1975). Analysis of *Drosophila* mRNA by *in situ* hybridization: sequences transcribed in normal and heat shocked cultured cells. *Cell* **4,** 395–404.

Spradling, A., Pardue, M. L., and Penman, S. (1977). Messenger RNA in heat shocked *Drosophila* cells. *J. Mol. Biol.* **109,** 559–587.

Storti, R. V., Scott, M. P., Rich, A., and Pardue, M. L. (1980). Translational control of protein synthesis in response to heat shock in *D. melanogaster* cells. *Cell* **22,** 825–834.

Thomas, G. P., Welch, W. J., Mathews, M. P., and Feramisco, J. R. (1982). Molecular and cellular effects of heat shock and related treatments on mammalian tissue culture cells. *Cold Spring Harb. Symp. Quant. Biol.* **46,** 985–996.

Tissières, A., Mitchell, H. K., and Tracy, U. M. (1974). Protein synthesis in salivary glands of *Drosophila melanogaster:* relation to chromosomal puffs. *J. Mol. Biol.* **84,** 389–398.

4

Coordinate and Noncoordinate Gene Expression during Heat Shock: A Model for Regulation

SUSAN LINDQUIST AND BETH DIDOMENICO

Changes in Eukaryotic Gene Expression
in Response to Environmental Stress

ISBN 0-12-066290-6

I. INTRODUCTION

Eukaryotic cells respond to mild heat treatments in a very similar fashion; they induce the synthesis of a small set of highly conserved proteins (see Cold Spring Harbor monograph 1982 for survey, Schlesinger *et al.*).

This so-called heat shock response was initially discovered in the fruit fly *Drosophila melanogaster* and has been studied more intensively in this organism than in any other. The independent findings of several laboratories fit together to provide a remarkable example of gene regulation in which controls acting at several different levels are coordinated to produce heat shock proteins (hsp's) at the fastest possible rate. We now have a clear impression of what kinds of control mechanisms are operating during heat shock; we are still in the process of determining how they work. In this chapter we focus on an issue which has been the source of some confusion in the heat shock literature, the fact that certain aspects of the response are coordinately regulated while others are not. We also draw together a variety of recent findings from our laboratory based on several different experimental approaches. Taken together with findings from other laboratories, we believe the data provide a strong argument for the existence of feedback control in the heat shock response and point to a model for post-transcriptional regulation that explains certain aspects of coordinate and noncoordinate regulation.

II. BASIC FEATURES OF THE HEAT SHOCK RESPONSE

In *Drosophila,* transcriptional activation of genetic loci can be observed directly under the light microscope by monitoring the formation of puffs on the giant polytene chromosomes of certain larval tissues. It was through the observation of such puffs that the heat shock response was first discovered (Ritossa, 1962). Exposure to elevated temperatures results in the simultaneous appearance of new puffs at a few unlinked chromosomal sites and the regression of most preexisting puffs (see Ashburner and Bonner, 1979, for review). Analogous changes occur in the specificity of translation; a rapid increase in the synthesis of heat shock proteins is accompanied by a sharp decrease in the synthesis of other cellular proteins. With extremely few exceptions, the activation of heat shock gene expression and the repression of preexisting gene expression are independent of developmental stage or tissue type (Tissières *et al.,* 1974; Lewis *et al.*, 1975; McKenzie *et al.*, 1975; Chomyn *et al.*, 1979; Mirault *et al.*, 1978).

As a result of these types of observations, the heat shock response has often been described as a coordinate system of gene regulation, and in this very general and important sense it is. However, more detailed investigations have revealed that certain aspects of the response are not tightly coordinated. Drawing such distinctions is more than a sterile descriptive exercise. As we shall show, it leads to important insights into the nature of gene regulation in this system. Before delineating the distinctions between coordinate and noncoordinate aspects of regulation, it seems most appropriate to present a brief overview of the basic features of regulation in the heat shock response of *Drosophila*.

III. MAJOR CONTROL POINTS OF HEAT SHOCK GENE REGULATION

The induction of hsp's in *Drosophila* is accomplished by mechanisms which act at three major control points: RNA synthesis, processing, and translation. The net effect is to ensure that hsp's will be produced at the maximum possible rate, sacrificing the synthesis of normal cellular proteins and their messenger RNA's to that end.

A. Transcription

When *Drosophila* cells are shifted from their normal growing temperature of 25°C to 36 or 37°C a change in transcription can be detected almost immediately. In salivary glands [^{3}H]uridine incorporation is observed at heat shock (hs) puff sites within 1–3 min (Ritossa, 1964; Ellgaard and Clever, 1971; Belyaeva and Zimulev, 1976; Bonner and Pardue, 1976); in tissue culture cells the first transcripts for hsp70, the major heat-induced protein, are completed within 4 min (Lindquist, 1980a). Over the next 60–90 min at 37°C hs mRNA's are continuously synthesized and accumulate in a rapid and linear fashion. Under optimal conditions, the message for hsp70 reaches a concentration of 10,000–15,000 molecules per haploid genome within the first hour of heat treatment (Lindquist, 1980b).

As demonstrated by a variety of techniques, the induction of hs synthesis is accompanied by the repression of preexisting transcription. At a morphological level, preexisting polytene chromosome puffs are markedly reduced in size after 40 min of heat treatment (Ashburner, 1970). RNA polymerase II (visualized by immunofluorescent staining) gradually disappears from these puff sites and after 45 min is detectable only over heat shock puffs (Jamrich *et al.*, 1977; Greenleaf *et al.*, 1978). Transcriptional repression almost certainly precedes these gross morphological changes, although a detailed temporal study has not been made. At any rate, autoradiographic analysis indicates that the incorporation of RNA precursors is greatly reduced over most of the chromosomes during heat shock (Belyaeva and Zhimulev, 1976; Bonner and Pardue, 1976; Lewis *et al.*, 1975; Spradling *et al.*, 1975), and biochemical analysis of RNA's isolated from heat-treated cells shows a marked reduction in synthesis of those RNA species which are most abundantly transcribed at normal temperatures (Spradling *et al.*, 1977; Lengyel *et al.*, 1980; Lindquist, 1981).

B. Processing

A second level of control appears to operate in heat-shocked cells to prevent the 25°C transcripts produced just prior to heat shock, as well as any which manage to escape the general transcriptional repression, from entering the cytoplasm. It has been known for some time that both ribosomal RNA (Ellgaard and Clever, 1971) and 5 S RNA (Rubin and Hogness, 1975) are not processed properly during heat shock. More recent evidence suggests that hnRNP assembly is altered, presumably leading to abortive processing of mRNA precursors (Mayrand and Pederson, 1983).

While the effect of heat shock on hnRNA processing is still a matter of some controversy (Kloetzel and Bautz, 1983), it is intriguing that hs gene transcripts are unusual in structure. They do not contain intervening sequences and thus do not require splicing (Holmgren *et al.*, 1979). Their transport to the cytoplasm would, therefore, remain unaffected by a block in RNA processing. (An exception to this rule is hsp83 mRNA, which contains an intervening sequence, but notably, it has the lowest temperature maxima of all the hsp's and is not strongly induced at high temperatures.)

C. Translation

Within 30 min of temperature elevation hsp's are the major products of protein synthesis (Lindquist, 1980b). This is due not only to the rapid synthesis of hs mRNA's but also to the activation of a specific cytoplasmic mechanism which discriminates against the translation of preexisting mRNA's. Notably, these mRNA's are not degraded but are sequestered in an inactive state, ready to be reused in translation when cells have recovered from heat shock (Mirault *et al.*, 1978; Lindquist, 1980b; 1981; Storti *et al.*, 1980; Petersen and Mitchell, 1981).

It is important to realize that the change in protein synthesis is not due to an adverse effect of temperature on the translational machinery itself. Heat shock mRNA's are translated with an extremely high efficiency both during the initial induction period and for several hours thereafter (Lindquist, 1980b).

The change in translational specificity is also not a simple consequence of temperature per se. Cell-free lysates made from heat-shocked cells continue to display translational discrimination during incubation at lower temperatures (Scott and Pardue, 1981; Krüger and Benecke, 1981). Furthermore, when hs and normal cellular mRNA's are mixed and translated in heterologous cell-free systems, the temperature of incubation has only negligible effects on the pattern of protein synthesis (Lindquist, 1981).

IV. DISTINCTION BETWEEN COORDINATE AND NONCOORDINATE ASPECTS OF REGULATION

A. Heat Shock Proteins

Heat shock proteins show a strong capacity for independent regulation in three aspects of their expression: (1) basal levels of synthesis at normal temperatures, (2) rates and levels of induction at different temperatures, (3) rates of repression during recovery from heat shock.

1. Heat Shock Synthesis at Normal Growing Temperatures

In *Drosophila melanogaster* there are seven major heat-induced proteins, designated hsp83, hsp70, hsp68, hsp27, hsp26, hsp23, and hsp22 in Fig. 1. (The proteins are named according to their apparent molecular masses on SDS–polyacrylamide gels, i.e., hsp70 = 70,000 daltons.) They show very different patterns of expression during normal growth and development. One protein, hsp83, appears to be constitutively synthesized in tissue culture cells at normal temperatures (Lindquist, 1980a) and can be readily detected as a stainable band on SDS–polyacrylamide gels. This protein is also found in substantial quantities in ovaries and developing egg chambers and in early embryogenesis in animals which have never been exposed to heat (Zimmerman *et al.*, 1983; Chomyn *et al.*, 1979; Loyd *et al.*, 1981).

On the other hand, hsp70 is found only in exquisitely small quantities in tissue culture cells growing at normal temperatures and can be detected only when immunological procedures are pushed to the limits of their resolution (Velazquez *et al.*, 1983). Hsp70 mRNA is not synthesized in oocytes, nurse cells, or preblastoderm embryos even with a heat shock (Zimmerman *et al.*, 1983). A group of proteins closely related to hsp70 (called hs cognates) are constitutively synthesized at normal temperatures (Ingolia and Craig, 1981a; Loyd *et al.*, 1981; Wadsworth, 1982). In the earlier literature, their synthesis was often confused with that of the heat-induced protein.

The four small heat shock proteins display still another pattern of expression. In tissue culture cells and during most of development, their basal level of synthesis is very low (Lindquist, 1980a). However, messenger RNA's for all four of these proteins are induced by ecdysone in isolated imaginal disks as well as in tissue culture cells (Ireland and Berger, 1982). At least one of the proteins (hsp23) is produced in substantial quantities in late third instar larvae and prepupae coincident with the rise in ecdysone titer (Cheney and Shearn, 1983). Finally, hsp28 and hsp26 mRNA's (as well as hsp83 mRNA) are induced in adult ovaries, passed into the developing oocyte, and retained in the developing embryo until blastoderm formation (Zimmerman *et al.*, 1983).

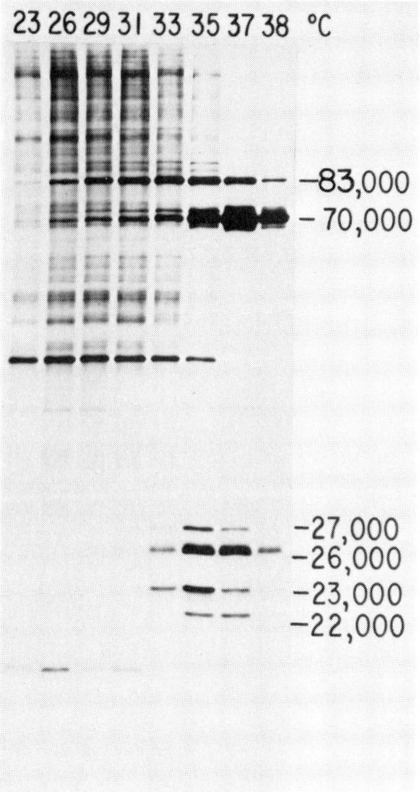

Fig. 1. Distribution of *Drosophila* proteins labeled with [^{3}H]leucine at various temperatures. Tissue culture cells were grown at 23°C and incubated at the indicated temperatures for 2 hr with [^{3}H]leucine added during the final hour of incubation. TCA-precipitable proteins were separated on SDS–polyacrylamide gels (Laemmli, 1970). Labeled proteins were visualized by fluorography (Laskey and Mills, 1975). The apparent molecular weights of the hs proteins are indicated on the right. Individual hsp's reach maximal levels of synthesis at different temperatures.

2. *Differential Induction of Individual hsp's*

During heat shock each hsp has its own unique induction characteristics with regard to the temperature at which it is maximally induced, the range of temperatures over which it is produced, and rate of its induction (Lindquist, 1980a). Heat shock protein 83 is synthesized over the broadest temperature range and is maximally induced at the lowest temperature. Heat shock protein 70 and its close relative hsp68 display the sharpest induction profiles at the highest temperatures (see Fig. 2). The four small hsp's also display a narrower induction range than hsp83 but it is centered 1–2°C below that of hsp70. The different effects of temperature on the synthesis of these proteins is primarily regulated at the level of RNA accumulation. In one of the earliest studies of the heat shock response, it was noted that individual hs puffs on polytene chromosomes display markedly different responses to temperature (Ashburner, 1970). Biochemical studies of mRNA concentrations have confirmed this basic observation (Spradling *et al.*, 1977, Lindquist, 1980a). More recently, some success has been achieved in fractionating components from tissue culture cells which induce different puffs to different levels (Bonner, 1982).

3. *Differential Rates of Repression After Heat Shock*

When returned to normal temperatures, *Drosophila* cells do not have a translational control mechanism to discriminate against hs mRNA's as a single class in the same manner that normal cellular messages are discriminated against during heat shock. What we find is that the repression of hsp's is always distinctly asynchronous. Figure 3 displays time courses of recovery following four different types of heat treatment. Synthesis of hsp70 increases briefly after return to 25°C and then declines. As this protein is being repressed, hsp26 and hsp83 are still being induced. Synthesis of hsp26 reaches maximal levels and begins to decline while synthesis of hsp83 is maintained at a high level. Although the rate of repression vary enormously depending on the severity of heat treatment, a specific temporal order is followed; hsp70 is the first protein to be repressed and hsp83 is the last (see also Fig. 5B).

The repression of heat shock synthesis is accomplished in an interesting way. One might expect that, as hsp synthesis declines and normal protein synthesis

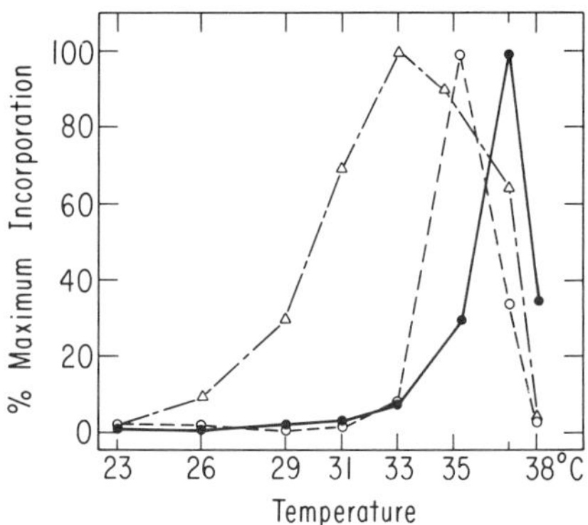

Fig. 2. Percentage of maximum incorporation versus temperature. The quantity of protein synthesized at each temperature is plotted as the percentage of the maximal incorporation rate for that protein: hsp83 (△), hsp70 (●), hsp26 (○). The hsp's have distinct induction characteristics with respect to both the temperature at which maximal synthesis occurs and the range of temperatures over which induction is strongest.

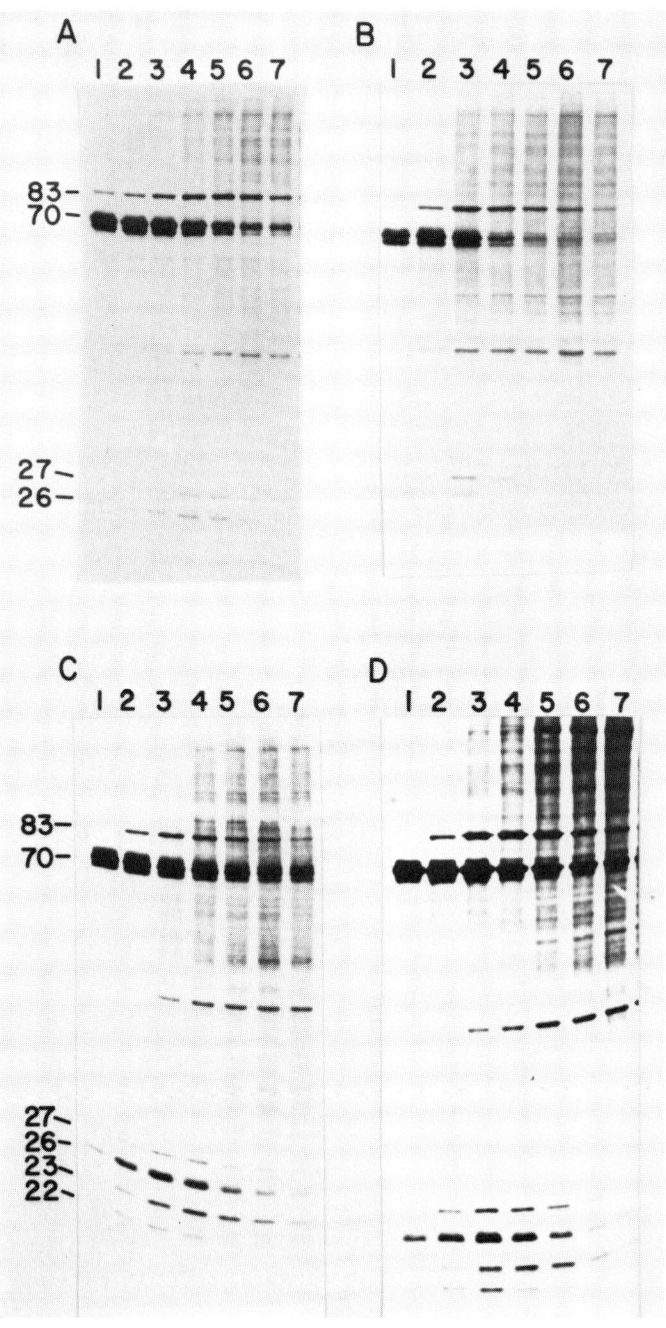
A
1 2 3 4 5 6 7
83-
70-
27-
26-
B
1 2 3 4 5 6 7
C
1 2 3 4 5 6 7
83-
70-
27-
26-
23-
22-
D
1 2 3 4 5 6 7

returns, hs mRNA's would be translated with gradually lower and lower efficiencies. This is not the case. In studying polysome profiles during recovery from heat shock we have found that it is the number of hs mRNA's in polysomes that gradually declines while the number of ribosomes per hsp70 mRNA appears to remain the same (DiDomenico *et al.*, 1982a). In tissue culture cells, the decline in hsp synthesis is closely paralleled by the degradation of hs messages (DiDomenico *et al.*, 1982a). It is not yet possible to determine whether it is a block in the translation of individual mRNA's which leads to their degradation or whether degradation is the primary control point. It is clear, however, that as individual hs mRNA's are withdrawn from translation those which remain in the active pool retain full activity.

B. Normal Cellular Proteins

While heat shock proteins show markedly different patterns of expression as the temperature increases or decreases, normal cellular proteins are more uniform in their response. The rate and extent of their repression vary with the severity of the heat shock, but, overall, they appear to be regulated together.

Figure 4 shows a time course of protein synthesis for *Drosophila* tissue culture cells shifted from 25 to 35 or 37°C. At 37°C virtually all proteins are repressed faster and to a greater extent than at 35°C. Considering the extraordinary diversity of proteins produced at 25°C, the response to temperature elevation is remarkably uniform with regard to the extent and rate of repression of these proteins.

We do not wish to overstate the case. Individual messages are certainly affected by heat to different extents. Some proteins will be repressed to 25% of their normal synthetic rates at 35°C and 10% at 37°C. Other proteins may have corresponding values of 5 and 2% (Lindquist, 1980a). There are even a few messenger RNA's (histones and certain viral RNA's) which appear to escape the general repression altogether and are translated at nearly normal rates during heat shock (Sanders, 1981; Camato and Tanguay, 1982; Scott *et al.*, 1980). Nev-

---

Fig. 3. Patterns of protein synthesis during recovery from different heat treatments. Cells were heat treated and returned to 25°C. At regular intervals individual aliquots were pulse labeled with [^{3}H]leucine. (A) 36.5°C for 30 min, pulse labeled in consecutive 30-min intervals during recovery; (B) 37°C for 30 min, pulse labeled in consecutive 1-hr intervals during recovery; (C) 36.5°C for 60 min, pulse labeled in consecutive 60-min intervals during recovery; (D) 39°C for 30 min after a gradual increase in temperature (i.e., 2°C/15 min), pulse labeled in 60 min-intervals during recovery. The hsp's are repressed at different times and with different speeds. Film exposure times are maximized to show the asynchrony of hsp70 and hsp83 (A) and (B) and the asynchrony of the four small hsp's (C) and (D).

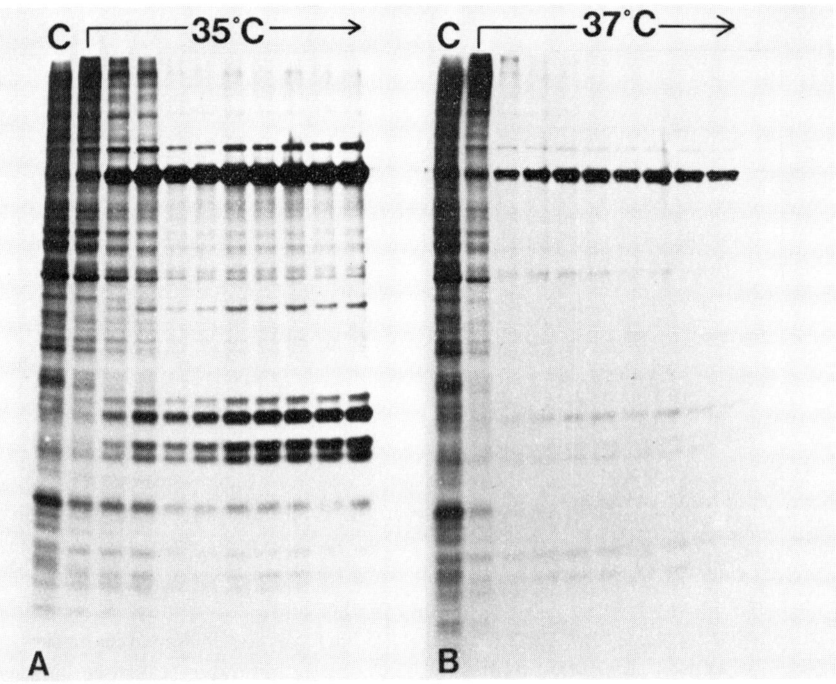

Fig. 4. Induction of heat shock proteins. Cells were shifted to (A) 35°C or (B) 37°C. Protein synthesis was monitored by the addition of [^{3}H]leucine to individual aliquots in 15-min consecutive intervals. The first lane in each figure (denoted C) was labeled at 25°C and serves as a control for the pattern of protein synthesis at normal temperatures. Lanes 2–9 display the pattern of protein synthesis after a temperature shift. Although 25°C protein synthesis is repressed at different rates and to different extents at the two temperatures, in each case most 25°C mRNA's appear to be affected by heat in the same manner.

ertheless, as a quick glance at Figs. 1 and 4 will demonstrate, most preexisting mRNA's appear to be affected by heat in a similar fashion.

This similarity in behavior applies not only to repression during heat shock but also to reactivation during recovery. Referring again to Fig. 3, the return to normal protein synthesis proceeds at different rates in cells which have been exposed to heat shocks of differing severity. In all cases, however, messenger RNA's for most normal cellular proteins are reactivated at the same rate. This uniformity in behavior is in sharp contrast to the asynchronous repression of hs mRNA's occurring in the same cells at the same time. The point is illustrated graphically in Fig. 5. The rates of synthesis for a randomly chosen series of normal cellular proteins are plotted next to the rates of synthesis for three of the major heat shock proteins during recovery from a 30 min heat shock at 36.5°C.

C. Repression of Heat Shock Protein 70 and Reactivation of Normal Cellular Proteins Are Coordinate

Given that most normal cellular mRNA's are reactivated at the same time, while heat shock mRNA's are repressed asynchronously, it might be assumed that no special relationship would exist between the two processes. However, when we plotted the rates of normal message reactivation against the rates of hsp repression, an interesting correlation was revealed. The repression of hsp70 always appeared inversely proportional to the reactivation of normal protein synthesis (see Fig. 6). (No particular relationship was observed between synthesis of other heat shock proteins and reactivation of normal protein synthesis.) A reciprocal relationship between hsp70 and normal cellular protein synthesis was observed in every recovery experiment we performed following heat treatments over a wide range of temperatures and times (DiDomenico *et al.*, 1982a). This relationship was also observed when the normal process of recovery was altered by the action of various drugs and inhibitors (DiDomenico *et al.*, 1982b). For example, when actinomycin was used to limit the production of hsp70 mRNA, slowing the rate of hsp70 accumulation, a longer period of recovery at 25°C was required for hsp70 repression. A corresponding delay was observed for the reactivation of normal protein synthesis (Fig. 6D).

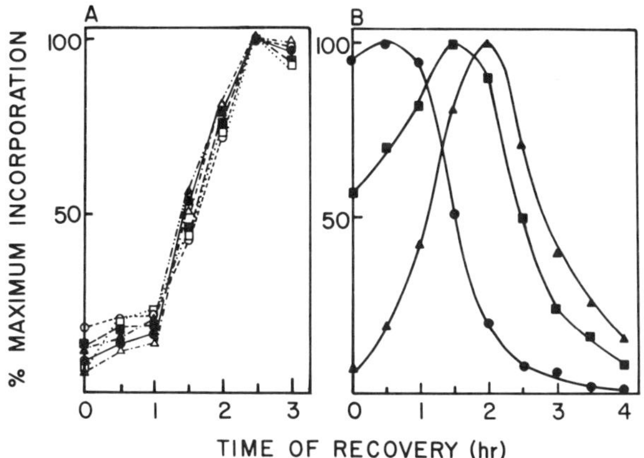

Fig. 5. Restoration of normal protein synthesis and repression of hsp synthesis during recovery. Cells were heat treated at 36.5°C for 30 min and then returned to 25°C for recovery. Protein synthesis was monitored by pulse labeling with [^{3}H]isoleucine, and the resulting fluorograms were quantified by densitometry. (A) Profiles for actin and five randomly chosen 25°C proteins. (B) Profiles for three hsp's: hsp70 (●), hsp26 (■), and hsp83 (▲). 25°C mRNA's are synchronously reactivated but hs mRNA's are asynchronously repressed.

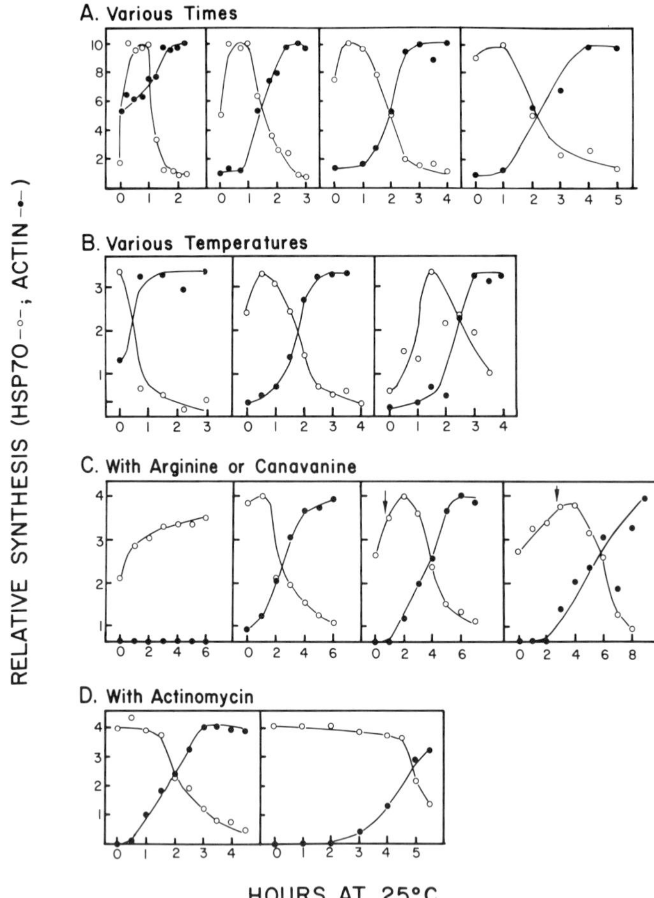

Fig. 6. Kinetics of hsp70 repression and actin recovery. Cells were heat shocked under a variety of conditions and then returned to 25°C. Proteins were pulse labeled with [^{3}H]leucine or [^{3}H]isoleucine. Electrophoretically separated proteins were quantified by densitometry of fluorograms. Incorporation into hsp70 (○) and actin (●) is plotted for cells: (A) heat shocked at 36.5°C for 10, 30, 60, or 120 min; (B) heat shocked for 60 min at 35, 36.5, or 39°C; (C) heat treated for 60 min at 36.5°C in medium containing canavanine for the entire experiment, heat treated in medium containing arginine for the entire experiment, initially heat treated in medium with arginine and then shifted to medium containing canavanine (arrow), or initially heat treated in medium containing canavanine and then shifted to medium containing arginine (arrow); (D) heat treated for 60 min at 36.5°C with actinomycin D added (to block further hsp mRNA synthesis) 15 or 5 min after the temperature shift up. In all cases, restoration of 25°C protein synthesis is coordinate with repression of hsp70 synthesis.

V. OTHER RECENT FINDINGS RELEVANT TO REGULATION

A. Heat Shock Protein 70 Moves Back and Forth between Nucleus and Cytoplasm

Until recently, there has been considerable confusion in the literature concerning the distribution of heat shock proteins in *Drosophila* cells, especially with regard to the major protein hsp70. Autoradiographic analysis suggested the protein concentrated in nuclei (Velazquez *et al.*, 1980), and cell fractionation experiments suggested it concentrated in the cytoplasm or was freely distributed throughout the cell (Arrigo *et al.*, 1980; Tanguay and Vincent, 1982; Levinger and Varshavsky, 1981; Sinibaldi and Morris, 1981).

To overcome technical problems inherent in the latter types of experiments, we developed monoclonal antibodies against hsp70 (Velazquez *et al.*, 1983) and used them to determine the protein's intracellular distribution by immunofluorescence. During heat shock hsp70 does concentrate in nuclei, with only a small amount remaining in the cytoplasm. This nuclear localization is characteristic of all cells tested, including both polytene and diploid tissues. Nuclear concentration is apparently stress specific, not temperature specific, since it also occurs in larvae recovering from anoxia without heat treatment. More germane to the issue of regulation, however, is the remarkable redistribution this protein undergoes during recovery from and subsequent reexposure to heat.

When cells are returned to normal temperatures, hsp70 remains localized over nuclei for a period of time that is proportional to the severity of the preceding heat treatment. With continued incubation at 25°C, the protein begins to leave the nucleus. By the time cells have returned to normal patterns of protein synthesis, most of the protein has migrated to the cytoplasm. The protein is stable in the cytoplasm for several hours at normal growing temperatures, but, if cells are reexposed to high temperatures, it rapidly returns to the nucleus (Velazquez and Lindquist, 1984).

B. Heat Shock Protein 70 Is an RNA Binding Protein

By several different criteria, hsp70 has been shown to be an RNA binding protein. First, when the protein is analyzed on two-dimensional gels it forms a broad smear across the first dimension unless the sample is treated with ribonuclease. This has been observed both for proteins produced *in vivo* and for proteins produced by translation of mRNA's in reticulocyte cell-free extracts (Storti *et al.*, 1980; DiDomenico *et al.*, 1982b). The RNA binding capacity of hsp70 has also been demonstrated *in vivo* through UV-induced cross-linking studies (Kloetzel and Bautz, 1983) and *in vitro* by filter binding assays (E. Sirkin

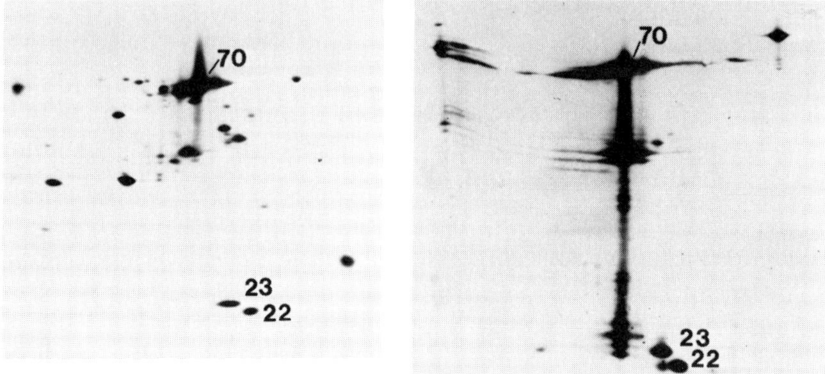

Fig. 7. Electrophoretic mobility of hsp's on two-dimensional gels. Identical aliquots of tissue culture cells were heat treated at 37°C for 3 hr with [^{3}H]isoleucine added after the first 30 min. Heat shock proteins were analyzed on two-dimensional gels as described by O'Farrell (1975). Samples were prepared in the absence of PMSF, permitting partial degradation by endogenous proteases. Left, with RNase and DNase; right, without nucleases.

and S. Lindquist, unpublished observations). The protein binds to single-stranded RNA homopolymers, but not to double-stranded RNA's or to RNA's with a high degree of secondary structure (E. Sirkin, unpublished observations).

In order to determine whether hsp70 forms a true, stable complex with RNA or merely an adventitious association, we analyzed proteins prepared in the absence of the protease inhibitor, phenylmethylsulfonyl fluoride (PMSF). (The major proteolytic activity of *Drosophila* lysates is PMSF sensitive.) Partially degraded proteins were analyzed on two-dimensional gels with and without nuclease treatment. As can be seen in Fig. 7, the presence of RNA in the sample holds the protein fragments together during electrophoresis in the first dimension. This complex is then dissociated by SDS in the second dimension so that all the protein fragments appear to fall on the same vertical line. When the same partially degraded sample was treated with nucleases prior to electrophoresis, the various fragments of hsp70 migrated independently of each other in the isoelectric focusing dimension and appeared scattered throughout the gel.

C. Synthesis of Heat Shock Proteins Is Autoregulated

Recently we presented evidence that heat shock proteins are involved in regulating their own synthesis (DiDomenico *et al.*, 1982b). The proteins appear to act at two different levels, repressing further transcription of their own genes, as well as translation of their own messengers. When the synthesis of functional heat shock proteins is blocked by a variety of different methods the synthesis of hs mRNA's continues much longer than it normally would. Under these conditions hs mRNA's are very stable and accumulate in enormous quantities. When the block in protein synthesis is released, transcription of hs mRNA's continues briefly, but then, as the proteins accumulate in substantial quantities, further transcription is repressed. Some time later, when a specific quantity of heat shock protein has been produced, hs protein synthesis is repressed and hs mRNA's are degraded. It should be noted that in *Drosophila* cells hsp's act to repress the transcription of their own mRNA's whether cells are maintained at high temperatures or returned to normal temperatures. Repression of their own translation, however, requires a return to normal temperatures.

D. Heat Shock Leader Sequences Are Unusual

The hs genes of *Drosophila* have all been cloned and sequenced in other laboratories (Ingolia *et al.*, 1980; Ingolia and Craig, 1981b; Holmgren *et al.*, 1981). The analysis has shown that hs mRNA's have distinctive structural features. In comparison with other *Drosophila* RNA's, hs mRNA's have very long 5′ leader sequences containing from 111 to 253 nucleotides. These leaders have an unusual nucleotide composition ranging from 46 to 51% adenine. The hs mRNA's may also be distinctive at their 3′ ends. Although the evidence is indirect, we and others have suspected that hs mRNA's may have unusually short poly(A) tails; their efficiency of binding to oligo(dT)cellulose is low compared to other mRNA's (S. Lindquist, unpublished observations).

VI. A MODEL FOR REGULATION

Before proposing a model for posttranscriptional regulation of the heat shock response, let us briefly recap: At the time when translation is shifting from hs proteins to normal cellular proteins hsp70 undergoes a dramatic redistribution by moving from the nucleus to the cytoplasm; hsp70 is an RNA binding protein; hsp70 synthesis is autoregulated; the hsp70 message has an unusually long 5′ leader sequence. The evidence leads us to propose that hsp70 modulates its own synthesis (and perhaps that of other hsp's) by direct interaction with the 5′ end of its own messenger RNA. This need not be a sequence-specific interaction, but may depend on a structural feature such as long stretches of unpaired nucleotides. If this model is correct, several otherwise puzzling features of the response would be explained.

For example, since hsp70 continues to concentrate in nuclei of *Drosophila* cells as long as the temperature is elevated, it is easy to understand why the repression of hs protein synthesis requires a return to normal temperatures. In other organisms, such as *Tetrahymena* and yeast, normal patterns of protein synthesis are restored even when cells are maintained at high temperatures. One might expect that in these organisms hsp70 would initially concentrate in nuclei and then appear in the cytoplasm with continued high-temperature incubation.

The model would also explain the intriguing manner in which hs mRNA's are repressed during heat shock. Recall that the number of hsp70 mRNA's in polysomes gradually declines but the number of ribosomes per message remains the same. Individual hs mRNA's are withdrawn from translation while others remain fully active. With an expected ribosome transit time of less than 3 min, a block in initiation would rapidly lead to disappearance of that message from the polysome profile and produce exactly this result. We would further postulate that in *Drosophila* tissue culture cells, with their characteristically high endogenous nuclease activities, the putative hsp70–hs mRNA complex would be unstable and the message rapidly degraded. It may well be the case that in other cell types, with lower nucleolytic activities, hs mRNA–protein complexes might be stable at low temperatures. Indeed, it is tempting to speculate that this may be the mechanism for the storage of translationally inactive hs mRNA's in *Xenopus* oocytes (Bienz and Gurdon, 1982).

Finally, this model provides an explanation for the asynchronous repression of hs proteins during recovery. Since each hs mRNA has a different leader sequence, each may have a different affinity for hsp70. Alternatively, the repression of each message may require the binding of its own protein, individually or as a complex with other proteins.

But what about the regulation of preexisting messenger RNA's? Although hsp70 concentrates in nuclei during heat shock, the proteins do not seem to be involved in repressing preexisting synthesis. This follows from the simple fact that repression is evident before hsp's have begun to accumulate in substantial quantities and occurs even when hsp synthesis is blocked by cycloheximide. Rather, we believe hsp70 is transported to the nucleus during heat shock in order to protect nuclear structure and function (Velazquez and Lindquist, 1984).

Heat shock proteins are also not involved in establishing the initial block in preexisting translation. Again, this block occurs before hsp's have been produced and occurs even when hs synthesis is inhibited with actinomycin.

On the other hand, three lines of evidence lead us to suggest that hsp70 is involved in reestablishing the translation of normal cellular mRNA's after heat shock. First, after all but the mildest heat shocks, reactivation appears to require the synthesis of a particular quantity of proteins. When the synthesis of functional hsp's is blocked, normal protein synthesis is not restored when cells are returned to normal temperatures. Normal protein synthesis resumes only when the block is released and a specific quantity of hs protein has accumulated (DiDomenico *et al.*, 1982b). Second, in all of the recovery experiments we have performed, reactivation of preexisting messages is coordinate with repression of hsp70. None of the other hsp's displays a special relationship to 25°C synthesis

(DiDomenico *et al.*, 1982a). Third, the gradual movement of hsp70 from nucleus to cytoplasm during recovery closely coincides with the gradual reactivation of normal protein synthesis (Velazquez and Lindquist, 1984).

We do not know how hsp70 facilitates the reactivation of preexisting mRNA's. Yet it must do so by a mechanism which affects most of these sequences in a similar way without regard to their coding capacity. We have already proposed that hsp70 interacts with its own A-rich leader sequence. An obvious and tempting possibility is that the protein may interact with the poly(A) tails of preexisting mRNA's facilitating their movement from an inactive to an active compartment. Alternatively, it might interact with proteins which are responsible for blocking their translation, thereby releasing them from inhibition.

The mechanism of posttranscriptional repression we propose for hsp70 is similar to mechanisms which have been proposed for other nucleic acid binding proteins in bacteria and viruses (for example, gene 32 protein; Gold *et al.*, 1981). It is attractive to think that control circuits for coping with similar regulatory problems may display this degree of conservation. In addition to its role in protecting cells from the toxic effects of stress, we suggest that hsp70 is involved in repressing its own synthesis and in reactivating the expression or preexisting mRNA's. This multiplicity of function may, in part, account for the extremely high level of coding sequence conservation recently reported for this protein (Craig *et al.*, 1982).

REFERENCES

Arrigo, A. P., Fakan, S., and Tissières, A. (1980). Localization of the heat shock induced protein in *Drosophila melanogaster* tissue culture cells. *Dev. Biol.* **78,** 86–103.

Ashburner, M. (1970). Patterns of puffing activity in salivary gland chromosomes of *Drosophila*. V. Responses to environmental treatment. *Chromosoma* **31,** 356–376.

Ashburner, M., and Bonner, J. J. (1979). The induction of gene activity in *Drosophila* by heat shock. *Cell* **17,** 241–254.

Belyaeva, E. S., and Zhimulev, I. F. (1976). RNA synthesis in the *Drosophila melanogaster* puffs. *Cell Differ.* **4,** 415–427.

Bienz, M., and Gurdon, J. B. (1982). The heat shock response in *Xenopus* oocytes is controlled at the translational level. *Cell* **29,** 811–819.

Bonner, J. J. (1982). Regulation of the *Drosophila* heat shock response. *In* "Heat Shock from Bacteria to Man" (M. J. Schlesinger, M. Ashburner, and A. Tissières, eds.), pp. 147–154. Cold Spring Harbor Lab., Cold Spring Harbor, New York.

Bonner, J. J., and Pardue, M. L. (1976). The effect of heat shock on RNA synthesis in *Drosophila* tissues. *Cell* **8,** 43–50.

Camato, R., and Tanguay, R. M. (1982). Changes in the methylation pattern of core histones during heat shock in *Drosophila* cells. *EMBO J.* **1,** 1529–1532.

Cheney, C. M., and Shearn, A. (1983). Developmental regulation of *Drosophila* disc proteins: synthesis of a heat shock protein under non-heat shock conditions. *Dev. Biol.* **95,** 325–330.

Chomyn, A., Moller, G., and Mitchell, H. K. (1979). Patterns of protein synthesis following heat shock in pupae of *D. melanogaster*. *Dev. Genet.* **1,** 77–95.

Craig, E. A., Ingolia, T. D., Slater, M., Manseau, L. J., and Bardwell, J. (1982). *Drosophila*, yeast, and *E. coli* genes related to the *Drosophila* heat-shock genes. *In* "Heat Shock, from Bacteria to Man" (M. J. Schlesinger, M. Ashburner, and A. Tissières, eds.), pp. 11–18. Cold Spring Harbor Lab., Cold Spring Harbor, New York.

DiDomenico, B. J., Bugaisky, G. E., and Lindquist, S. (1982a). Heat shock and recovery are mediated by different translational mechanisms. *Proc. Natl. Acad. Sci. U.S.A.* **79,** 6181–6185.

DiDomenico, B. J., Bugaisky, G. E., and Lindquist, S. (1982b). The heat shock response is self-regulated at both the transcriptional and posttranscriptional levels. *Cell* **31,** 593–603.

Ellgaard, E. G., and Clever, U. (1971). RNA metabolism during puff induction in *D. melanogaster. Chromosoma* **36,** 60–78.

Gold, L., Pribnow, D., Schneider, T., Shinedling, S., Singer, B. S., and Stormo, G. (1981). Translational initiation in prokaryotes. *Annu. Rev. Microbiol.* **35,** 365–403.

Greenleaf, A. L., Plagens, U., Jamrich, M., and Bautz, E. R. F. (1978). RNA polymerase B (or II) in heat-induced puffs on *Drosophila* polytene chromosomes. *Chromosoma* **65,** 127–136.

Holmgren, R., Livak, K., Morimoto, R., Freund, R., and Meselson, M. (1979). Studies of cloned sequences from four *Drosophila* heat shock loci. *Cell* **18,** 135–137.

Holmgren, R., Corces, V., Morimoto, R., Blackman, R., and Meselson, M. (1981). Sequence homologies in the 5′ regions of four *Drosophila* heat shock genes. *Proc. Natl. Acad. Sci. U.S.A.* **78,** 3775–3778.

Ingolia, T. D., and Craig, E. A. (1981a). *Drosophila* gene related to the major heat shock-induced gene is transcribed at normal temperatures and not induced by heat shock. *Proc. Natl. Acad. Sci. U.S.A.* **79,** 525–529.

Ingolia, T. D., and Craig, E. A. (1981b). Primary sequence of the 5′ flanking regions of the *Drosophila* heat shock genes in chromosome subdivision 67B. *Nucleic Acids Res.* **9,** 1627–1642.

Ingolia, T. D., Craig, E. A., and McCarthy, B. J. (1980). Sequence of three copies of the gene for the major *Drosophila* heat shock induced protein and their flanking regions. *Cell* **21,** 669–679.

Ireland, R. C., and Berger, E. M. (1982). Synthesis of low molecular weight heat shock peptides stimulated by ecdysterone in a cultured *Drosophila* cell line. *Proc. Natl. Acad. Sci. U.S.A.* **79,** 855–859.

Jamrich, M., Greenleaf, A. L., and Bautz, E. K. F. (1977). Localization of RNA polymerase in polytene chromosomes of *Drosophila melanogaster. Proc. Natl. Acad. Sci. U.S.A.* **74,** 2079–2083.

Kloetzel, P. M., and Bautz, E. K. F. (1983). Heat shock proteins are associated with hnRNA in *Drosophila melanogaster* tissue culture cells. *EMBO J.* **2,** 705–710.

Krüger, C., and Benecke, B. J. (1981). *In vitro* translation of *Drosophila* heat shock and non-heat shock mRNA's in heterologous and homologous cell-free systems. *Cell* **23,** 595–604.

Laemmli, U. K. (1970). Cleavage of structural proteins during the assembly of the head of the bacteriophage T4. *Nature (London)* **227,** 680–685.

Laskey, R. A., and Mills, A. D. (1975). Quantitative film detection of ^{3}H- and ^{14}C in polyacrylamide gels by fluorography. *Eur. J. Biochem.* **56,** 335–341.

Lengyel, J. A., Ransom, L., Graham, M., and Pardue, M. L. (1980). Comparison of transcription from three major *D. melanogaster* heat shock puff sites. *Chromosoma* **80,** 237–252.

Levinger, L., and Varshavsky, A. (1981). Heat shock proteins of *Drosophila* are associated with nuclease-resistant, high salt resistant nuclear structures. *J. Cell Biol.* **90,** 793–796.

Lewis, M., Helmsing, P. J., and Ashburner, M. (1975). Parallel changes in puffin activity and patterns of protein synthesis in salivary glands of *Drosophila. Proc. Natl. Acad. Sci. U.S.A.* **72,** 3604–3608.

Lindquist, S. (1980a). Varying patterns of protein synthesis in *Drosophila* during heat shock: implications for regulation. *Dev. Biol.* **77,** 463–479.

Lindquist, S. (1980b). Translational efficiency of heat-induced messages in *D. melanogaster* cells. *J. Mol. Biol.* **137,** 151–158.

Lindquist, S. (1981). Regulation of protein synthesis during heat shock. *Nature (London)* **293,** 311–314.

Loyd, J. E., Raff, E. C., and Raff, R. A., (1981). Site and timing of synthesis of tubulin and other proteins during oogenesis in *Drosophila melanogaster. Dev. Biol.* **86,** 272–284.

McKenzie, S. L., Henikoff, S., and Meselson, M. (1975). Localization of RNA from heat-induced polysomes at puff sites in *Drosophila melanogaster. Proc. Natl. Acad. Sci. U.S.A.* **72,** 1117–1121.

Mayrand, S., and Pederson, T. (1983). Heat shock alters nuclear ribonucleoprotein assembly in *Drosophila* cells. *Mol. Cell Biol.* **3,** 161–171.

Mirault, M.-E., Goldschmidt-Clermont, M., Moran, L., Arrigo, A. P., and Tissières, A. (1978). The effect of heat shock on gene expression in *Drosophila melanogaster. Cold Spring Harbor Symp. Quant. Biol.* **42,** 819–827.

O'Farrell, P. H. (1975). High resolution two-dimensional electrophoresis of proteins. *J. Biol. Chem.* **250,** 4007–4021.

Petersen, N. S., and Mitchell, H. K. (1981). Recovery of protein synthesis following heat shock: preheat treatment affects mRNA translation. *Proc. Natl. Acad. Sci. U.S.A.* **78,** 1708–1711.

Ritossa, F. M. (1962). A new puffing pattern induced by heat shock and DNP in *Drosophila. Experientia* **18,** 571–573.

Ritossa, F. M. (1964). Behavior of RNA and DNA synthesis at the puff level in salivary gland chromosomes of *Drosophila. Exp. Cell Res.* **36,** 515–523.

Rubin, G. M., and Hogness, D. S. (1975). Effect of heat shock on the synthesis of low molecular weight RNAs in *Drosophila*: accumulation of a novel form of 5S RNA. *Cell* **6,** 207–213.

Sanders, M. M. (1981). Identification of histone H2b as a heat shock protein in *Drosophila. J. Cell Biol.* **91,** 69a.

Schlesinger, M. J., Ashburner, M., and Tissières, A., eds. (1982). "Heat Shock, from Bacteria to Man." Cold Spring Harbor Lab., Cold Spring Harbor, New York.

Scott, M. P., and Pardue, M. L. (1981). Translational control in lysates of *Drosophila melanogaster* cells. *Proc. Natl. Acad. Sci. U.S.A.* **78,** 3353–3357.

Scott, M. P., Fostel, J. M., and Pardue, M. L. (1980). A new type of virus from cultured *Drosophila* cells: characterization and use in studies of the heat-shock response. *Cell* **22,** 929–941.

Sinibaldi, R. M., and Morris, P. W. (1981). Putative function of *Drosophila melanogaster* heat shock proteins in the nucleoskeleton. *J. Biol. Chem.* **256,** 10,735–10,738.

Spradling, A., Penman, S., and Pardue, M. L. (1975). Analysis of *Drosophila* mRNA by *in situ* hybridization: sequences transcribed in normal and heat shocked cultured cells. *Cell* **4,** 395–404.

Spradling, A., Pardue, M. L., and Penman, S. (1977). Messenger RNA in heat-shocked *Drosophila* cells. *J. Mol. Biol.* 109,559–109,587.

Storti, R. V., Scott, M. P., Rich, A., and Pardue, M. L. (1980). Translational control of protein synthesis in response to heat shock in *D. melanogaster* cells. *Cell* **22,** 825–834.

Tanguay, R. M., and Vincent, M. (1982). Intracellular translocation of cellular and heat shock induced proteins upon heat shock in *Drosophila* Kc cells. *Can. J. Biochem.* **60,** 306–315.

Tissières, A., Mitchell, H. K., and Tracy, U. M. (1974). Protein synthesis in salivary glands of *D. melanogaster* cells. *J. Mol. Biol.* **84,** 389–398.

Velazquez, J. M., and Lindquist, S. (1984). hsp70: Nuclear concentration during environmental stress and cytoplasmic storage during recovery. *Cell,* **36,** 655–662.

Velazquez, J. M., DiDomenico, B. J., and Lindquist, S. (1980). Intracellular localization of heat shock proteins in *Drosophila*. *Cell* **20,** 679–689.

Velazquez, J. M., Sonoda, S., Bugaisky, G., and Lindquist, S. (1983). Is the major *Drosophila* heat shock protein present in cells that have not been heat shocked? *J. Cell Biol.* **96,** 286–290.

Wadsworth, S. C. (1982). A family of related proteins is encoded by the major *Drosophila* heat shock family. *Mol. Cell. Biol.* **2,** 286–292.

Zimmerman, J. L., Petri, W. L., and Meselson, M. (1983). Accumulation of specific subsets of *D. melanogaster* heat shock mRNAs in normal development without heat shock. *Cell* **32,** 1161–1170.

5

Intracellular Localization and Possible Functions of Heat Shock Proteins

ROBERT M. TANGUAY

I. INTRODUCTION

While data on the structure and regulation of genes induced by heat shock have rapidly accumulated in recent years, our understanding of the functional significance of this response is still considerably limited due to our ignorance of the nature and function of the induced proteins, the hsp's. Various enzymatic functions for hsp's have been proposed on the basis of correlative evidence between increased enzyme activities and hsp induction during heat shock (reviewed in Leenders *et al.*, 1974, and in Ashburner and Bonner, 1979).

Changes in Eukaryotic Gene Expression
in Response to Environmental Stress

ISBN 0-12-066290-6

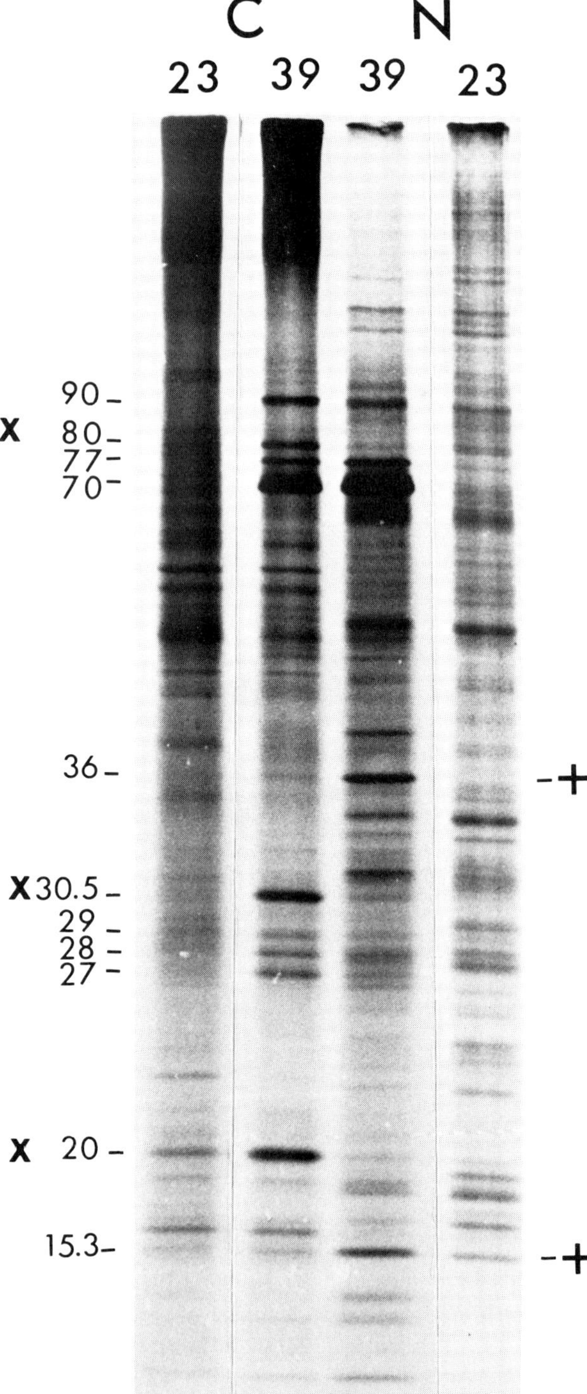
C
N
23
39
39
23
90
X
80
77
70
36
X 30.5
29
28
27
X
20
15.3
+
+

A rather elementary but nevertheless potentially useful approach to searching for the function of proteins is to study their localization within the cell. Such studies could provide important clues for the elucidation of the function of these ubiquitous proteins. Previous reviews on the heat shock response have dealt only briefly with the questions of hsp localization and function (Ashburner and Bonner, 1979; Schlesinger *et al.*, 1982; Adams and Rinne, 1982; Tanguay, 1983). This chapter reviews biochemical studies on the cellular distribution of hsp's and presents some new data on the use of specific antibodies for localizing these proteins in dipteran cells.

II. BIOCHEMICAL STUDIES ON HEAT SHOCK PROTEIN LOCALIZATION

A. Presence of Heat Shock Proteins in Nuclear Preparations

Early studies by Holt (1971) had shown the accumulation of nonhistone proteins in nuclei as well as in certain puffed regions of *Drosophila hydei* salivary gland cells following heat shock. However, the nature of the proteins accumulating at these sites was unknown. Following the observation of Tissières *et al.*, (1974) that heat shock induced the synthesis of new proteins, Mitchell and Lipps (1975) originally reported the presence of hsp's in isolated salivary gland nuclei of *Drosophila melanogaster* following heat shock. By autoradiography, these authors also observed the accumulation of labeled proteins in some heat shock puffs (87B) but not in others (93D, 95D). The nature of the labeled proteins accumulating at these sites remained unknown. Vincent and Tanguay (1979) similarly observed the presence of hsp's in microdissected salivary gland nuclei of another dipteran *Chironomus tentans*. In this case, furthermore, some hsp's were shown to have a preferential if not exclusive nuclear or cytoplasmic localization, whereas others were present in both of these cellular compartments. This is illustrated in Fig. 1, which compares nuclear and cytoplasmic labeled proteins in normal and heat-shocked cells. The nuclei are enriched with hsp36 and hsp15.3, while hsp 80, 30, and 20 are exclusively found in the cytoplasm of those cells. On the other hand, some hsp's are found in both cellular compartments. When nuclei were microdissected into chromosomes, nucleoli, and nu-

Fig. 1. Heat shock proteins in microdissected nuclei and cytoplasms of *Chironomus tentans* salivary gland cells. 4th instar larvae were heat shocked at 39°C for 20 min. Glands were explanted in [^{35}S]methionine-supplemented medium and labeled for 6 hr at 20°C. Proteins from microdissected nuclei (N,75) and cytoplasms (C,3) were separated in 7.5–15% SDS–polyacrylamide gradient gels. All procedures are described in Vincent and Tanguay (1979). Nucleus-predominant hsp's (+); cytoplasm-specific hsp's (x).

clear sap, the two major nuclear hsp's were present in all three fractions (Vincent and Tanguay, 1979).

These two early reports on the presence of hsp's in the nucleus of polytene cells tended to overestimate the putative functional importance of hsp's in this compartment. Thus in both cases, equal amounts of radioactively labeled proteins were loaded on gels rather than equivalent numbers of nuclei and cytoplasms; the amounts of nuclei loaded were 25- to 75-fold higher than cytoplasms. If this is corrected to a single-cell basis, approximately 2–5% of the cellular labeled hsp's are found in the nucleus of these large polytene cells. This is markedly lower than what has been found in *Drosophila* tissue culture cells (see later).

The presence of hsp's in the nucleus of heat-shocked cells was further substantiated by an autoradiographic approach with *Drosophila* salivary gland and tissue culture cells (Velazquez *et al.*, 1980; Arrigo *et al.*, 1980). Both studies also provided direct microscopic evidence that in *Drosophila,* mitochondria were certainly not the major cellular target for hsp's. This observation was particularly important since a model favored until then proposed that the main function of the heat shock response was to restore normal mitochondrial function (Leenders *et al.*, 1974). This model was based mainly on the heat shock-mimicking effects of various inhibitors of electron transfer or oxidative phosphorylation. In *Chironomus* the mitochondria are distributed in a thin cortical zone at the gland periphery. Analysis of proteins from this microdissected mitochondria-rich fraction also failed to show any enrichment of hsp's (Tanguay and Vincent, 1981).

The increased availability and use of *Drosophila*-cultured cells for studying the heat shock response facilitated cell fractionation studies, and a differential localization of the various hsp's was soon reported in various laboratories (Arrigo *et al.*, 1980; Tanguay and Vincent, 1980; Sinibaldi and Morris, 1981; Levinger and Varshavsky, 1981; Tanguay and Vincent, 1982). Thus, hsp's could

TABLE I

Intracellular Localization of Heat Shock Proteins in Diptera

hsp	Organism	Localization	Comment
hsp83	*Drosophila*	Cytosol	
hsp80	*Chironomus*	Cytosol	
hsp 68–70	*Drosophila*	Nucleus or cytoplasm	Dependent on temperature or nature of stress
hsp 34–36	*Drosophila*	Nucleus	Soluble in low salt
hsp 22–28	*Drosophila*	Nucleus and cytoplasm	Dependent on temperature and nature of stress Insoluble in high salt Nuclease resistant

be subdivided into various groups with respect to their cellular localization. This is summarized in Table I. The high-M_r hsp83 of *Drosophila* has an exclusively cytoplasmic distribution as evidenced by cell fractionation and by immunochemical localization studies (see later and Section III,A). This is similar to the cellular localization of the homologous high-molecular-weight hsp's found in various vertebrate cells (Welch and Feramisco, 1982; Kelley and Schlesinger, 1982; Voellmy *et al.*, 1983). It is also compatible with a report from Hughes and August (1982) suggesting that in mouse 3T3 cells hsp's mediate an association between the plasma membrane and the cell cytoskeleton. On the other hand, the less well characterized minor hsp 34K–36K species, found in both *Drosophila* and *Chironomus,* has an almost exclusively nuclear localization with solubility properties differing from those of the hsp 22–28 family (Vincent and Tanguay, 1979; Tanguay and Vincent, 1982).

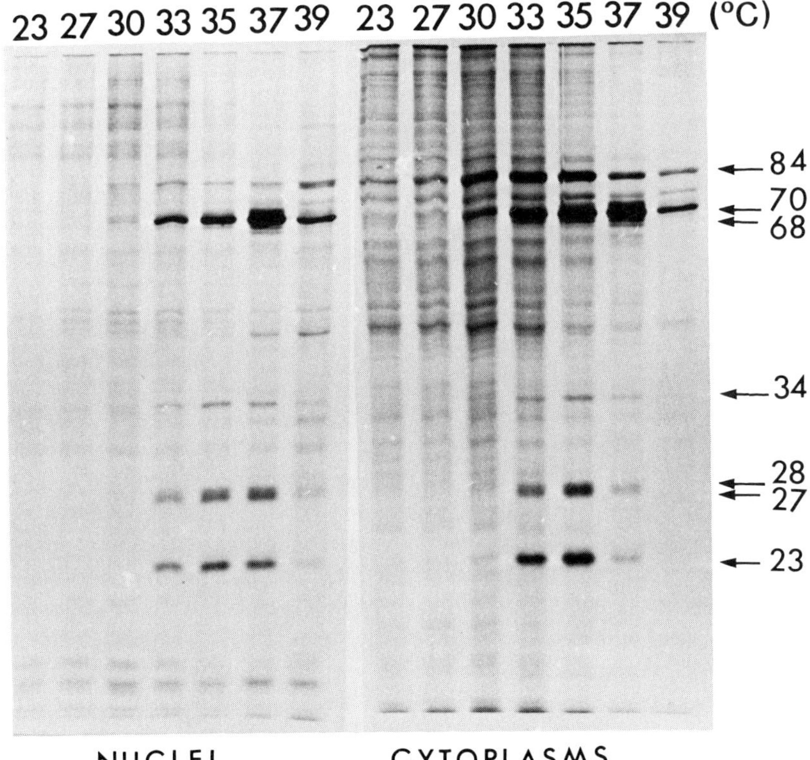

Fig. 2. Temperature dependence of the intracellular distribution of hsp's. *Drosophila* Kc cells were preincubated for 1 hr and labeled for 2 hr at the indicated temperature. Nuclear and cytoplasmic proteins were analyzed on SDS–PAGE [From Vincent and Tanguay (1982) with permission from *Journal of Molecular Biology* **162,** 365–378. Copyright: Academic Press Inc. (London) Ltd.]

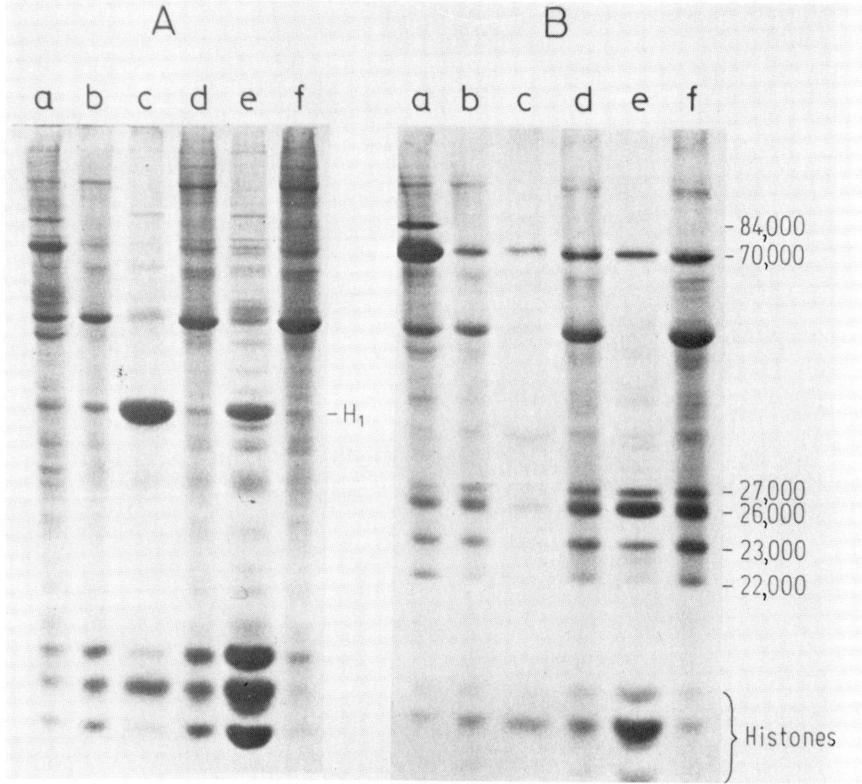

Fig. 3. Acid extraction of chromosomal and heat shock proteins. (A) Coomassie blue stain, (B) autoradiogram. Labeled chromatin from heat-shocked cells was extracted with various acids and the proteins analyzed. Lane a, total cells; lane b, nontreated chromatin; lane c, proteins extracted in 2% trichloroacetic acid; lane d, pellet after extraction; lane e, proteins extracted in 0.2 N H_2SO_4, and lane f, final pellet. Details can be found in Arrigo (1980). (Courtesy of A. P. Arrigo.)

The major 68,000–70,000 hsp family as well as the low-M_r 22,000–28,000 hsp family are found in both the nuclear and the cytoplasmic fractions of heat-shocked cells. Both can shuttle between cellular compartments, as shown by recovery studies following heat shock (Arrigo *et al.*, 1980). However, they are not necessarily associated with the nuclear fraction, as shown in studies where inducers other than heat were used. Thus, when induced by arsenite, hsp70 has

---

Fig. 4. Indirect immunofluorescent localization of hsp83 in *Drosophila*-cultured cells. Control cells: (A) immunofluorescence; (B) phase contrast. Heat shock protein 83 shows a diffuse staining in the whole cytoplasm. Heat-shocked cells: (C) immunofluorescence; (D) phase contrast. After 1 hr at 37°C, the staining with hsp83 antibody concentrates at the cell periphery. All details are described in Duband *et al.* (1985b).

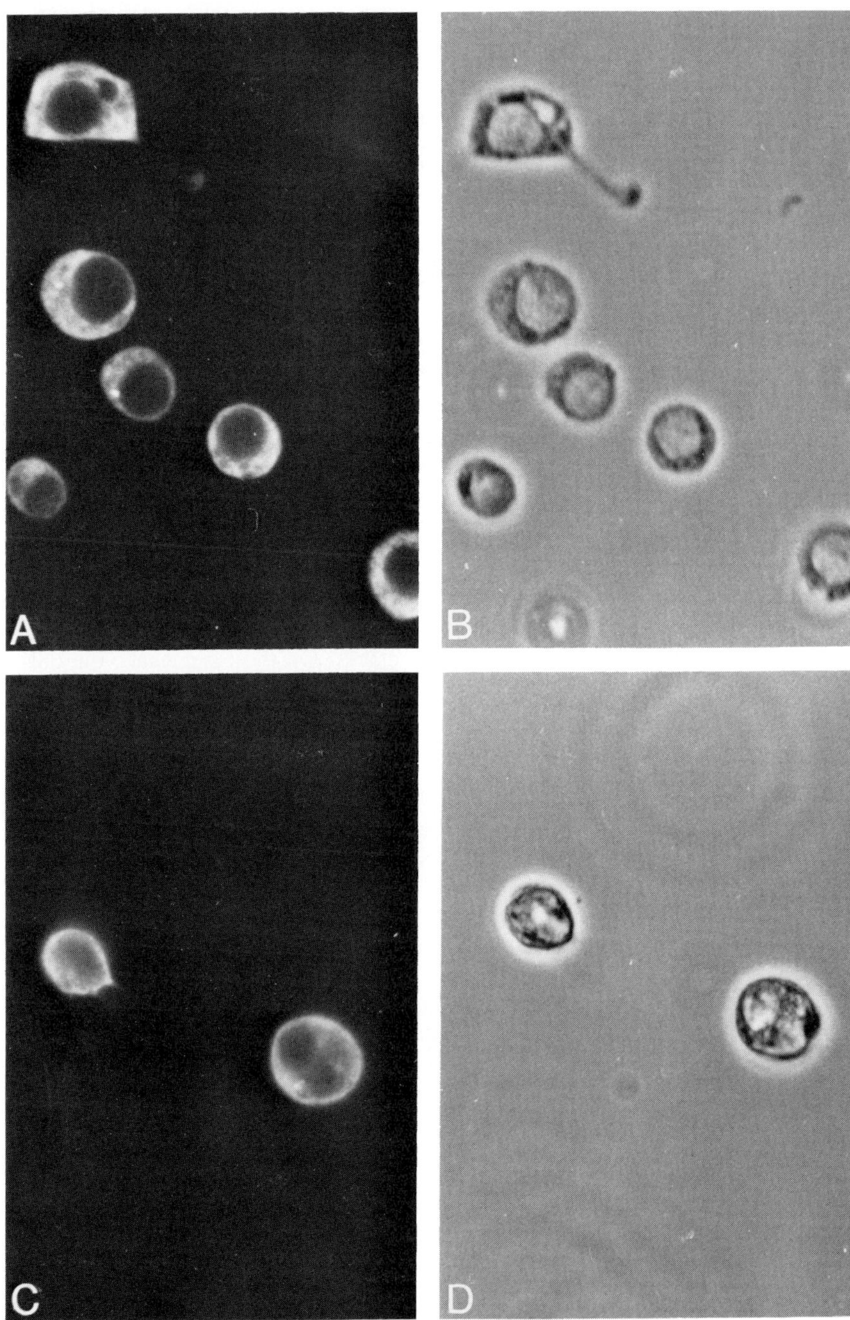
A
B
C
D

an exclusive cytoplasmic localization. The presence of these two hsp families in the nuclear fraction is temperature dependent, as illustrated in Fig. 2. Both families are mainly cytoplasmic at lower temperatures (33°C) and nuclear at higher temperatures (37°C). In this respect the report of Ireland *et al.* (1982) that the low-M_r hsp's induced by ecdysterone are located in the cytoplasmic fraction is of considerable importance in our interpretation of the relationship between cellular localization and function. Finally, we have recently observed that these classes of hsp's can have a cytoplasmic or nuclear localization depending on the percentage of ethanol used during induction of hsp's by this substance (R. M. Tanguay *et al.*, in preparation).

There is still very little published information on the cellular localization of hsp's in other eukaryotic cells. Although similarities to diptera can be found in certain organisms such as *Dictyostelium* (Loomis and Wheeler, 1982), important discrepancies are found in others in terms of both cellular localization and properties (see Tanguay, 1983, for short review).

B. Properties of Nuclear Heat Shock Proteins

The hsp's which are found associated with the nuclear pellet in heat-treated cells show interesting and intriguing properties. Heat shock protein 34 is soluble in low-salt washes and, thus, easily extractable. On the other hand, the 68K–70K hsp and 22K–28K hsp families show high resistance to salt extraction as well as to nuclease treatments.

These properties are supportive of a structural function for hsp's either as intranuclear nucleoskeletal elements (Sinibaldi and Morris, 1981; Levinger and Varshavsky, 1981) or as extranuclear cytoskeletal elements tightly bound to the nuclear structure (Tanguay and Vincent, 1982).

Early data from microscopic as well as microdissection studies (see Section II,A) indicated the presence of hsp's in chromosomes and nucleoli. However, in certain aspects, the intranuclear distribution of hsp's remains controversial, as evidenced by contradictory localization in heterochromatin or euchromatin (Arrigo *et al.*, 1980; Velazquez *et al.*, 1980).

A second controversy deals with the association of hsp's with chromatin or with RNA. The low-M_r hsp's can be released from chromatin by DNase I but not by RNase (Arrigo, 1980). They also seem to be associated with polynucleosomes in micrococcal nuclease-digested chromatin, suggesting an internucleosomal localization. Nuclear hsp's are not solubilized by 2% trichloroacetic acid solutions and are present in a residual fraction after histone extraction (see Fig. 3). They show the properties of nonhistone proteins. In CsCl gradients hsp's can be found in particles of 1.40 g/cm^3, indicating their association with nucleic acids (Arrigo *et al.*, 1980). A linkage of hsp's to nuclear RNA has recently been suggested by UV cross-linking studies on particles of 1.38–1.43 g/cm^3 from normal or heat-

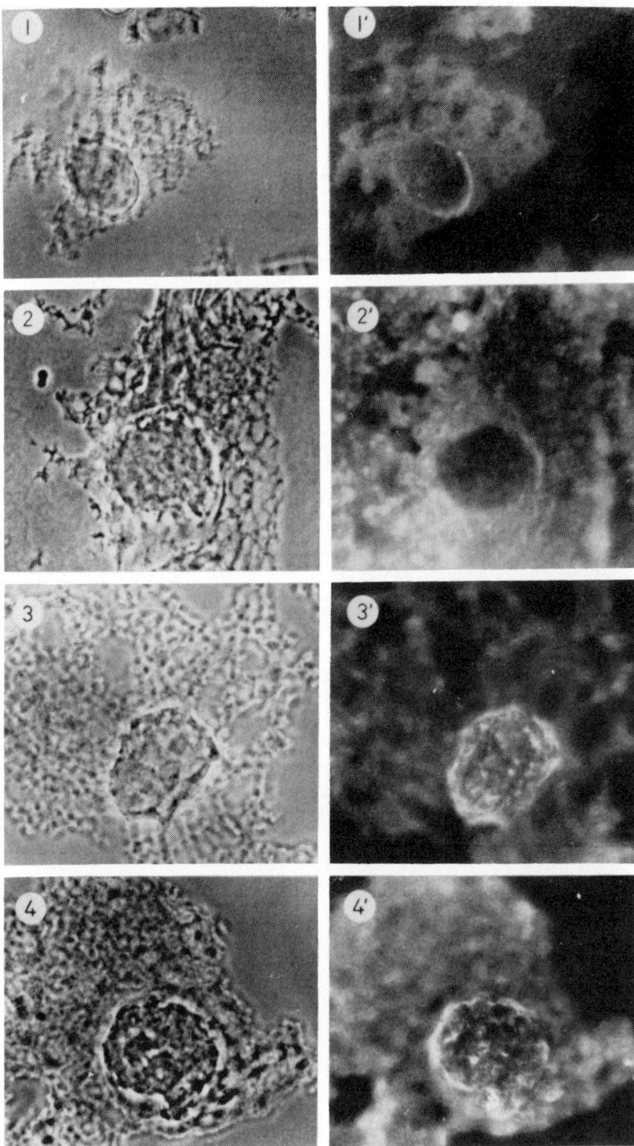

Fig. 5. Cellular localization of hsp23 by immunofluorescence in salivary gland cells: 1 and 1′, cells from larvae heated for 1 hr at 37°C, preimmune serum; 2 and 2′, cells from larvae kept at 23°C, anti-hsp23 antiserum; 3 and 3′ cells from larvae heated for 1 hr at 37°C, anti-hsp23 antiserum; 4 and 4′, cells from larvae heated for 1 hr at 37°C followed by 3 hr of recovery at 25°C, anti-hsp23 antiserum. (1–4) Phase Contrast (1–4) indirect immunofluorescence. (From Arrigo and Ahmad-Zadeh, 1981.)

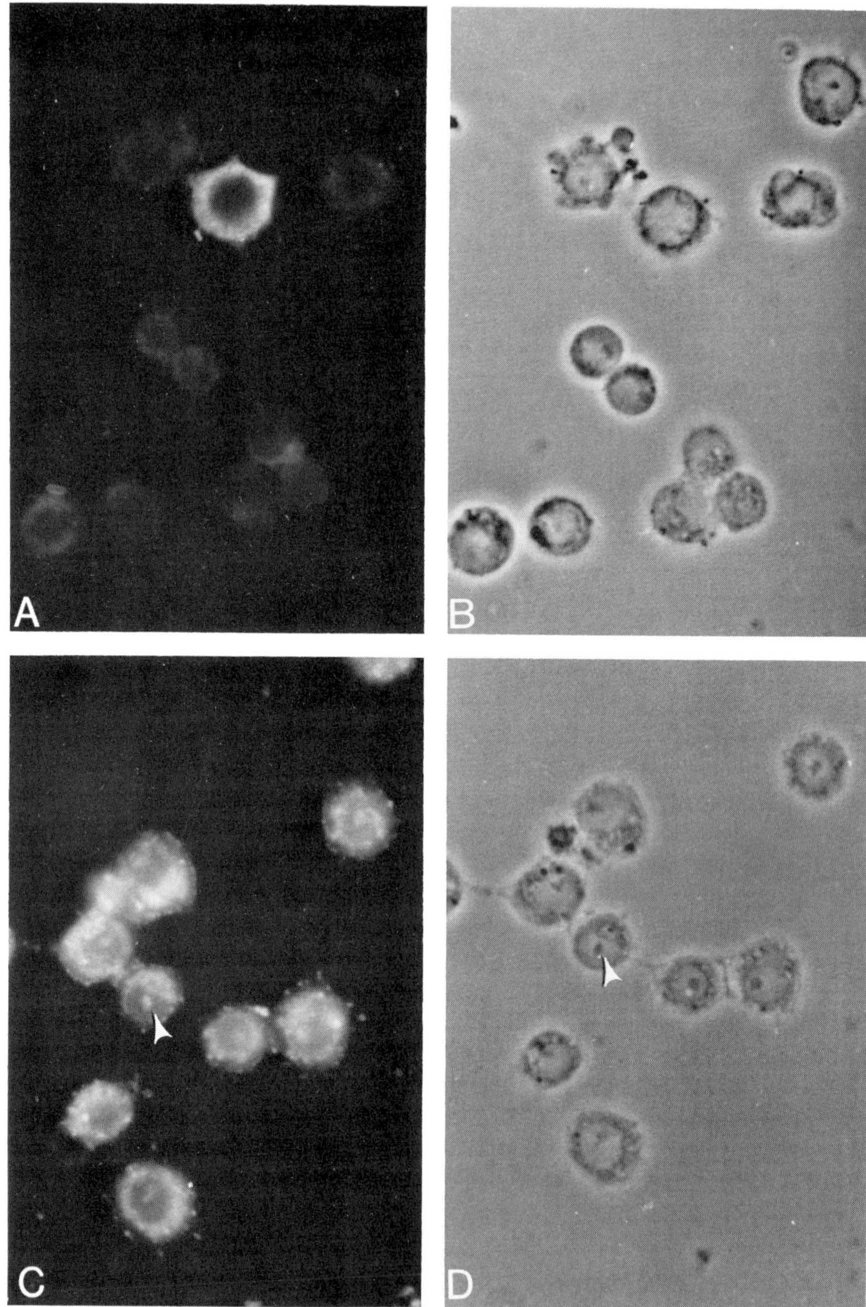
A
B
C
D

shocked Kc cells (Kloetzel and Bautz, 1983). On the other hand, the different hnRNP density of heat shock cells (1.63 g/cm^3) reported by Mayrand and Pederson (1983) remains unexplained. Finally, at the cytoplasmic level the low-M_r hsp have been shown to exist as 30 S aggregates associated with RNA of unknown function (Arrigo, 1980; Arrigo and Ahmad-Zadeh, 1981).

On the other hand, hsp's have been reported to have properties similar to cytoskeletal elements (Tanguay and Vincent, 1982). Since heat shock is accompanied by extensive cytoskeletal rearrangement (Falkner *et al.*, 1981), and since these changes seem to correlate with the intracellular localization of hsp's in cultured cells (Tanguay and Vincent, 1982; Vincent and Tanguay, 1982), an extranuclear cytoskeletal function for hsp's has also been suggested. Cytoskeletal reorganization has also been reported during heat shock in mammalian cells (Thomas *et al.*, 1982; Voellmy *et al.*, 1983).

III. IMMUNOCYTOCHEMICAL LOCALIZATION OF HEAT SHOCK PROTEINS

As seen in the previous section, the biochemical fractionation studies do not give unambiguous results on the localization of hsp's. For example, it cannot be dismissed that the hsp's associated with the nuclear pellet may be outside elements collapsed on the nuclear surface. This has been shown in the case of the intermediate filament network in vertebrate cells (Staufenbiel and Deppert, 1982). This distribution can also result from foreign structures cosedimenting with nuclei, as described by Sanders *et al.* (1982), in *Drosophila*-cultured cells. Localization of hsp's by indirect immunofluorescence techniques with specific antibodies could help solve these dilemmas and perhaps provide important clues about the function of hsp's. Such antibodies have been produced and used in localization studies (Duband *et al.*, 1985a,b).

A. Heat Shock Protein 83

The cellular distribution of heat shock protein 83, which is synthesized at 23°C and further induced at 37°C, is shown in Fig. 4. The immunocytochemical data confirm the cell fractionation studies showing an exclusively cytoplasmic lo-

---

Fig. 6. Indirect immunofluorescent localization of hsp23 in *Drosophila*-cultured cells. Control cells: (A) immunofluorescence; (B) phase contrast. Heat-shocked cells: (C) immunofluorescence; (D) phase contrast. Same conditions as in Fig. 4. The anti-hsp23 was kindly supplied by A. P. Arrigo. While most cells are negative at 23°C, hsp23 is mainly localized in the perinuclear region as bright granules after heat shock. Arrowheads point to nucleoli.

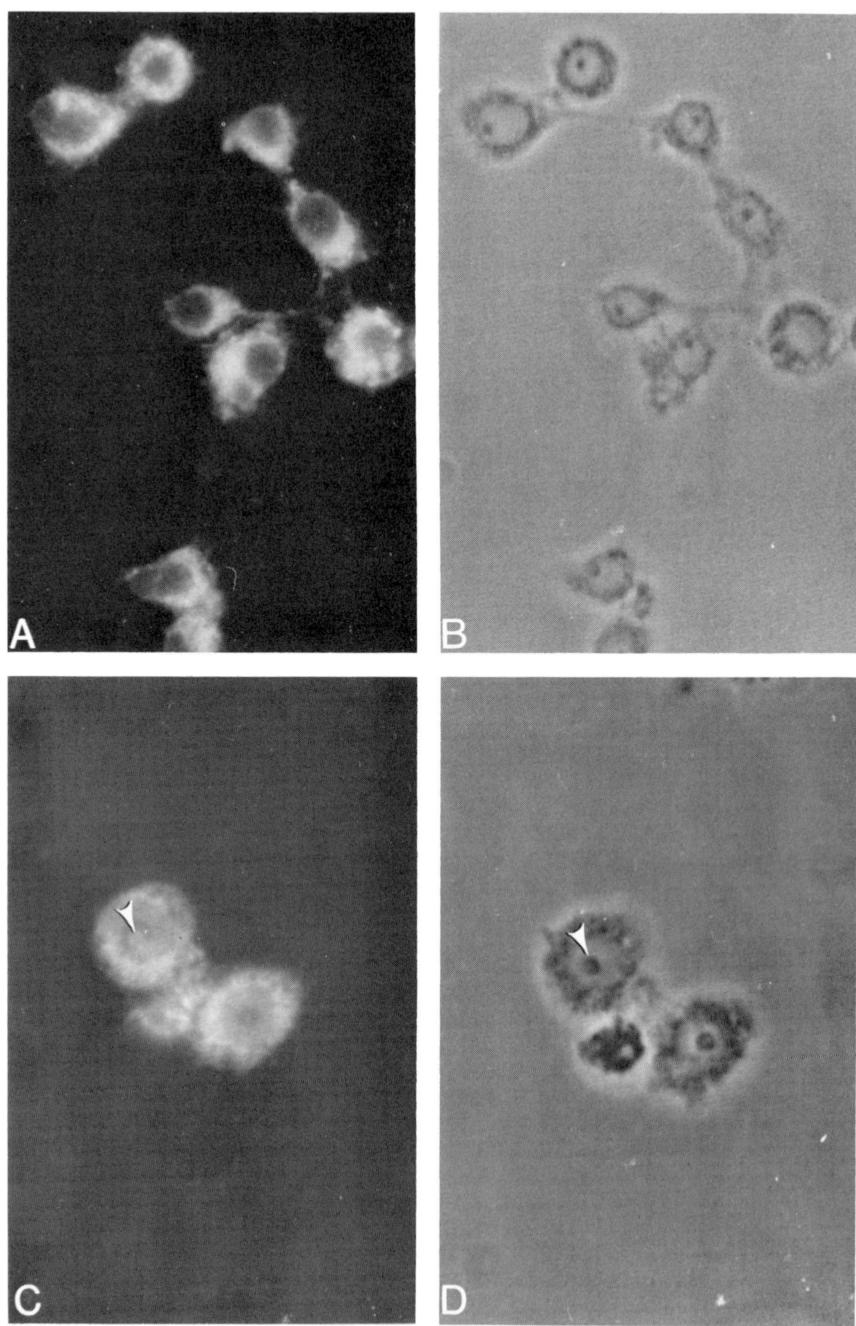
A
B
C
D

calization at either temperature. However, during heat shock the staining seems to concentrate at the cell periphery (Fig. 4C,D). A generalized cytoplasmic distribution pattern is seen within 2 hr of cell recovery at 23°C (Duband *et al.*, 1985b). This localization of hsp83 at or close to the plasma membrane is compatible with a previous suggestion of the involvement of a corresponding hsp in the modulation of plasma membrane ATPase in human cells (Burdon and Cutmore, 1982).

B. Heat Shock Protein 23

Arrigo and Ahmad-Zadeh (1981) prepared an antibody against hsp23 using the 20–30 S cytoplasmic aggregate from *Drosophila* heat-shocked cells. From immunofluorescent studies they concluded that the majority of hsp23 was in the nucleus of salivary gland cells after heat shock (Fig. 5). However, from the data presented it cannot be absolutely ascertained that the antigen is intranuclear rather than being on the surface of the nucleus (see Fig. 5, parts 3′, 4′). When assayed on *Drosophila*-cultured Kc cells (Fig. 6), this hsp23 antibody shows that the antigen has a granular perinuclear localization. There is a faint nuclear staining and the nucleolus is clearly positive (Fig. 6C,D). These granules, which are also seen in the region of the plasma membrane, disappear during recovery. A generally diffuse cytoplasmic staining pattern is then observed.

C. Heat Shock Protein 68–70 Group

An antiserum recognizing hsp68 and hsp70 shows cytoplasmic staining of *Drosophila* Kc cells at 23°C (Fig. 7A,B). Following heat shock there is an increase of fluorescent material in both the cytoplasm and the nucleus (Fig. 7C,D). It should also be observed that the perinucleolar region is negative. During recovery there is a slow return of this group of antigens to the cytoplasm. An affinity-purified antibody against hsp68 exhibits a perinuclear diffuse cytoplasmic staining at 23°C (Fig. 8A,B). After 1 hr at 37°C, the antigen retains a cytoplasmic localization with the exception of the nucleolus, which then shows intense staining while the rest of the nucleus is negative (Fig. 8C,D). Nuclear staining is seen after 1 hr of recovery at 23°C, and then the antigen returns to the cytoplasm (Duband *et al.*, 1985b).

The hsp 68–70 group is composed of a high number of subspecies as reported by Buzin and Petersen (1982). Some of these referred to as cognate proteins are

---

Fig. 7. Indirect immunofluorescent localization of the hsp68–70 complex in *Drosophila*-cultured cells. Control cells: (A) immunofluorescence; (B) phase contrast. Heat-shocked cells: (C) immunofluorescence; (D) phase contrast. Same conditions as in Fig. 4. The arrowheads show the unstained perinucleolar region.

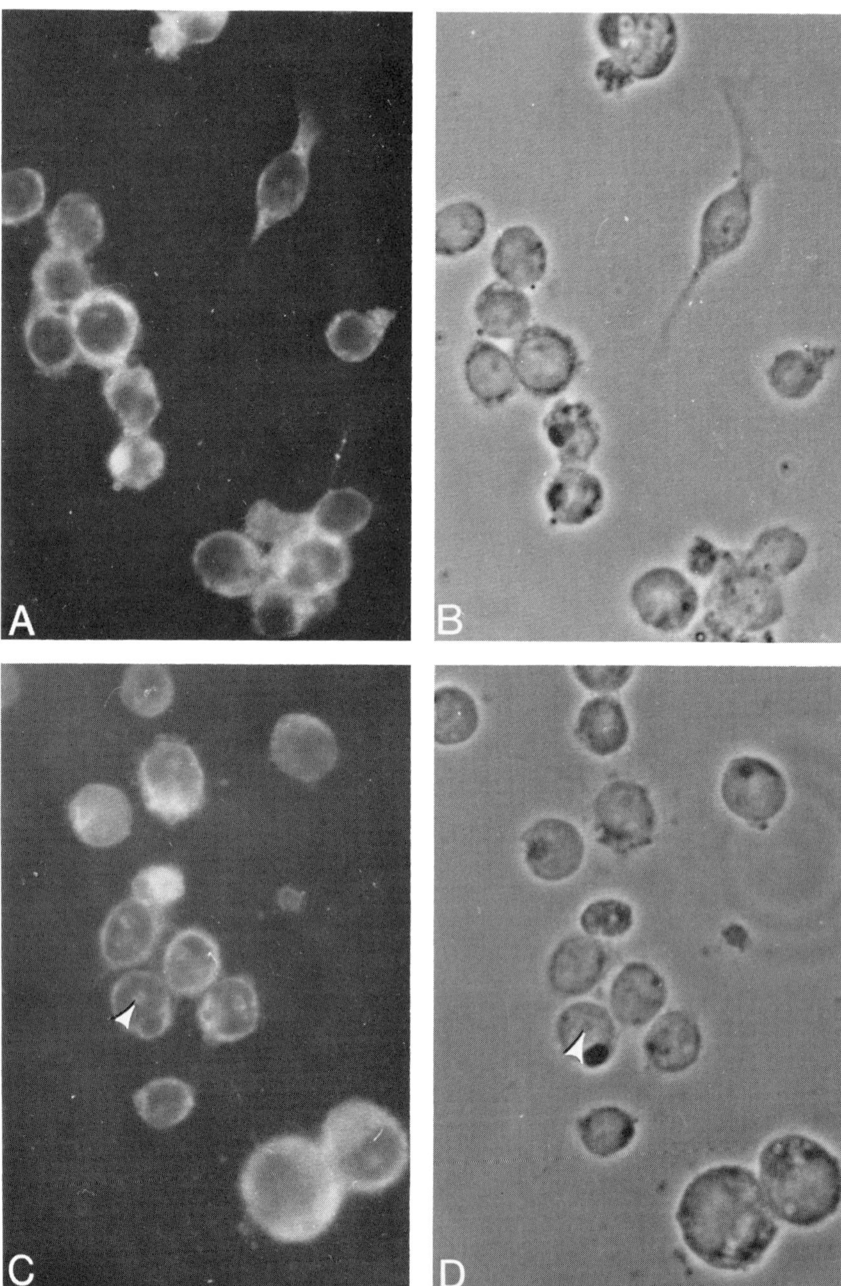

Fig. 8. Indirect immunofluorescent localization of anti-hsp68 antibody in *Drosophila*-cultured cells. Control cells: (A) immunofluorescence; (B) phase contrast. Heat-shocked cells: (C) immunofluorescence; (D) phase contrast. Same conditions as in Fig. 4. The antibody was affinity purified and its specificity tested by protein blotting. Note the positive nucleolar staining after heat shock

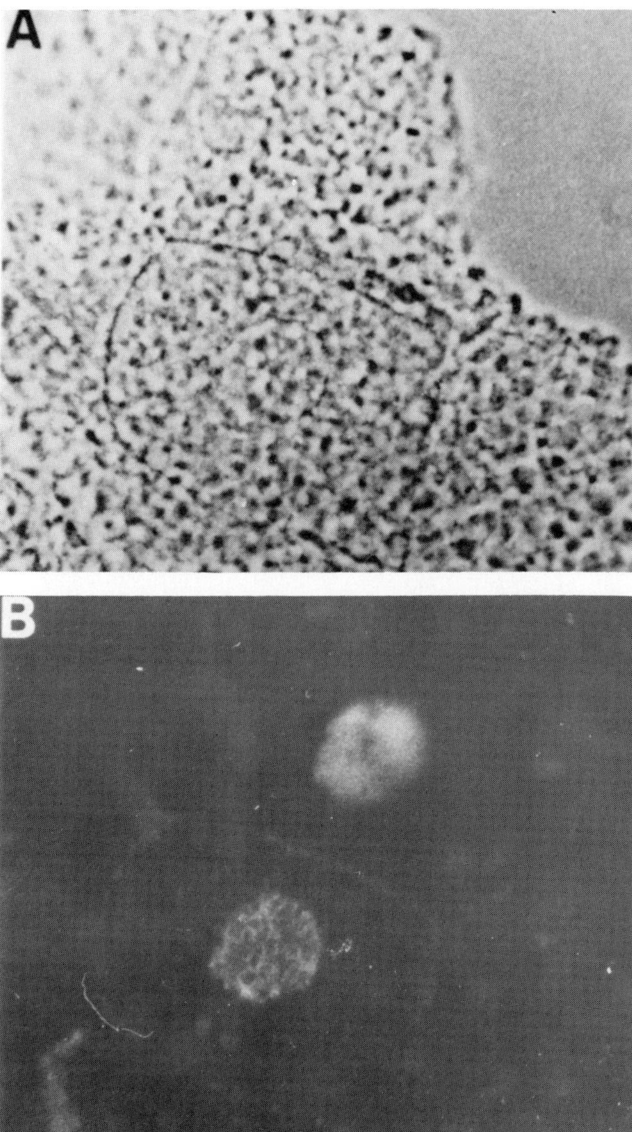

Fig. 9. Indirect immunofluorescence localization of hsp70 in *Drosophila* salivary gland cells with a monoclonal antibody. (A) Phase contrast; (B) immunofluorescence. Larvae were heat shocked for 1 hr at 36.5°C. (Courtesy of S. Lindquist.)

(arrowheads). The preparation of the anti-hsp68 antibody and the immunofluorescence staining are described in Duband *et al.* (1985a,b).

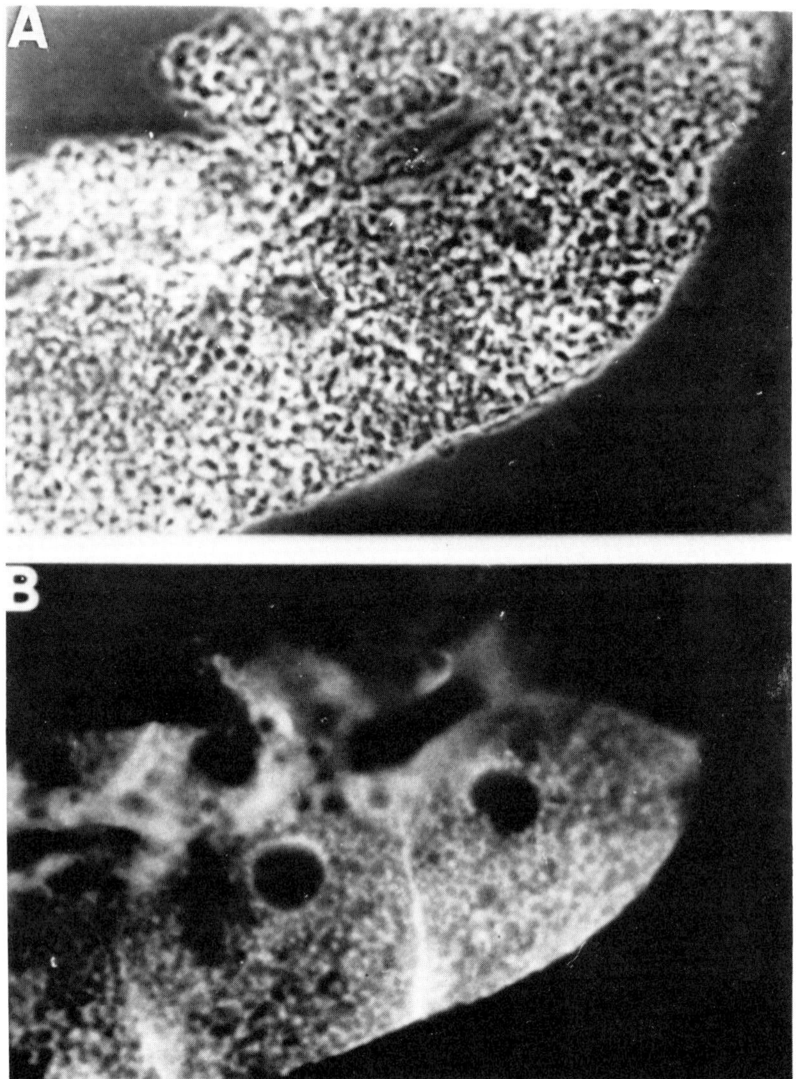

Fig. 10. Localization of hsp70 during recovery. (A) Phase contrast; (B) immunofluorescence. Larvae were allowed 5 hr of recovery at 25°C after a 1-hr heat shock at 36.5°C. (Reproduced with permission from Velazquez and Lindquist, 1984. Copyright 1984 M.I.T. Press.)

also present at normal temperature and do not show induction by heat (Ingolia and Craig, 1982a). Furthermore these various species are not uniformly distributed in the nucleus and the cytoplasm (R. M. Tanguay *et al.*, in preparation). The availability of monoclonal antibodies can help in localizing these variants. The localization of hsp70 with a monoclonal antibody following a 1-hr heat shock to salivary gland cells of *Drosophila* is shown in Fig. 9. The staining is clearly nuclear. After 5 hr of recovery at 25°C, this antigen shows a diffuse cytoplasmic localization (Fig. 10).

In summary, the immunological localization studies are generally in agreement with the biochemical fractionation data. Furthermore they support the existence of distinct classes of hsp's in terms of cellular localization.

IV. PUTATIVE FUNCTION OF HEAT SHOCK PROTEINS

A. Heat Shock Proteins; Normal or Abnormal Proteins

While hsp's were originally defined as new proteins, further studies in various laboratories have led to a widened definition of hsp's as proteins whose synthesis is induced or increased at supraoptimal temperatures or under various conditions of cellular stress. Thus, in *Drosophila,* the number of hsp's now seems considerably larger than the original number of seven (reviewed in Petersen and Mitchell, 1984). For example, even one of the core histones, H2B, has been reported to be induced by heat shock (Sanders, 1981; Tanguay *et al.,* 1983).

If one concentrates on the major heat shock proteins, the question of their expression under normal conditions, that is, without stress, is of considerable importance for the elucidation of their function. While their presence in cultured cells at normal temperatures can be criticized on the basis of culture conditions, recent data on the expression of these genes in animals during development suggest that most hsp's are expressed at one time or another during normal development in *Drosophila*. Thus, the low-M_r group of hsp's can be independently induced by the molting hormone ecdysterone, in both cultured cells and imaginal disks (Ireland and Berger, 1982; Ireland *et al.,* 1982). Further evidence for the differential transcription of these genes during normal development has been reported by Zimmerman *et al.,* (1983) in the early embryo and by Sirotkin and Davidson (1982) during puparium formation.

The presence of hsp83 under normal conditions has been reported by numerous laboratories as an indication of its function under these physiological conditions (Mirault *et al.,* 1977; Arrigo, 1980; Storti *et al.,* 1980; Chomyn and Mitchell, 1982; Lindquist, 1980). The presence of hsp70 in normal cells is somewhat more controversial (Sanders, 1981; Findly and Pederson, 1981; Kloetzel and Bautz, 1983; Velazquez *et al.,* 1983). The answer to this problem is further complicated by the presence of different proteins and variants of this group and by the existence of cognate hsp70 proteins (see Section III,C). Further studies with specific probes will be needed to assess the function of this protein class.

Studies in vertebrate cells also show the presence, although at a reduced amount, of hsp's in normal cells and in unstressed animal tissues (Welch and Feramisco, 1982; Welch *et al.*, 1983; Currie and White, 1983). But, again in this case, the presence of variants and perhaps of cognate proteins similar to those described in *Drosophila* does not permit one to decide whether those genes are expressed or not in unstressed cells.

B. Heat Shock Proteins, Thermotolerance, and Cellular Recovery

An increased resistance to supraoptimal temperatures has been correlated with the induction of hsp synthesis in various organisms such as yeast (McAlister and Finkelstein, 1980), *Dictyostelium* (Loomis and Wheeler, 1982), insects (Mitchell *et al.*, 1979; Dean and Atkinson, 1983; Berger and Woodward, 1983), and in various mammalian cell lines (Landry *et al.*, 1982a,b; Li *et al.*, 1982). While the correlation seems strong, a direct causal effect of hsp's on the development of thermotolerance remains to be shown. Landry and Chrétien (1983) have recently reported the induction of hsp's without the development of thermotolerance in Morris hepatoma cells. Furthermore, when the increase in temperature is slow, thermotolerance develops without an increase in the rate of hsp synthesis (Tomasovic *et al.*, 1983). Finally, it is rather difficult to identify which hsp could be directly involved in this process. While the ubiquitous well-conserved major hsp70 protein seems a likely candidate, the low-M_r group of hsp's, which seem less conserved, could also be involved. In *Dictyostelium*, for example, a mutant lacking expression of the low-M_r hsp's fails to develop thermotolerance (Loomis and Wheeler, 1982). Similarly in *Drosophila*, ecdysterone, which induces only the low-M_r hsp's, can nonetheless induce thermotolerance (Berger and Woodward, 1983). Also, it cannot be excluded that the various hsp's could perform a similar function through a general mechanism such as the nonspecific stabilization of proteins as proposed by Minton *et al.* (1982). The presence of hsp's in the nuclear fraction and the fact that the response seems to be self-regulated at both the transcriptional and translational levels (DiDomenico *et al.*, 1982) could be suggestive of a function at the chromatin level. However, when thermotolerance is induced by ecdysterone in *Drosophila* (Berger and Woodward, 1983), the hsp's do not have a nuclear localization (Ireland *et al.*, 1982). Thus, while there seems to be a correlation between the induction of hsp's and thermotolerance, this relation seems quite complex and the molecular mechanism underlying it remains obscure.

Finally, in *Drosophila*, the synthesis of hsp's has been shown to be essential for normal cell recovery following heat shock (Petersen and Mitchell, 1981; DiDomenico *et al.*, 1982). The involvement of hsp's in translation could be equally important in processes such as protection from phenocopy induction and protection from various teratogens following sublethal mild heat shocks (Buzin and Bournias-Vardiabasis, 1982).

V. SUMMARY

While there is a high conservation of heat shock proteins from bacteria to man, it is still too early to assign them functions based on subcellular localization studies. For example, the observed localization of the various hsp's varies depending on the organism and on the nature of the stress. At this time, it cannot be excluded that the apparently contradictory localization of hsp's in various systems could result from the varying intensity of the stress given.

On the other hand, these localization studies indicate that the various hsp's probably have different cellular functions. This is supported by observations that, in *Drosophila* at least, gene expression of the various hsp's can be uncoupled. During oogenesis, for example, mRNA's for hsp 83, 28, and 26 accumulate while no RNA's for hsp68 or hsp70 are produced (Zimmerman *et al.*, 1983). The uncoupling of hsp83 from the low-M_r hsp's can also be observed, since the latter are already abundant when the former appears. The early synthesis of these hsp's is suggestive of their function in early embryonic development. Furthermore, the fact that hsp70 cannot be induced at that time (Dura, 1981) suggests that its function probably differs from that of hsp 83, 28, and 26. Another indication for a variety of functions comes from observations on the rate of synthesis of the various hsp's during cellular recovery. Thus, while hsp70 is maximally synthesized under heavy stress conditions, its synthesis is rapidly turned off during recovery in contrast to the low-M_r hsp's, which are made at a maximum rate during this period.

Both biochemical and immunofluorescent data indicate a structural function for hsp's. The putative structural nature of the low-M_r hsp's has already been suggested by sequence analyses of these genes, which show regional homology with α-crystallin (Ingolia and Craig, 1982b). This vertebrate lens protein forms high-molecular-weight aggregates. The hsp 22–28 group also shows a tendency to form aggregates (Arrigo, 1980).

The observation that hsp's can show a different intracellular localization depending on the nature of the inducer stresses the importance of taking this factor into consideration in any attempt to derive function from cellular localization data. With this reservation, further biochemical studies, coupled with a more precise localization of the various hsp's with specific antibodies and at the electron microscopic level, should eventually lead to a better understanding of the function of these proteins at both the cellular and the molecular level.

Acknowledgments

I am particularly grateful to Dr. A. P. Arrigo and Dr. S. Lindquist for communicating unpublished material. The hsp23 antibody was kindly provided by Dr. A. P. Arrigo. I would also like to thank Mr. J. L. Duband for helpful discussions and reviewing of the manuscript and Mrs. M. A. Godbout for secretarial assistance. The author's work described here has been supported by grants from The Medical Research Council of Canada and l'Université Laval (Budget Special Recherche).

REFERENCES

Adams, C., and Rinne, R. W. (1982). Stress protein formation: gene expression and environmental interaction with evolutionary significance. *Int. Rev. Cytol.* **79,** 305–315.

Arrigo, A. P. (1980). Etude des protéines induites par le choc thermique chez *Drosophila melanogaster*. Ph.D. thesis, Univ. of Geneva.

Arrigo, A. P., and Ahmad-Zadeh, C. (1981). Immunofluorescence localization of a small heat shock protein (hsp23) in salivary gland cells of *Drosophila melanogaster. Mol. Gen. Genet.* **184,** 73–79.

Arrigo, A. P., Fakan, S., and Tissières, A. (1980). Localization of the heat shock proteins in *Drosophila melanogaster* tissue culture cells. *Dev. Biol.* **78,** 86–103.

Ashburner, M., and Bonner, J. J. (1979). The induction of gene activity in *Drosophila* by heat shock. *Cell* **17,** 241–254.

Berger, E. M., and Woodward, M. P. (1983). Small heat shock proteins in *Drosophila* may confer thermal tolerance. *Exp. Cell Res.* **147,** 437–442.

Burdon, R. H., and Cutmore, C. M. M. (1982). Human heat shock gene expression and the modulation of plasma membrane Na^+ K^+ ATPase activity. *FEBS Lett.* **140,** 45–48.

Buzin, C. H., and Bournias-Vardiabasis, N. (1982). The induction of heat shock proteins by drugs that inhibit differentiation in *Drosophila* embryonic cell cultures. *In* "Heat Shock, from Bacteria to Man" (M. J. Schlesinger, M. Ashburner, and A. Tissières, eds.), pp. 387–394. Cold Spring Harbor Lab., Cold Spring Harbor, New York.

Buzin, C. H., and Petersen, N. S. (1982). A comparison of the multiple *Drosophila* heat shock proteins in cell lines and larval salivary glands by two-dimensional gel electrophoresis. *J. Mol. Biol.* **158,** 181–201.

Chomyn, A., and Mitchell, H. K. (1982). Synthesis of 84,000 dalton protein in normal and heat shocked *Drosophila melanogaster* cells as detected by specific antibody. *Insect Biochem.* **12,** 105–114.

Currie, R. W., and White, F. P. (1983). Characterization of a 71-kilodalton protein induced in rat tissues after hyperthermia. *Can. J. Biochem. Cell Biol.* **61,** 438–446.

Dean, R. L., and Atkinson, B. G. (1983). The acquisition of thermal tolerance in larvae of *Calpodes ethlius* (Lipidoptera) and the *in situ* and *in vitro* synthesis of heat shock proteins. *Can. J. Biochem. Cell Biol.* **61,** 472–479.

DiDomenico, B. J., Bugaisky, G. E., and Lindquist, S. (1982). The heat shock response is self-regulated at both the transcriptional and post transcriptional levels. *Cell* **31,** 593–603.

Duband, J. L., Valet, J. P., and Tanguay, R. M. (1985a). Production and characterization of antibodies specific to three major heat shock proteins of *Drosophila melanogaster* cultured cells. Submitted for publication.

Duband, J. L., Lettre, F., Arrigo, A. P., and Tanguay, R. M. (1985b). Immunofluorescence localization of the major heat shock proteins (hsp's 83, 70, 68, 23) in normal, heat shocked, and recovering *Drosophila* cultured cells. Submitted for publication.

Dura, J. M. (1981). Stage dependent synthesis of heat shock induced proteins in early embryos of *Drosophila melanogaster. Mol. Gen. Genet.* **184,** 381–385.

Falkner, F. G., Saumweber, H., and Biessmann, H. (1981). Two *Drosophila melanogaster* proteins related to intermediate filament proteins of vertebrate cells. *J. Cell. Biol.* **91,** 175–183.

Findly, R. C., and Pederson, T. (1981). Regulated transcription of the genes for actin and heat shock proteins in cultured *Drosophila* cells. *J. Cell Biol.* **88,** 323–328.

Holt, Th. K. H. (1971). Local protein accumulation during gene activation. II. Interferometric measurements of the amount of solid material in temperature induced puffs of *Drosophila hydei. Chromosoma* **32,** 428–435.

Hughes, E. N., and August, J. T. (1982). Coprecipitation of heat shock proteins with a cell surface glycoprotein. *Proc. Natl. Acad. Sci. U.S.A.* **79,** 2305–2309.

Ingolia, T. D., and Craig, E. A. (1982a). *Drosophila* gene related to the major heat shock-induced gene is transcribed at normal temperature and not induced by heat shock. *Proc. Natl. Acad. Sci. U.S.A.* **79,** 525–529.

Ingolia, T. D., and Craig, E. A. (1982b). Four small *Drosophila* heat shock proteins are related to each other and to mammalian α-crystallin. *Proc. Natl. Acad. Sci. U.S.A.* **79,** 2360–2364.

Ireland, R. C., and Berger, E. M. (1982). Synthesis of low molecular weight heat shock peptides stimulated by ecdysterone in a cultured *Drosophila* cell line. *Proc. Natl. Acad. Sci. U.S.A.* **79,** 855–859.

Ireland, R. C., Berger, E. M., Sirotkin, K., Yund, M. A., Osterbur, D., and Fristrom, J. (1982). Ecdysterone induces the transcription of four heat-shock genes in *Drosophila* S3 cells and imaginal discs. *Dev. Biol.* **93,** 498–507.

Kelley, P. M., and Schlesinger, M. J. (1982). Antibodies to two major chicken heat shock proteins cross-react with similar proteins in widely divergent species. *Mol. Cell. Biol.* **2,** 267–274.

Kloetzel, P. M., and Bautz, E. K. F. (1983). Heat-shock proteins are associated with hnRNA in *Drosophila melanogaster* tissue culture cells. *EMBO J.* **2,** 705–710.

Landry, J., and Chrétien, P. (1983). Relationship between hyperthermia-induced heat shock proteins and thermotolerance in Morris hepatoma cells. *Can. J. Biochem. Cell Biol.* **61,** 428–437.

Landry, J., Chrétien, P., Bernier, D., Nicole, L. M., Marceau, N., and Tanguay, R. M. (1982a). Thermotolerance and heat shock proteins induced by hyperthermia in rat liver cells. *Int. J. Radiat. Oncol. Biol. Phys.* **8,** 59–62.

Landry, J., Bernier, D., Chrétien, P., Nicole, L. M., Tanguay, R. M., and Marceau, N. (1982b). Synthesis and degradation of heat shock proteins during development and decay of thermotolerance. *Cancer Res.* **42,** 2457–2461.

Leenders, H. J., Berendes, H. D., Helmsing, P. J., Derksen, J., and Koninkx, J. F. J. G. (1974). Nuclear-mitochondrial interactions in the control of mitochondrial respiratory metabolism. *Subcell. Biochem.* **3,** 119–147.

Levinger, L., and Varshavsky, A. (1981). Heat shock proteins of *Drosophila* are associated with nuclease-resistant high salt resistant nuclear structures. *J. Cell Biol.* **90,** 793–796.

Li, G. C., Petersen, N. S., and Mitchell, H. K. (1982). Induced thermal tolerance and heat shock protein synthesis in CHO cells. *Int. J. Radiat. Oncol. Biol. Phys.* **8,** 63–67.

Lindquist, S. (1980). Varying patterns of protein synthesis in *Drosophila* during heat shock: implications for regulation. *Dev. Biol.* **77,** 463–479

Loomis, W. F., and Wheeler, S. A. (1982). Chromatin-associated heat shock proteins of *Dictyostelium. Dev. Biol.* **90,** 412–418.

McAlister, L., and Finkelstein, D. B. (1980). Heat shock proteins and thermal resistance in yeast. *Biochem. Biophys. Res. Commun.* **93,** 819–824.

Mayrand, S., and Pederson, T. (1983). Heat shock alters nuclear ribonucleoprotein assembly in *Drosophila* cells. *Mol. Cell. Biol.* **3,** 161–171.

Minton, K. W., Karmin, P., Hahn, G., and Minton, A. P. (1982). Nonspecific stabilization of stress-susceptible proteins by stress-resistant proteins: a model for the biological role of heat shock proteins. *Proc. Natl. Acad. Sci. U.S.A.* **79,** 7107–7111.

Mirault, M.-E., Goldschmidt-Clermont, M., Moran, L., Arrigo, A. P., and Tissières, A. (1977). The effect of heat shock on gene expression in *Drosophila melanogaster. Cold Spring Harbor Symp. Quant. Biol.* **42,** 819–827.

Mitchell, H. K., and Lipps, L. (1975). Rapidly labeled proteins on the salivary gland chromosomes of *Drosophila melanogaster. Biochem. Genet.* **13,** 585–602.

Mitchell, H. K., Moller, G., Petersen, N. S., and Lipps-Sarmiento, L. (1979). Specific protection from phenocopy induction by heat shock. *Dev. Genet.* (*Amsterdam*) **1,** 181–192.

Petersen, N. S., and Mitchell, H. K. (1981). Recovery of protein synthesis after heat shock: prior heat treatment affects the ability of cells to translate mRNA. *Proc. Natl. Acad. Sci. U.S.A.* **78,** 1708–1711.

Petersen, N. S., and Mitchell, H. K. (1984). Heat shock proteins. *In* "Comprehensive Insect Physiology, Biochemistry and Pharmacology" (G. A. Kerkut and L. A. Gilbert, eds.), Pergamon, New York (in press).

Sanders, M. M. (1981). Identification of histone H2b as heat shock protein in *Drosophila. J. Cell Biol.* **91,** 579–583.

Sanders, M. M., Fenney-Triemer, D., Olsen, A. S., and Farrell-Towt, J. (1982). Changes in protein phosphorylation and histone H2b disposition in heat shock in *Drosophila. In* "Heat Shock, from Bacteria to Man" (M. J. Schlesinger, M. Ashburner, and A. Tissières, eds.), pp. 235–242. Cold Spring Harbor Lab., Cold Spring Harbor, New York.

Schlesinger, M. J., Aliperti, G., and Kelley, P. M. (1982). The response of cells to heat shock. *Trends Biochem. Sci. (Pers. Ed.)* **7,** 222–225.

Sinibaldi, R. M., and Morris, P. W. (1981). Putative Function of *Drosophila melanogaster* heat shock proteins in the nucleoskeleton. *J. Biol. Chem.* **256,** 10,735–10,738.

Sirotkin, K., and Davidson, N. (1982). Developmentally regulated transcription from *Drosophila melanogaster* chromosomal site 67B. *Dev. Biol.* **89,** 196–210.

Staufenbiel, M., and Deppert, W. (1982). Intermediate filament systems are collapsed onto the nuclear surface after isolation of nuclei from tissue culture cells. *Exp. Cell Res.* **138,** 207–214.

Storti, R. V., Scott, M. P., Rich, A., and Pardue, M. L. (1980). Translational control of protein synthesis in response to heat shock in *D. melanogaster. Cell* **22,** 825–834.

Tanguay, R. M. (1983). Genetic regulation during heat shock and function of heat-shock proteins: a review. *Can. J. Biochem. Cell Biol.* **61,** 387–394.

Tanguay, R. M., and Vincent, M. (1980). Intracellular distribution of heat shock induced proteins in diptera. *Eur. J. Cell Biol.* **22,** 44.

Tanguay, R. M., and Vincent, M. (1981). Biosynthesis and characterization of heat shock proteins in *Chironomus tentans* salivary glands. *Can. J. Biochem.* **59,** 67–73.

Tanguay, R. M., and Vincent, M. (1982). Intracellular translocation of cellular and heat shock induced proteins upon heat shock in *Drosophila* Kc cells. *Can. J. Biochem.* **60,** 306–315.

Tanguay, R. M., Camato, R., Lettre, F., and Vincent, M. (1983). Expression of histone genes during heat shock and in arsenite-treated *Drosophila* Kc cells. *Can. J. Biochem. Cell Biol.* **61,** 414–420.

Thomas, G. P., Welch, W. J., Mathews, M. B., and Feramisco, J. R. (1982). Effects of heat shock and related treatments of mammalian tissue-culture cells. *Cold Spring Harbor Symp. Quant. Biol.* **46,** 985–996.

Tissières, A., Mitchell, H. K., and Tracy, U. M. (1974). Protein synthesis in salivary glands of *Drosophila melanogaster*: relation to chromosome puffs. *J. Mol. Biol.* **84,** 389–398.

Tomasovic, S. P., Steck, P. A., and Heitzman, D. (1983). Heat-stress proteins and thermal resistance in rat mammary tumor cells. *Radiat. Res.* **95,** 399–414.

Velazquez, J. M., and Lindquist, S. (1984). hsp70: Nuclear concentration during environmental stress and cytoplasmic storage during recovery. *Cell* **36,** 655–662.

Velazquez, J. M., DiDomenico, B. J., and Lindquist, S. (1980). Intracellular localization of heat shock proteins in *Drosophila. Cell* **20,** 679–689.

Velazquez, J. M., Sonoda, S., Bugaisky, G., and Lindquist, S. (1983). Is the major *Drosophila* heat shock protein present in cells that have not been heat shocked? *J. Cell Biol.* **96,** 286–290.

Vincent, M., and Tanguay, R. M. (1979). Heat-shock induced proteins present in the cell nucleus of *Chironomus tentans* salivary gland. *Nature (London)* **281,** 501–503.

Vincent, M., and Tanguay, R. M. (1982). Different intracellular distribution of heat shock and arsenite-induced proteins in *Drosophila* Kc cells. Possible relation with the phosphorylation and translocation of a major cytoskeletal protein. *J. Mol. Biol.* **162,** 365–378.

Voellmy, R., Bromley, P., and Kocher, H. P. (1983). Structural similarities between corresponding heat-shock proteins from different eucaryotic cells. *J. Biol. Chem.* **258,** 3516–3522.

Welch, W. J., and Feramisco, J. R. (1982). Purification of the major mammalian heat shock proteins. *J. Biol. Chem.* **257,** 14,949–14,959.

Welch, W. J., Garrels, J. I., Thomas, G. P., Lin, J. J. C., and Feramisco, J. R. (1983). Biochemical characterization of the mammalian stress proteins and identification of two stress proteins as glucose and Ca^{2+}-ionophore regulated proteins. *J. Biol. Chem.* **258,** 7102–7111.

Zimmerman, J. L., Petri, W., and Meselson, M. (1983). Accumulation of a specific subset of *D. melanogaster* mRNAs in normal development without heat shock. *Cell* **32,** 1161–1170.

6

Heat Shock Proteins in Sea Urchin Development

GIOVANNI GIUDICE

I. INTRODUCTION

In the last few years, the scientific literature has repeatedly stressed the important and intriguing phenomena involved with the production of heat shock proteins. Some fundamental questions have been asked: Are the phenomena restricted to special tissues like the salivary gland? How do entire organisms react to a rise in temperature? How does heat shock affect developmental processes?

To answer these and other questions dealt with in this chapter, the sea urchin embryo has proved a very convenient model for a number of reasons. It is very easy to raise the temperature of a culture medium in which millions of individuals develop synchronously and reach the larval stage in a couple of days.

Changes in Eukaryotic Gene Expression
in Response to Environmental Stress

ISBN 0-12-066290-6

Furthermore, a great deal of information is available on the morphological and biochemical features of its development (see Giudice, 1973, for a review).

II. HEAT TREATMENT OF EMBRYOS AT GASTRULA STAGE

Cultures of *Paracentrotus lividus* embryos are normally maintained at 20°C. We raised the temperature of cultures of embryos at the gastrula stage for 45 min, added [^{3}H]leucine, and maintained the cultures for an additional 15 min at the same temperature (Guidice *et al.*, 1980). Embryos were then lysed in buffered SDS, and their proteins were analyzed by electrofluorography according to Laskey and Mills (1975). When the temperature was raised to 25°C, little effect on protein synthesis was observed. At 37°C an apparently complete inhibition of protein synthesis occurred. However, at 31°C bulk protein synthesis was greatly inhibited (over 50%), while the synthesis of a few proteins, the heat shock proteins (hsp's), was greatly stimulated (Fig. 1). The most highly labeled and frequently encountered hsp's observed in sea urchin embryos have apparent molecular weights of 72,500 and 70,000, and it is these two proteins that we will

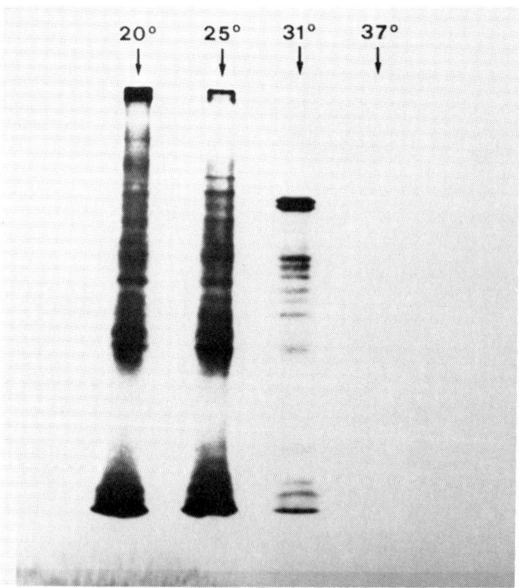

Fig. 1. Effect of heat treatment at different temperatures on the pattern of protein synthesis of *P. lividus* gastrulas. (From Roccheri *et al.*, 1981b.)

refer to as hsp's throughout this chapter. Other minor hsp's have molecular weights of 79,000, 50,000, 47,000, 44,000, and 42,000, respectively (Roccheri *et al.*, 1981a). This response is not unique to *Paracentrotus;* embryos of *Arbacia lixula* respond to heat shock in the same manner.

We raised the temperature of *Paracentrotus* embryos for various time periods. Only 15 min at the higher temperature was required to elicit the described response, that is, inhibition of bulk protein synthesis and stimulation of hsp synthesis. This response was greater when the length of treatment was increased up to about 1 hr, remained unchanged for about 3 more hours, and then began to decline. This apparently paradoxical behavior was explained, as is described later, by a study which showed that hsp synthesis occurs in a wave shortly after heating, and then, after about 4 hr, the pattern of protein synthesis starts to revert to normal even if the high temperature is maintained (Roccheri *et al.*, 1981b).

III. HEAT TREATMENT OF EMBRYOS AT DIFFERENT DEVELOPMENTAL STAGES

Up to now we have described the effect of heat treatment on embryos at the gastrula stage. What happens if embryos at other stages are exposed to heat shock? The results of such an experiment are reported in detail in Fig. 2 (Roccheri *et al.*, 1981b). If the embryos are heated at any stage between fertilization and early blastula, there is severe inhibition of bulk protein synthesis but no synthesis of heat shock proteins. When the embryos are heated at any stage from the hatched blastula onward, besides inhibition of bulk protein synthesis, a great stimulation of the synthesis of heat shock proteins is observed. There is, therefore, a turning point during development, the hatched blastula stage, at which embryos acquire the ability to respond to heat with the production of heat shock proteins.

This observation is intriguing because hatching has long been known to be a turning point for several aspects of sea urchin embryonic development (Giudice, 1973). For example, only after hatching are new messenger RNA's needed for development to proceed (Giudice *et al.*, 1968), paternal antigens are detectable only at hatching, and it is shortly after hatching that the development of lethal hybrids is arrested (reviewed in Giudice, 1973).

The next question, therefore, was: What happens to development when the embryos are heated for 45 min. at different developmental stages either before or after hatching? The answer to this question is detailed in Table I (Roccheri *et al.*, 1981b) and can be summarized in the interesting conclusion that follows: "embryos heated at any stage before hatching do not produce heat shock proteins and rapidly degenerate; embryos heated at any stage following hatching produce a wave of heat shock proteins and continue to develop normally.''

These observations suggest a role for heat shock proteins in protection from thermal shock. That development is arrested by heat shock at stages preceding hatching can be explained by the fact that bulk protein synthesis is severely inhibited and reversal is never observed, even if the embryos are returned to normal temperature, whereas such a reversal is observed when heat is applied at posthatching stages. A comparable phenomenon has been described by Graziosi *et al.* (1980) in *Drosophila* embryos. These, if heated at stages preceding the migration of nuclei to the periphery, produce small amounts of heat shock

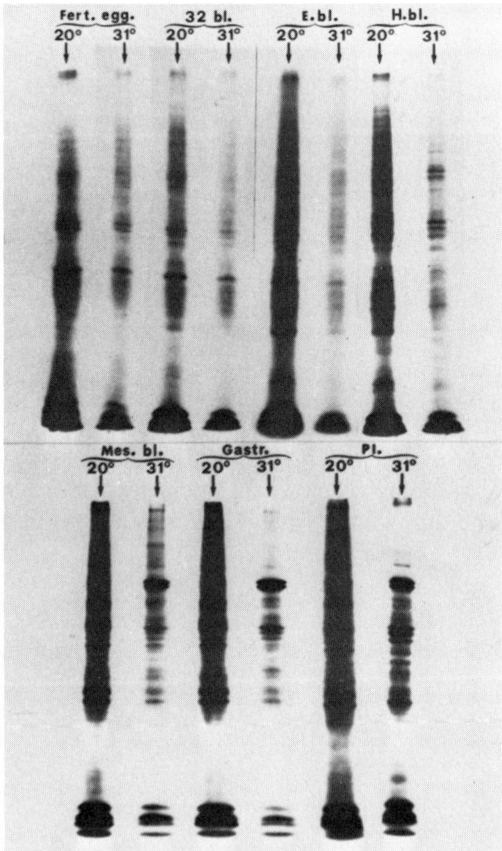

Fig. 2. Effect of heating at 31°C on the pattern of protein synthesis at different developmental stages. Fert.egg., fertilized eggs; 32 bl., 32 blastomeres; E.bl., early blastulas; H.bl., hatched blastulas; Mes.bl., mesenchyme blastulas; Gastr., gastrulas; Pl., plutei. (From Roccheri *et al.*, 1981b.)

TABLE I

Effect on Development of Heating at 31°C at Different Embryonic Stages

Stage of treatment	Duration of heating	After fertilization (36 hr)
Fertilized eggs	1 hr	50% Degeneration 50% Irregular morulas
2 Blastomeres	1 hr	50% Degeneration 25% Irregular morulas 25% Irregular blastulas
32 Blastomeres	1 hr	44% Degeneration 55% Irregular blastulas 1% Gastrulas
Early blastulas	1 hr	34% Degeneration 65% Irregular blastulas 1% Gastrulas
Hatched blastulas	1 hr	99% Plutei 1% Degeneration
Mesenchyme blastulas	1 hr	100% Plutei
Gastrulas	1 hr	100% Plutei
Prisms	1 hr	100% Plutei
Gastrulas	4 hr	100% Plutei
Gastrulas	8 hr	100% Plutei
Gastrulas	Continuous	100% Plutei
Control embryos	None	100% Plutei

proteins and degenerate; if heated at later stages, they produce larger amounts of heat shock proteins and continue developing.

It is of interest that sea urchin oocytes, when heated, are able to produce heat shock proteins. This ability, therefore, must be switched off after oogenesis in the egg and then switched on again after hatching.

As has already been mentioned, the production of the heat shock proteins at sensitive, that is, posthatching, stages occurs in a wave after heating, and then, irrespective of whether the embryos are returned to normal temperature, protein synthesis reverts to the usual complex pattern (see Fig. 3), and then heat shock proteins are no longer synthesized.

IV. FATE OF THE HEAT SHOCK PROTEINS AFTER REVERSAL TO NORMAL PROTEIN SYNTHESIS

What happens to the heat shock proteins which have already been synthesized? Are they destroyed when the reversal of protein synthesis occurs, or do they remain for some time within the embryo? To answer this question a pulse and chase experiment was performed: embryos at the mesenchyme blastula or at the gastrula stage were pulse labeled with [^{3}H]leucine for 15 min after 45 min of heating, as usual, and then an excess of cold leucine was added in order to prevent further isotope incorporation. The electrophoretic pattern of the labeled proteins, shown in Fig. 4, remained unchanged for at least an additional 23 hr, that is, until the pluteus stage, thus demonstrating that "the heat shock proteins, although synthesized for only a short period after heating, are not dismantled and remain within the embryo for a much longer time."

Are the hsp's still serving a function in the embryo? That this might be so is suggested by the observation that, if heat-shocked embryos are brought back to normal temperature and then heated again, they produce fewer heat shock proteins.

V. HEAT SHOCK PROTEIN SYNTHESIS IN DISSOCIATED CELLS

As we have described, the ability to respond to heat treatment by synthesis of heat shock proteins arises at a precise period of development, that is, after hatching. Is this because a certain embryonic architecture is needed for this property to be developed, or do the cells "know" that, after a certain time following egg fertilization, they have to develop such a property, irrespective of cell interactions? To answer this question we carried out three experiments (Sconzo *et al.*, 1984).

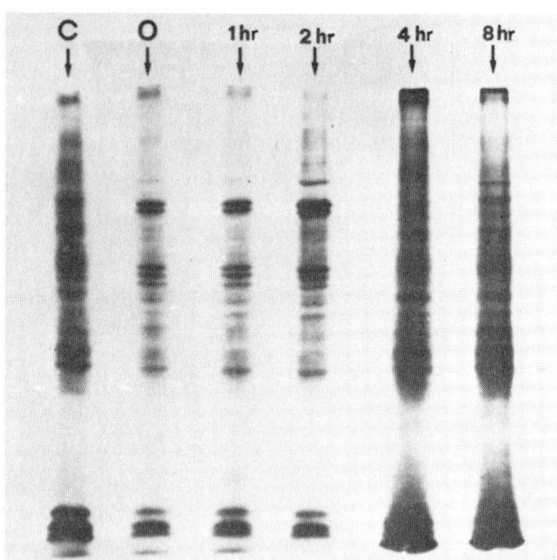

Fig. 3. Reversal of the pattern of protein synthesis at various time lengths after 1 hr of heating for *Paracentrotus* gastrulas. C, control, nonheated embryos; O, 1 hr of heating, and 1 hr of heating followed by 1, 2, 4, and 8 hr, respectively, at normal temperature. (From Roccheri *et al.*, 1981b.)

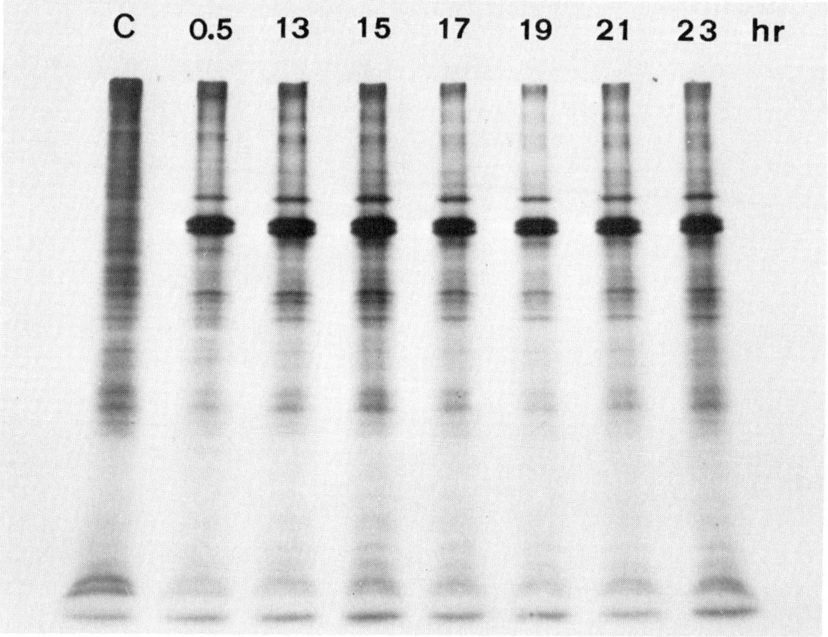

Fig. 4. Electrofluorograms of the proteins of embryos homogenized at various times after 1 hr of heating and which had been pulse labeled at the time of heating. C, nonheated embryos. (From Roccheri *et al.*, 1982.)

In the first one, embryos at different developmental stages were dissociated into cells and reaggregated, according to the procedure of Giudice (1973), or the procedure of Noll *et al.*, (1979) for preblastular stages, and then immediately tested for their response to heat shock. It was invariably found that the freshly dissociated cells retained the ability to respond to heat shock, characteristic of the embryonic stage of development; i.e., cells dissociated from preblastular stages, if heated, underwent a severe inhibition of bulk protein synthesis and did not synthesize heat shock proteins, whereas cells dissociated from postblastular stages and immediately heated underwent inhibition of bulk protein synthesis but at the same time synthesized heat shock proteins.

In the second experiment, the embryos were dissociated into cells at the 16-blastomere stage, that is, when they were still unable to synthesize heat shock proteins, and the isolated cells were allowed to reaggregate, according to the procedure of Giudice (1962). When the nondissociated control embryos reached the early gastrula stage, that is, a stage at which heat shock proteins can be synthesized, the aggregates, still in the form of blastulalike structures, were heated and tested for their ability to produce heat shock proteins. It was found that they too had acquired such an ability.

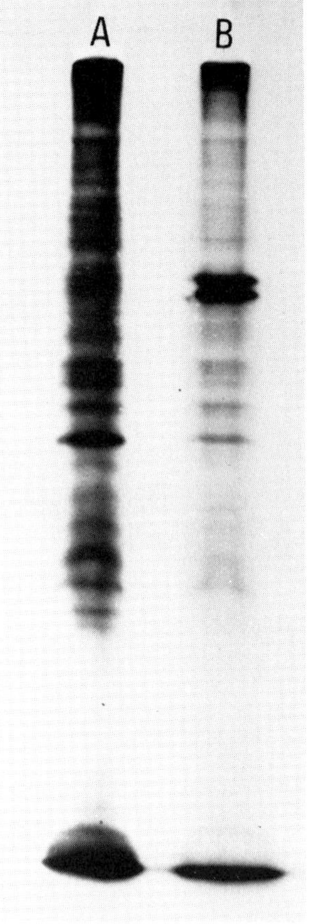
A
B

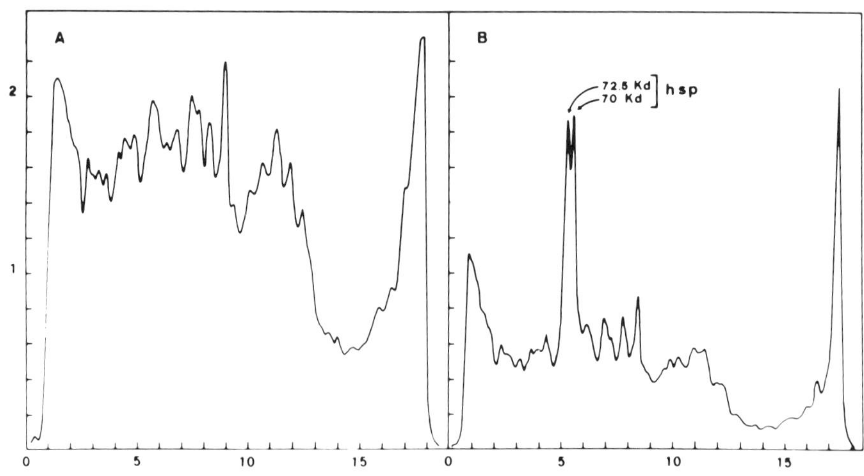
A
B
72.5 Kd
70 Kd
hsp
2
1
0
5
10
15

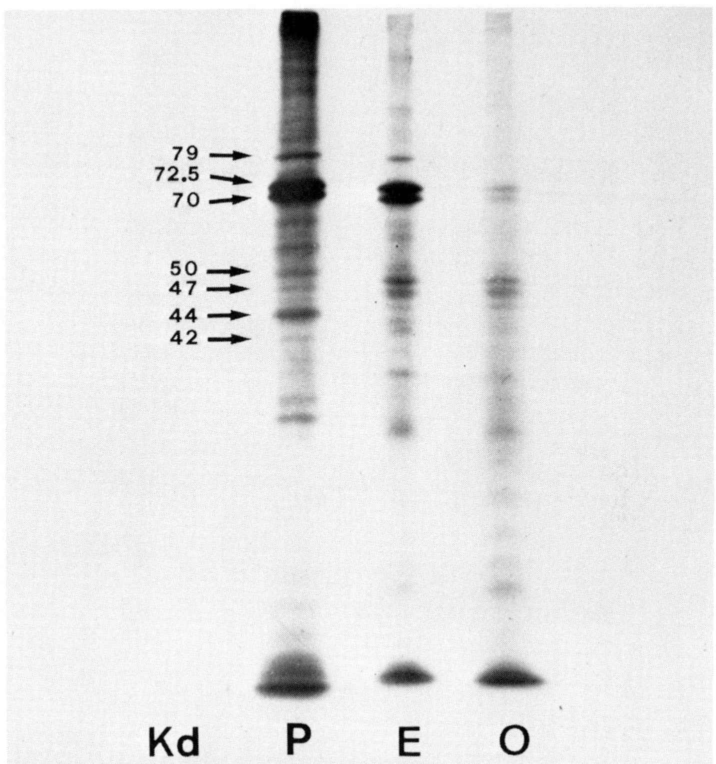

Fig. 6. Electrofluorogram of the proteins of plutei labeled after heat shock. P, entire plutei; E, ectoderm, and O, other tissues. (From Roccheri *et al.*, 1981a.)

In the third experiment, embryos were again dissociated into cells at the 16-blastomere stage, but this time the cells were not permitted to reaggregate by overdiluting them according to Vittorelli *et al.* (1973). When the undissociated control embryos reached the early gastrula stage, the dissociated, nonreaggregated cells were heated and tested for their ability to produce heat shock proteins. This time, also, the cells had developed such an ability (Fig. 5).

The conclusion of this set of experiments is, therefore, that "the sea urchin embryonic cells, after a certain time following egg fertilization, develop the ability to respond to heating with the production of heat shock proteins even in the complete absence of cell interactions."

What then triggers such a change in the cells? A simple hypothesis is that such

---

Fig. 5. Fluorograms (upper) and densitograms (lower) of the labeled proteins of cells dissociated at 16 blastomeres and kept unreaggregated for 18 hr. (A) Kept at 20°C; (B) labeled after heating. (From Sconzo *et al.*, 1983.)

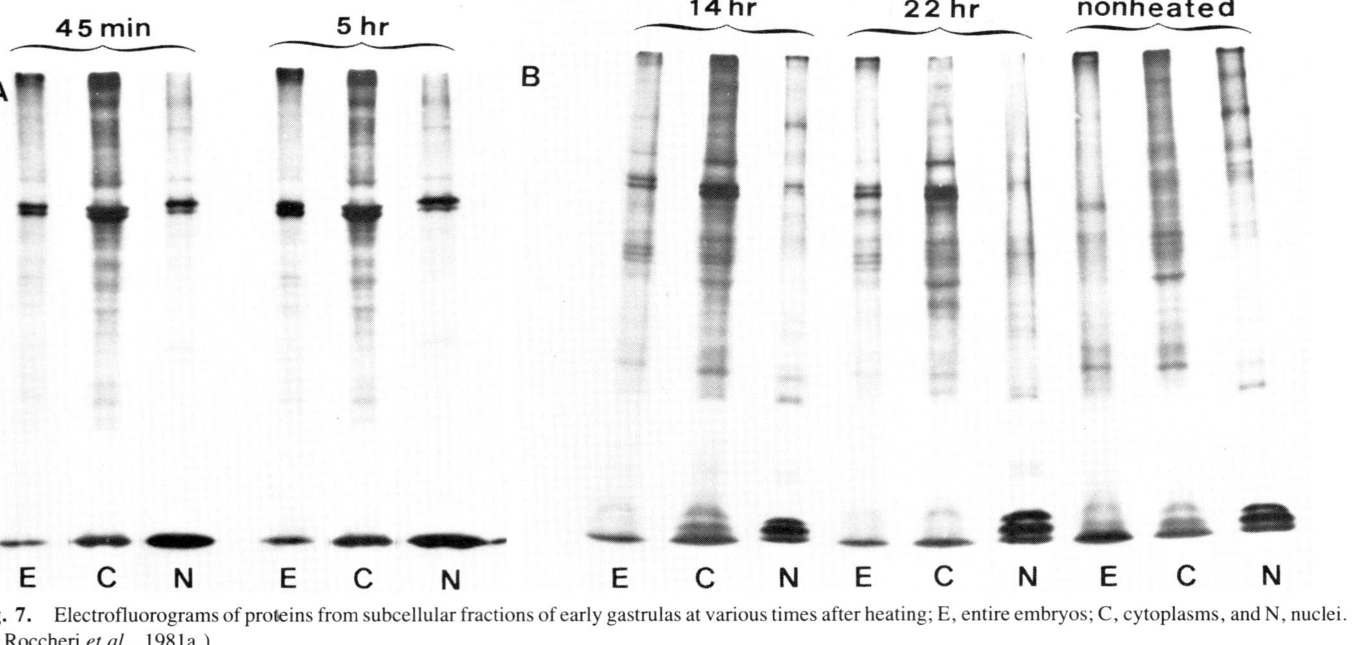

Fig. 7. Electrofluorograms of proteins from subcellular fractions of early gastrulas at various times after heating; E, entire embryos; C, cytoplasms, and N, nuclei. (From Roccheri *et al.*, 1981a.)

a property arises in the embryonic cells after a certain number of cell divisions. This hypothesis may be tested by the use of inhibitors of DNA synthesis, such as aphidicholin, known not to interfere with the synthesis of RNA and proteins.

VI. TISSUE SPECIFICITY IN THE PRODUCTION OF HEAT SHOCK PROTEINS

Although these embryonic cells do not need to interact to develop the ability to synthesize heat shock proteins, do all the cells synthesize hsp's, or only those cells belonging to special embryonic territories?

We approached this question (Roccheri *et al.*, 1981a) by subjecting embryos at the pluteus stage to heat shock and pulse labeling, and then separating and isolating the ectoderm from one side and the remaining tissues from the other, according to the method of McClay and Chambers (1978), which was slightly modified to investigate if both groups of tissues had synthesized the heat shock proteins. The results reported in Fig. 6 show that "the ectoderm is clearly able to

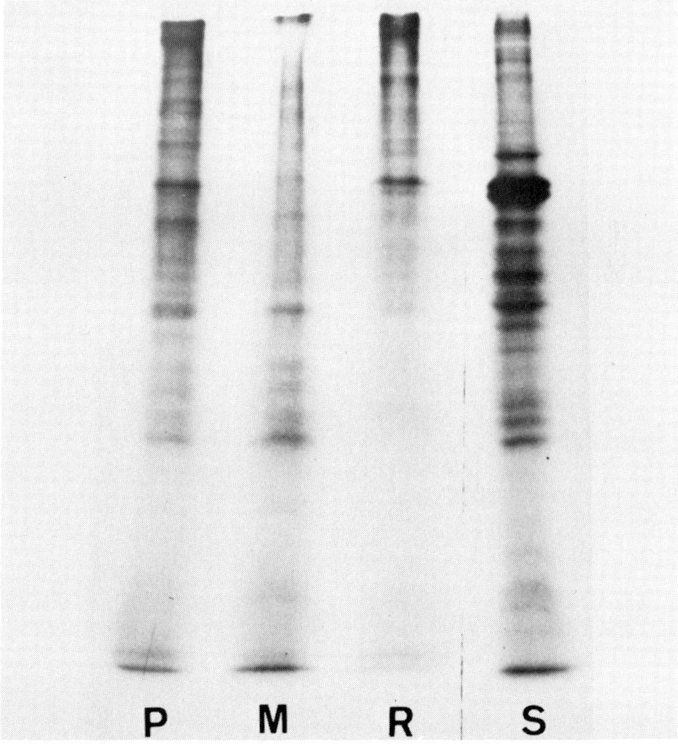

Fig. 8. Electrofluorograms of proteins of the cytoplasmic fractions from early gastrulas labeled after heat shock. P, polysomes; M, monosomes; R, RNP particles, and S, soluble cytoplasm. (From Roccheri *et al.*, 1981a.)

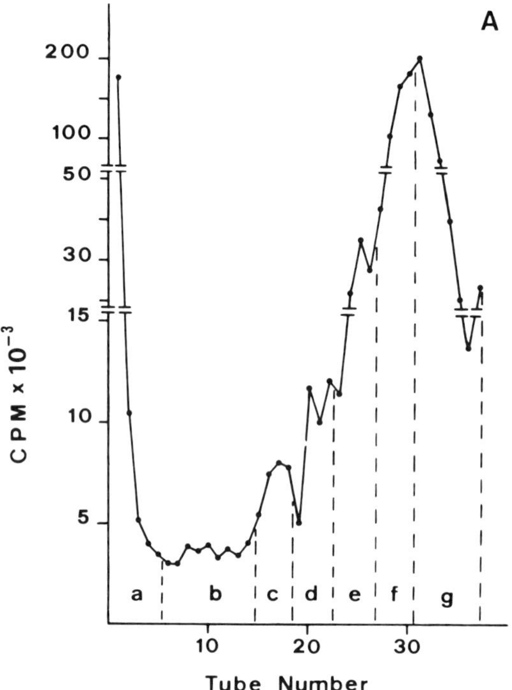
A
200
100
50
30
15
10
5
CPM x 10^{-3}
a
b
c
d
e
f
g
10
20
30
Tube Number

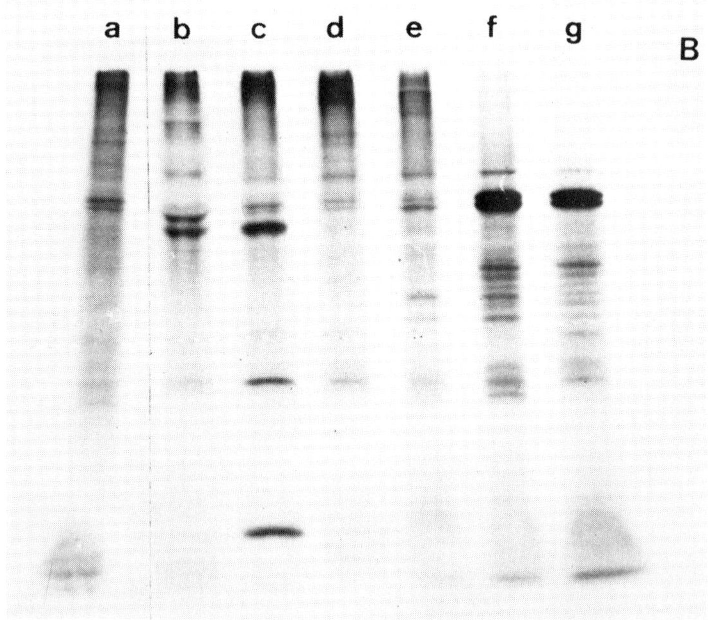
a
b
c
d
e
f
g
B

synthesize heat shock proteins, whereas the other tissues, although undergoing the inhibition of bulk protein synthesis following heating, synthesize much less, if any, heat shock protein.''

VII. INTRACELLULAR LOCATION OF HEAT SHOCK PROTEINS

Roccheri *et al.* (1981a) also investigated the intracellular location of heat shock proteins. Figure 7 shows that they are located mainly in the cytoplasm, either at short or at much longer times after heating. Two bands migrating closely, although not corresponding exactly to the two major heat shock proteins, are also seen in the nuclei at the earliest investigated times. The hsp's are, however, clearly visible in the nuclei shortly after heating and persist after a 5-hr chase. They tend to fade after a longer chase, so that they are barely visible after 14 hr and have almost completely disappeared after 22 hr. Do they have a function in the nucleus? We do not know but would like to recall here that proteins which bind to the DNA of their loci and repress the transcription of their own mRNA's have been described in bacteria (Putney and Schimmel, 1981) and that some proteins have been observed to bind to a locus for heat shock proteins (Jack *et al.*, 1981).

If the cytoplasm of heat-shocked embryos is resolved into its particulate and soluble components, most heat shock proteins are found in the soluble cytoplasm (Fig. 8), and, if this is further centrifuged to better purify it from contaminating RNP particles, it is still the lighter part of the gradient which contains most of the heat shock proteins (Fig. 9). The bands in zones 3 and 4 of the figure cannot be interpreted as heat shock proteins because they are also present with the same intensity in the nonheated embryos. One can, therefore, conclude that ``most of the two main heat shock proteins appear immediately in the soluble cytoplasm, where they accumulate and persist for at least 22 hr. A minor part also appears in the nuclei at a short time after heating, disappearing after 5 hr.''

VIII. BULK PROTEIN SYNTHESIS INHIBITION AFTER HEATING

The mechanism of inhibition of bulk protein synthesis following heating is still unknown. Some conclusions can, however, be drawn: The polysome-to-monosome ratio of the heated embryos is very much in favor of monosomes compared

Fig. 9. (A) Distribution in a sucrose gradient of TCA-insoluble radioactivity in the postribosomal region of the cytoplasm of early gastrulas labeled after heat shock; (B) electrofluorograms of aliquots from each fraction of A, which were adjusted to contain the same number of counts. (From Roccheri *et al.*, 1981a.)

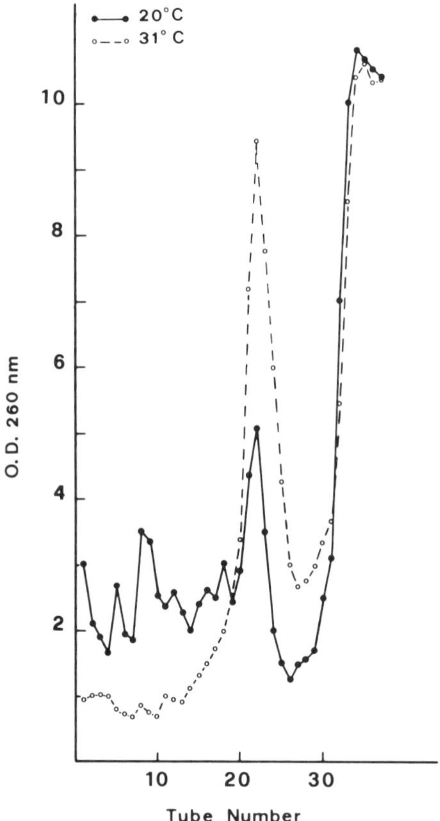

Fig. 10. Comparison of the polysomal profiles from embryos reared at normal temperature (●——●) or after 1 hr of heating (○- - -○). (From Roccheri *et al.*, 1982.)

to that of the control embryos (Fig. 10), that is, 0.56 (±0.09) and 1.22 (±0.14), respectively (Roccheri *et al.*, 1982).

A second observation is that the inhibition of bulk protein synthesis by heating does not depend on the production of heat shock proteins, since it also occurs in the absence of hsp's, for example, when the embryos are heated at stages preceding hatching.

The third observation is that the reversal of the inhibition of the overall protein synthesis, as can be observed 4 hr after the heating of hatched embryos, does not depend on the synthesis of new RNA's, because it also occurs in an apparently normal way in the presence of actinomycin D at concentrations which are known to inhibit over 70% of the RNA synthesis (Roccheri *et al.*, 1982). We can, therefore, conclude that "the inhibition of bulk protein synthesis following heating is due to disengagement of the ribosomes from the mRNA; it does not depend

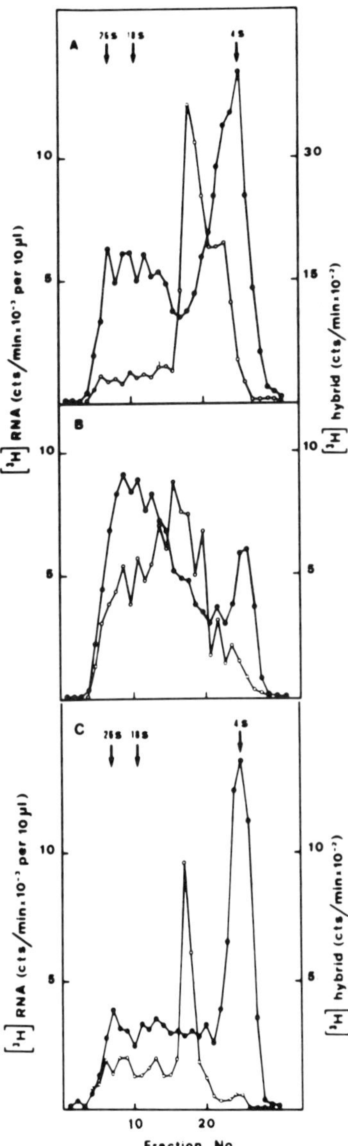

Fig. 11. Hybridization of histone DNA with nuclear RNA fractionated on glyoxal/agarose gels. Labeled nuclear RNA's, extracted from control (A), heat shocked (B), and chased (C) mesenchyme blastula embryos, were denatured with glyoxal and fractionated on horizontal 1.5% agarose gels in 10 m*M* phosphate buffer, pH 7.0. The dimensions of the gel were 1 cm × 18 cm × 21.5 cm. Electrophoresis was carried out at 70 V for 8.5 hr. Slices of 0.5 cm were cut and the RNA eluted. Each RNA fraction was hybridized with pPH70 DNA in 2 × NaCl/Citrate at 65°C. [^{3}H]RNA (●——●) and RNA–DNA hybrid (○——○). (From Spinelli *et al.*, 1982.)

upon the synthesis of heat shock proteins and can be reversed in the absence of transcription.''

This does not necessarily mean that heating has no effect on the bulk synthesis of RNA in sea urchin embryos, but only that they have a store of long-lived mRNA's (Giudice, 1973) which are not destroyed during the period following heating and preceding recovery. In fact, Spinelli *et al.* (1982) have directly investigated the effect of heat shock on the synthesis of RNA with special regard to that of the histone mRNA. If mesenchyme blastulas are submitted to the described heat shock procedure and then pulse labeled with [^{3}H]uridine, the

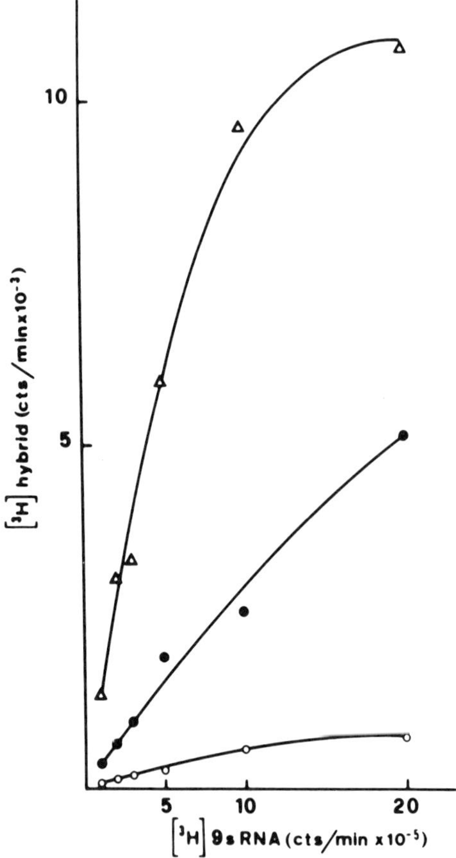

Fig. 12. Saturation of histone DNA with 9 S RNA. The cytoplasmic labeled RNA's, extracted from the same embryos used for the experiment described in Fig. 11, were fractionated on 15–30% sucrose density gradients and the 7 to 12 S RNA was hybridized with pPH70 DNA. RNA heat shock (○———○); RNA chase (●———●), and RNA control (△———△). (From Spinelli *et al.*, 1982.)

electrophoretic profile of their total RNA, under fully denaturing conditions, shows an accumulation of radioactivity in the large-sized RNA classes and a decrease in the small-sized RNA classes (Fig. 11A), where radioactivity, on the contrary, is predominant in the nonheated embryos (Fig. 11B). However, if these embryos are returned to normal temperature and a chase of 3 hr is given in the presence of an excess of cold uridine and cytidine, the radioactivity is observed to be chased from the heavy classes to the light ones (Fig. 11C). If histone RNA is identified by hybridization with a purified histone DNA probe, it is found that very little 9 S histone RNA is synthesized under heat shock conditions, whereas a larger RNA containing the histone sequences does appear. After the chase radioactivity is transferred from this large histone RNA to the 9 S histone mRNA. This suggests that heating has slowed down the maturation of a large precursor of the 9 S histone mRNA.

This inhibition of the maturation of the precursor of histone mRNA can be assumed also to slow down its transfer to the cytoplasm. That this is the case is demonstrated by the experiment reported in Fig. 12. A hybridization experiment using purified histone DNA at saturation with cytoplasmic RNA extracted from nonheated, heated, and heated and chased embryos shows, in fact, that histone mRNA is found in very small amounts in the heated embryos compared to the nonheated ones and that it reappears in substantial amounts in the cytoplasm after chasing.

The results of these experiments suggest that "heat shock probably inhibits maturation of the large RNA precursors into mature RNA's and their transport to cytoplasm. Such a hypothesis has been specifically verified for histone mRNA."

IX. DEPENDENCE OF HEAT SHOCK PROTEIN SYNTHESIS ON SYNTHESIS OF CORRESPONDING mRNA'S

Whereas the reversal of inhibition of bulk protein synthesis is regulated at a posttranscriptional level, the synthesis of heat shock proteins in sea urchins depends on the synthesis of the corresponding mRNA's. In fact, if hatched embryos are heated in the presence of various concentrations of actinomycin D, the synthesis of heat shock proteins is inhibited at the same time as that of RNA (Fig. 13). It is interesting to note that here again bulk protein synthesis is inhibited after hatching, although no heat shock proteins are produced.

In order to investigate whether the RNA whose synthesis is needed for the production of heat shock proteins is their own mRNA, Roccheri *et al.* (1982) extracted the total RNA from heated and nonheated hatched embryos, divided it into poly(A)$^+$ and poly(A)$^-$ fractions, and tested all the RNA's for their ability to stimulate the synthesis of heat shock proteins in a wheat germ cell-free system.

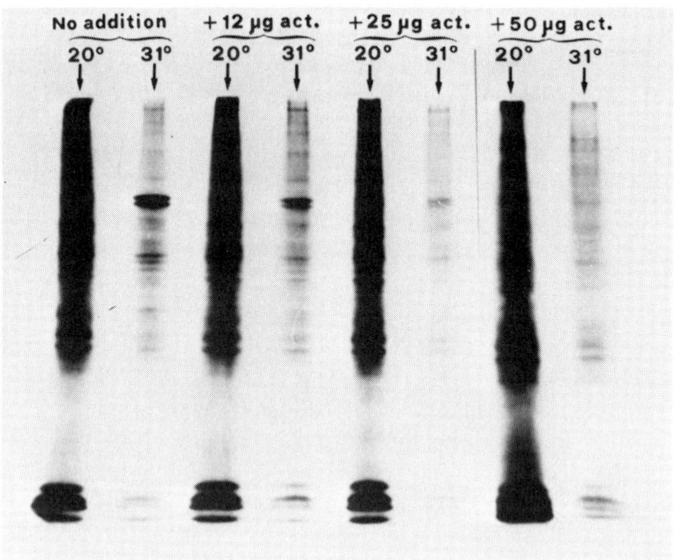

Fig. 13. Effect of actinomycin D on the production of heat shock proteins. (From Roccheri *et al.*, 1981b.)

The results showed that only the RNA's extracted from the heated embryos stimulated such a synthesis, the poly(A)$^+$ being more active than the poly(A)$^-$ fraction. One can, therefore, safely conclude that "the synthesis of heat shock protein in sea urchin embryos is regulated through the production of their mRNA's."

In a further attempt to clarify the mechanism through which this transcription is regulated, we digested the chromatin of heated or nonheated hatched embryos with pancreatic I DNase and demonstrated by hybridization with a labeled probe that the loci for heat shock proteins are exposed to DNase attack in only the heated embryos.

The fact that sea urchin heat shock proteins have the same molecular weights as those of *Drosophila* and other species suggests that their DNA has been highly conserved throughout evolution, as is to be expected if they have an important function for the survival of cells under various adverse conditions. In order to further investigate this point, we have cross-hybridized DNA from various species with a purified probe from *Drosophila* (56H8), corresponding to the 70-kilodalton heat shock protein. The probe was modified by removal of the repeated ends by treatment with the restriction enzyme *Sal,* which cuts the probe at its 5′ terminus and 160 nucleotides inside the 3′ terminus of its coding part. The results of such experiments show that DNA of the heat shock locus of *Drosophila* cross-hybridizes to that of sea urchins and of humans (G. Giudice, unpublished observations). This shows an extremely high conservation of these loci, thus, again, suggesting a very important function for them.

REFERENCES

Giudice, G. (1962). Restitution of whole larvae from disaggregated cells of sea urchin embryos. *Dev. Biol.* **5,** 402–411.

Giudice, G. (1973). "Developmental Biology of the Sea Urchin Embryo." Academic Press, New York.

Giudice, G., Mutolo, V., and Donatuti, G. (1968). Gene expression in sea urchin development. *Whilhelm Roux's Arch. Dev. Biol.* **161,** 118–128.

Giudice, G., Roccheri, M. C., and Di Bernardo, M. G. (1980). Synthesis of "heat shock" proteins in sea urchin embryos. *Cell Biol. Int. Rep.* **4,** 69–74.

Graziosi, G., Miceli, F., Marzari, R., De Cristini, F., and Savoini, A. (1980). Variability of response of early *Drosophila* embryos to heat shock. *J. Exp. Zool.* **214,** 141–145.

Jack, R. S., Gehring, W. J., and Brack, C. (1981). Protein component from *Drosophila* larval nuclei showing sequence specificity for a short region near a major heat shock protein gene. *Cell* **24,** 321–331.

Laskey, R. A., and Mills, A. D. (1975). Quantitative film detection of ^{3}H and ^{14}C in polyacrylamide gels by fluorography. *Eur. J. Biochem.* **56,** 335–341.

McClay, D. R., and Chambers, A. F. (1978). Identification of four classes of cell surface antigens appearing at gastrulation in sea urchin embryos. *Dev. Biol.* **63,** 179–186.

Noll, H., Matranga, V., Cascino, D., and Vittorelli, M. L. (1979). Reconsitution of membranes and embryonic development in dissociated blastula cells of the sea urchin by reinsertion of aggregation-promoting membrane proteins extracted with butanol. *Proc. Natl. Acad. Sci. U.S.A.* **76,** 288–292.

Putney, S. D., and Schimmel, P. (1981). An aminoacyl tRNA synthetase binds to a specific DNA sequence and regulates its gene transcription. *Nature (London)* **291,** 632–635.

Roccheri, M. C., Sconzo, G., Di Bernardo, M. G., Albanese, I., Di Carlo, M., and Giudice, G. (1981a). Heat shock proteins in sea urchin embryos. Territorial and intracellular location. *Acta Embryol. Morphol. Exp.* **2,** 91–99.

Roccheri, M. C., Di Bernardo, M. G., and Giudice, G. (1981b). Synthesis of heat shock proteins in developing sea urchins. *Dev. Biol.* **83,** 173–177.

Roccheri, M. C., Sconzo, G., Di Carlo, M., Di Bernardo, M. G., Pirrone, A., Gambino, R., and Giudice, G. (1982). Heat shock proteins in sea urchin embryos. Transcriptional and post-transcriptional regulation. *Differentiation (Berlin)* **22,** 175–178.

Sconzo, G., Roccheri, M. C., Di Carlo, M., Di Bernardo, M. G., and Giudice, G. (1983). Synthesis of heat shock proteins in dissociated sea urchin embryonic cells. *Cell Differ.* **12,** 317–320.

Spinelli, G., Casano, C., Gianguzza, F., Ciaccio, M., and Palla, F. (1982). Transcription of sea urchin mesenchyme blastula histone genes after heat shock. *Eur. J. Biochem.* **128,** 509–513.

Vittorelli, M. L., Cannizzaro, G., and Giudice, G. (1973). Trypsin treatment of cells dissociated from sea urchin embryos elicits DNA synthesis. *Cell Differ.* **2,** 279–284.

7

Heat Shock Gene Expression during Early Animal Development

JOHN J. HEIKKILA, JOHN G. O. MILLER, GILBERT A. SCHULTZ, MALGORZATA KLOC, AND LEON W. BROWDER

I. INTRODUCTION

Exposure of prokaryotic and eukaryotic cells to heat shock (a sudden increase in temperature) or other environmental stresses, such as sodium arsenite treatment or metal ion exposure, results in the production of a small number of new heat shock proteins (hsp's) (Ashburner and Bonner, 1979; Schlesinger *et al.*, 1982). In terms of gene expression, control of this induction phenomenon has been shown to occur at the level of selective synthesis of specific mRNA's coding for hsp's (Ashburner and Bonner, 1979) and at the level of protein synthesis (Storti *et al.*, 1980; Krüger and Benecke, 1981). Heat shock proteins are found in many diverse organisms (Schlesinger *et al.*, 1982) and may be structurally related since antibodies raised against two hsp's from chicken embryo fibroblasts cross-react with heat shock extracts from a wide range of organisms (Kelley and Schlesinger, 1982). The apparently ubiquitous nature of this temperature stress phenomenon raises the question of whether these hsp's have a

Changes in Eukaryotic Gene Expression
in Response to Environmental Stress

ISBN 0-12-066290-6

possibly homeostatic function. It has been reported that the induction of hsp's in yeast (McAlistar and Finkelstein, 1980), *Dictyostelium* (Loomis and Wheeler, 1980), *Drosophila* (Ashburner and Bonner, 1979), and Chinese hamster fibroblasts (Li and Werb, 1982) is correlated with a dramatic increase in capacity for thermotolerance.

The ability to induce the expression of a set of genes by heat shock provides a useful means by which one can study the expression of an inducible gene family during animal development. Such studies can address a number of fundamental questions: Is there a stage specificity in heat shock gene expression? What levels of gene expression are involved in regulation of the heat shock response in embryos? Does the capability for hsp synthesis confer thermotolerance on the embryo? In this chapter we review results demonstrating that the ability to synthesize hsp's and accumulate hsp mRNA in response to heat shock during early development in *Arbacia punctulata, Xenopus laevis,* rabbits, and mice is a stage-dependent phenomenon. Furthermore, heat shock treatment decreases the amount of preexisting mRNA (actin mRNA) in early cleavage-stage embryos but has no detectable effect on mRNA in later developmental stages (i.e., blastula and later stages of *Xenopus*). Finally, we have noted a correlation between the onset of hsp synthesis and the acquisition of thermotolerance during early animal development.

II. SEA URCHIN

In Chapter 6 (this volume), Giudice reviews work from his laboratory examining the effect of elevated incubation temperature on hsp synthesis and on the subsequent development of early sea urchin embryos. In this section, we present data that confirm and extend these earlier observations with respect to heat shock gene expression during the early development of *A. punctulata*. In agreement with Roccheri *et al.* (1981), we have found that the capability of early sea urchin embryos to express a set of hsp's in response to thermal stress is a stage-dependent process (Howlett *et al.*, 1983). For example, elevation of the incubation temperature from 20 to 31°C for 1 hr produces an increase in the relative labeling of two major hsp's (M_r = 80,000 and 70,000) in blastula and gastrula stages (Fig. 1A, lanes 5 to 8) but not in 2- to 4-cell or 8- to 16-cell stages (Fig. 1A, lanes 1 to 4) of sea urchin development. In early cleavage-stage embryos (i.e., 2- to 16-cell stages) we have also used selected temperatures (26–37°C) as well as selected durations of heat shock (5 min to 2 hr) but could not detect hsp synthesis under any of these experimental conditions.

A similar stage-dependent synthesis of hsp's has been reported in *Drosophila*; an increase in the synthesis of hsp70 and hsp68 was observed at the blastoderm and later stages but not in preblastoderm stages of development (Graziosi *et al.*,

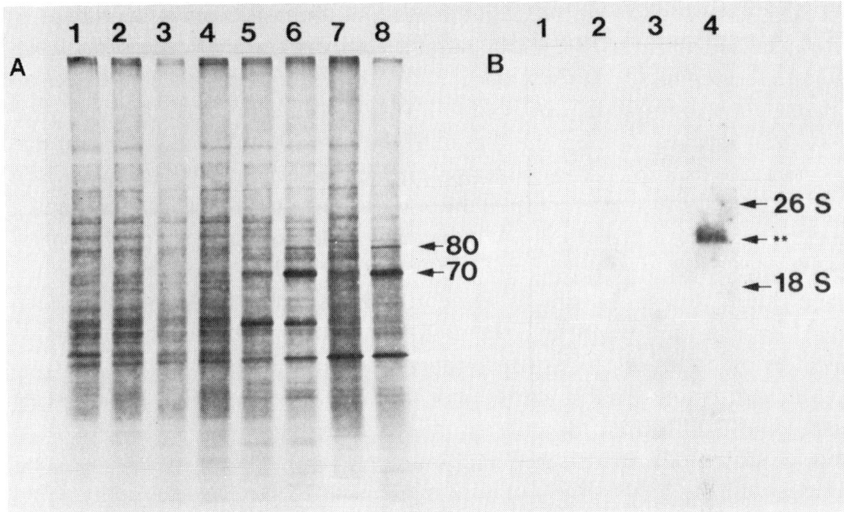

Fig. 1. Stage-dependent induction of hsp's and hsp mRNA during the early development of *A. punctulata*. (A) Sea urchin embryos (2-cell to gastrula stage) were incubated at either 20 or 31°C for 1 hr. Newly synthesized proteins were labeled with [^{35}S]methionine during the last 30 min of incubation. The labeled proteins were separated on a 7–17% gradient acrylamide slab gel and detected by fluorography. Equivalent acid-precipitable radioactivity (40,000 cpm) was loaded for each sample. Molecular mass of polypeptides identified by arrows was determined according to electrophoretic migration of standards and is expressed in daltons $\times 10^{-3}$. Lane 1, 2- to 4-cell stages, 20°C; lane 2, 2- to 4-cell stages, 31°C; lane 3, 8- to 16-cell stages, 20°C; lane 4, 8- to 16-cell stages, 31°C; lane 5, blastula, 20°C; lane 6, blastula, 31°C; lane 7, gastrula, 20°C; lane 8, gastrula, 31°C. (B) Polyadenylated RNA isolated from control and heat-shocked sea urchin embryos was resolved on a formaldehyde–agarose gel (Maniatis *et al.*, 1982), transferred to nitrocellulose, and hybridized against a recombinant DNA probe encoding the 5′ end of the *Drosophila* hsp70 gene (subclone 232.1; Livak *et al.*, 1978). Lane 1, 16-cell-stage RNA, 20°C; lane 2, 16-cell-stage RNA, 31°C; lane 3, blastula-stage RNA, 20°C; lane 4, blastula-stage RNA, 31°C.

1980; Dura, 1981). Whereas heat shock does not induce an enhanced synthesis of hsp's in early cleavage-stage sea urchin embryos, it is possible that hsp mRNA accumulates but is not translated. In order to examine this possibility, a cloned genomic fragment from *Drosophila* comprising the 5′ end of the gene coding for hsp70 (Livak *et al.*, 1978; Heikkila *et al.*, 1982) was utilized to examine by RNA blot analysis (Northern hybridization) whether hsp70 mRNA sequences had accumulated in sea urchin 16-cell embryos. In this experiment, poly(A)$^+$ mRNA from control and heat-shocked 16-cell-stage and blastula embryos was subjected to electrophoresis, transferred to nitrocellulose, and hybridized with a ^{32}P-labeled hsp70 probe. A typical autoradiogram resulting from this type of experiment is shown in Fig. 1B. As shown in lanes 1 and 2, we found no detectable hybridization of the hsp70 probe to RNA isolated from either control or heat-shocked 16-cell-stage embryos. On the contrary, the hsp70 probe hybridized to RNA isolated from heat-shocked blastula embryos (lane 4) but not to RNA isolated from control embryos (lane 3). We have not been able to detect an accumulation of hsp70 mRNA before the 64- to 128-cell stage.

In heat-shocked sea urchin blastulae the size of the RNA species that hybridized to the hsp70 probe was approximately 3.4 kb in length in comparison to 2.6 kb for the *Drosophila* hsp70 mRNA that was run in parallel. The finding that the sea urchin hsp70 mRNA is larger than the corresponding *Drosophila* hsp70 mRNA suggests the presence of an extensive untranslated region on the sea urchin messenger. A similar situation with respect to mRNA size has been found for a class of sea urchin actin mRNA (2.2 kb), which contains an untranslated region of nearly 1.0 kb (Bushman and Crain, 1983).

Although zygote genome transcription can be detected in early cleavage-stage embryos of sea urchin, the bulk of mRNA in eggs and early embryos is of oogenic origin (Humphreys, 1971). Therefore, the impact of embryonic transcription on the total mRNA pool is minimal in early cleavage embryos but increases with development as oogenic mRNA is degraded and replaced by newly synthesized mRNA (Brandhorst and Humphreys, 1972; Galau *et al.*, 1977). In early cleavage-stage embryos it is possible that either (1) heat shock gene transcription is not activated or (2) if these gene(s) are transcribed the newly synthesized transcripts are rapidly degraded and do not accumulate.

Roccheri *et al.* (1981) have examined the effect of heat shock on the subsequent development of early sea urchin embryos. They observed that survival of heat-shocked embryos is correlated with their ability to synthesize hsp's. In agreement with these studies, we have found that heat shock treatment (31°C for 1 hr) of early cleavage embryos of *A. punctulata,* which do not synthesize hsp's, arrests further development or leads to abnormalities. A similar correlation between hsp synthesis and thermotolerance during early animal development has been reported in *Drosophila* (Graziosi *et al.*, 1980) and in *X. laevis* (Heikkila *et al.*, 1984; see Section III).

III. *XENOPUS LAEVIS*

Exposure of *Xenopus* embryos at cleavage and early blastula stages to temperatures exceeding 30°C for 90 min results in a reduction in the level of total protein synthesis, relative to control that becomes more pronounced as temperature is increased. An electrophoretic pattern of the proteins synthesized after heat shock at the different developmental stages is displayed in Fig. 2. Thermal stress does not induce a detectable change in the spectrum of proteins synthesized in cleavage (Fig. 2; compare lanes 1 and 2) and early blastula stages. In contrast, enhanced synthesis of a 68–70K hsp is seen occasionally in medium-cell

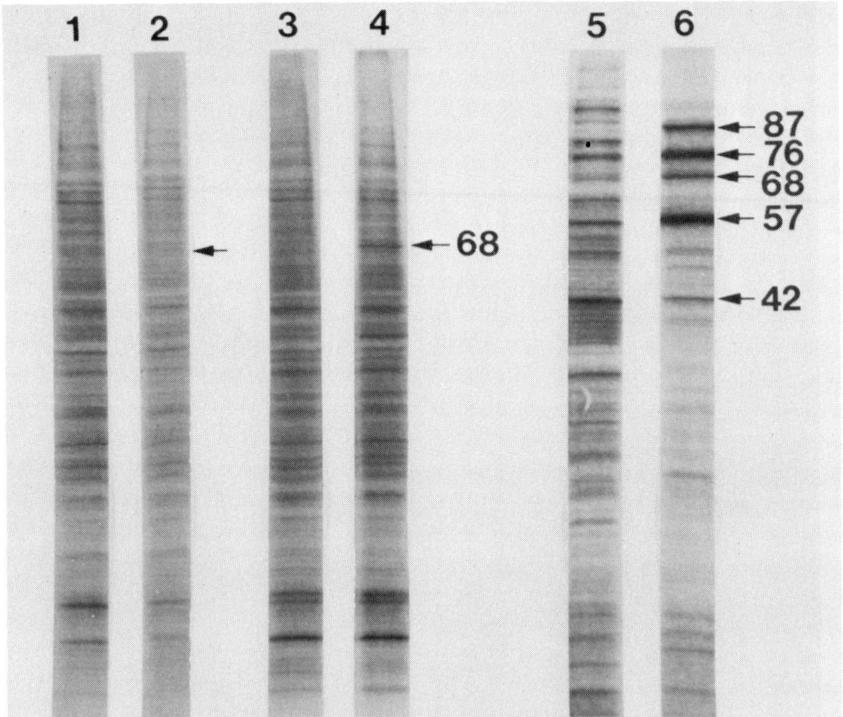

Fig. 2. Effect of heat shock on the pattern of protein synthesis during early *X. laevis* development. *Xenopus laevis* eggs were fertilized *in vitro,* degelled, and allowed to develop at 22°C. At the appropriate stages, embryos (groups of 10) were heat shocked for 20 min at either 33 or 37°C, followed by an additional 70 min of incubation in the presence of [^{35}S]methionine at the heat shock temperature. Control embryos were maintained and labeled at 22°C. The labeled proteins were resolved on a 7–17% gradient acrylamide slab gel and detected by fluorography. Equivalent amounts of acid-precipitable radioactivity (20,000 cpm) were loaded for each sample. Molecular mass of polypeptides identified by arrows was determined according to electrophoretic migration of standards and is expressed in daltons $\times$ 10^{-3}. Lane 1, 32-cell stage, 22°C; lane 2, 32-cell stage, 33°C; lane 3, fine-cell blastula, 22°C; lane 4, fine-cell blastula, 33°C; lane 5, neurula, 22°C; lane 6, neurula, 37°C. (From Heikkila *et al.,* 1984.)

blastulae (stage 8) and consistently in fine-cell blastulae (stage 9) and in embryos of all later stages when heat shocked (Fig. 2, lanes 3 and 4). Stage 9 and older embryos sometimes produced a more complex heat shock response, exhibiting enhanced synthesis of hsp87, hsp76, hsp57, and, less frequently, hsp42 (Fig. 2, lanes 5 and 6). At this time we are uncertain whether the variability in the detection of enhanced synthesis of these hsp's has a biological explanation or is due to experimental variation.

To determine whether the increase in hsp70 reflects an accumulation of hsp70 mRNA, the techniques of *in vitro* translation and Northern hybridization were utilized. In the first procedure, poly(A)$^+$ mRNA was isolated from control (22°C) and heat-shocked (33°C) cleavage- and neurula-stage embryos and translated in a rabbit reticulocyte lysate cell-free system. The fluorograms of the [^{35}S]methionine-labeled proteins synthesized in this system are shown in Fig. 3. An increase in the synthesis of hsp68, hsp69, and hsp70 was not detectable in the spectrum of translation products derived from the RNA of either control or heat-

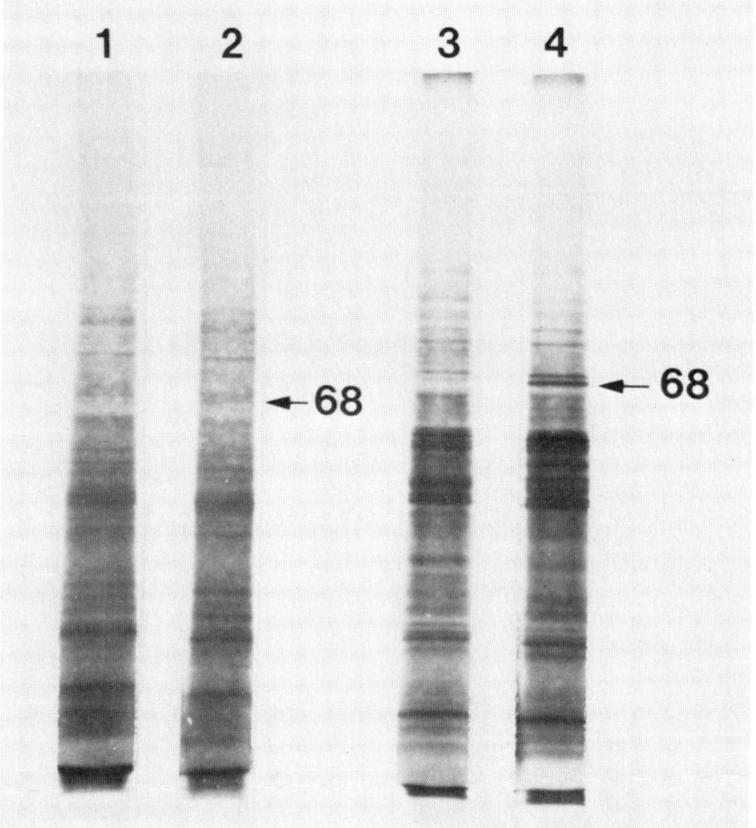

Fig. 3. Comparison of the cell-free translation products of RNA isolated from control and heat-shocked cleavage-stage and neurula-stage *X. laevis* embryos. Embryos at the 32-cell and neurula stages were maintained at either 22 or 35°C for 90 min. Polyadenylated RNA isolated from each sample was translated in a rabbit reticulocyte lysate cell-free system. The [^{35}S]methionine-labeled proteins were analyzed by electrophoresis and fluorography. Equivalent amounts of acid-insoluble radioactivity (40,000 cpm) were loaded for each sample. Lane 1, 32-cell stage, 22°C; lane 2, 32-cell stage, 35°C; lane 3, neurula, 22°C; lane 4, neurula, 35°C. (From Heikkila *et al.*, 1984.)

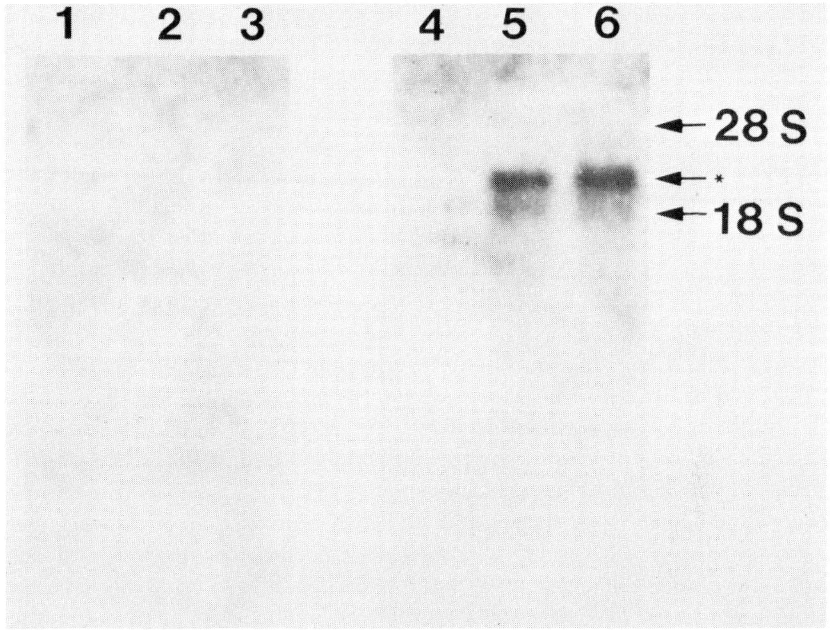

Fig. 4. Stage-dependent accumulation of the hsp70 mRNA during early *X. laevis* embryo development. Poly(A)$^+$ mRNA isolated from control and heat-shocked 32-cell- and neurula-stage embryos was resolved on a methylmercury hydroxide–agarose gel (Bailey and Davidson, 1976), transferred to DBM paper (Alwine *et al.*, 1977), and hybridized with a ^{32}P-labeled genomic subclone containing the coding region for the 5′ end of mouse hsp70 mRNA. (Courtesy of Dr. R. Morimoto, Northwestern University.) After posthybridization washes, the blot was exposed to preflashed Kodak XAR-5 X-ray film using Chronex intensifying screens. Lane 1, 32-cell stage, 22°C; lane 2, 32-cell stage, 33°C; lane 3, 32-cell stage, 35°C; lane 4, neurula, 22°C; lane 5, neurula, 33°C; lane 6, neurula, 35°C. (From Heikkila *et al.*, 1984.)

shocked cleavage-stage embryos (Fig. 3, lanes 1 and 2). However, an increase in the relative labeling of a 68–70K hsp was observed in the cell-free products of heat shock neurula-stage RNA but not in control RNA (Fig. 3, lanes 3 and 4).

The results of the Northern hybridization experiments are shown in Fig. 4. Poly(A)$^+$ mRNA from control and heat-shocked cleavage- and neurula-stage embryos was resolved on a methylmercury hydroxide–agarose gel, transferred to diazobenzyloxymethyl (DBM) paper, and then hybridized with a ^{32}P-labeled genomic subclone complementary to the 5′ end of mouse hsp70 mRNA. This probe shares homology with *Xenopus* hsp70 DNA sequences and cross-hybridizes with *Xenopus* hsp70 mRNA. As shown in Fig. 4, the hsp70 probe hybridized to neurula heat shock mRNA but not to RNA isolated from cleavage-stage embryos. In RNA blot analyses utilizing either polyadenylated mRNA or total RNA we have not been able to detect hsp70 mRNA accumulation until stage

9 (fine-cell blastula stage). These results are consistent with those obtained by *in vitro* translation presented above and indicate that the stage-dependent synthesis of the 68–70K hsp is correlated with an increase in the abundance of its corresponding mRNA.

The effect of heat shock on the steady-state levels of preexisting mRNA's has also been examined. As shown in Fig. 3, the synthesis of a broad spectrum of proteins by RNA isolated from heat-shocked cleavage- and neurula-stage embryos is indicative of the presence of intact mRNA. In order to examine the levels of a specific mRNA species we employed a *Drosophila* actin probe (Fyrberg *et al.*, 1980; Giebelhaus *et al.*, 1983), which has homology to and cross-hybridizes with *Xenopus* actin mRNA. In a typical experiment, the RNA blot shown in Fig. 4 was treated with 90% formamide at 75°C, which removed the ^{32}P-labeled mouse hsp70 probe but left the RNA covalently bound to the filter. The Northern blot was rehybridized to a ^{32}P-labeled subclone of a *Drosophila* actin gene. The *Drosophila* actin probe hybridized to 2.2-kb and 1.8-kb RNA species in both control cleavage- and neurula-stage RNA samples (Fig. 5). The level of the 1.8-kb RNA species was higher in neurula- than in cleavage-stage embryos. Heat shock (30–35°C) produced a decrease in the levels of actin mRNA in cleavage-stage embryos but had no degradative effect on RNA prepared from heat-shocked neurulae (Fig. 5). This finding is similar to observations we have made in heat-shocked sea urchin cleavage-stage RNA samples, which had a decrease in the level of actin mRNA relative to the control. However, heat shock had little effect on the levels of actin mRNA in blastula and gastrula sea urchin RNA samples. Therefore, whereas *in vitro* translation experiments suggest retention of total mRNA in the cleavage-stage embryos after heat shock, Northern hybridization experiments with an actin probe suggest a decrease in the levels of at least one specific mRNA species. Perhaps other specific transcripts are also affected. Since the level of a specific mRNA within the cell is dependent on its rate of transcription and rate of degradation, further experiments are required to determine the underlying mechanism for the decline in actin mRNA levels of cleavage-stage embryos after heat shock. As will be shown later, heat shock arrests further development of *X. laevis* cleavage-stage embryos. Thus, we believe that heat shock treatment, which initiates a decrease in the viability of the cleavage-stage embryo, may also lead to an increase in mRNA degradation.

The inability of early cleavage-stage embryos to accumulate hsp70 mRNA is consistent with previous data showing that the level of RNA synthesis during this developmental period in *Xenopus* is very low (Bachvarova and Davidson, 1966; Woodland and Gurdon, 1969; Newport and Kirschner, 1982a). Generalized RNA synthesis is activated at stage 8 or midblastula (Shiokawa *et al.*, 1981; Newport and Kirschner, 1982a). The mechanism for this activation at stage 8 is not fully understood. Newport and Kirschner (1982b) have demonstrated that this event depends on an increased DNA-to-cytoplasm ratio. They have postu-

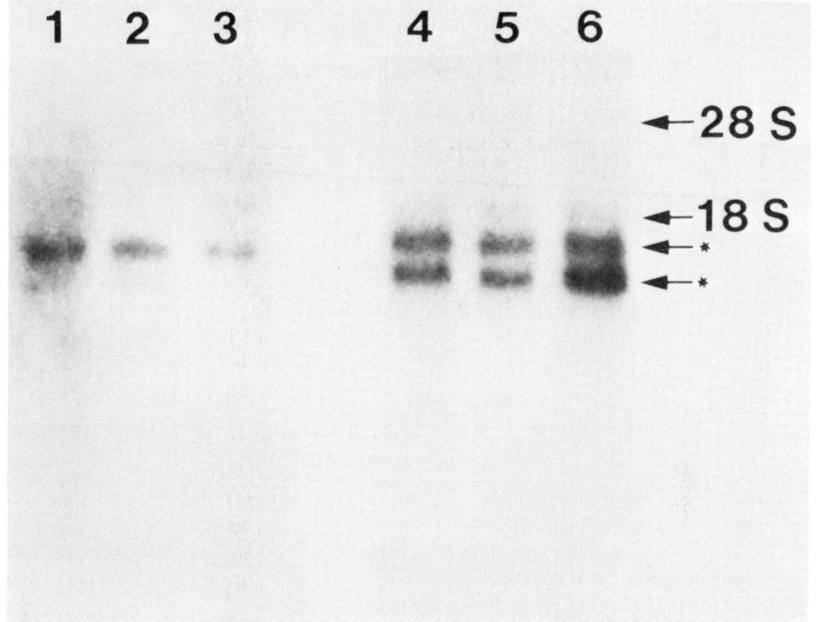

Fig. 5. Hybridization of a genomic clone of a *Drosophila* actin gene to RNA isolated from control and heat-shocked *Xenopus* embryos. The DBM paper shown in Fig. 4 was treated with 90% formamide to remove the ^{32}P-labeled mouse hsp70 probe. The covalently bound RNA on the DBM paper was rehybridized to the ^{32}P-labeled subclone of the *Drosophila* actin gene. Lane 1, 32-cell stage, 22°C; lane 2, 32-cell stage, 33°C; lane 3, 32-cell stage, 35°C; lane 4, neurula, 22°C; lane 5, neurula, 33°C; lane 6, neurula, 35°C. (From Heikkila *et al.*, 1984.)

lated that some substance present in the unfertilized egg suppresses transcription. During development the ratio of DNA to cytoplasm increases to a critical level. Beyond this level, the amount of preexisting suppressor is hypothesized to be limiting and, thus, unable to saturate the DNA template allowing transcription to ensue. It is tenable that heat shock gene expression may be blocked during cleavage and inducible after stage 8 in this way.

The *Xenopus* oocyte contains abundant amounts of stored oogenic messengers (Perlman and Rosbash, 1978), some of which are utilized to support protein synthesis in the cleavage-stage embryo until *de novo* RNA synthesis provides the transcripts that direct protein synthesis during later stages of development (for review, see Browder, 1984). The absence of hsp70 synthesis during cleavage and the failure to detect hsp70 mRNA in these embryos indicate that cleavage-stage *Xenopus* embryos lack oogenic hsp70 mRNA. However, Bienz and Gurdon (1982) have provided evidence that the *Xenopus* oocyte contains hsp70 mRNA and that the translation of this mRNA is induced by heat shock. The

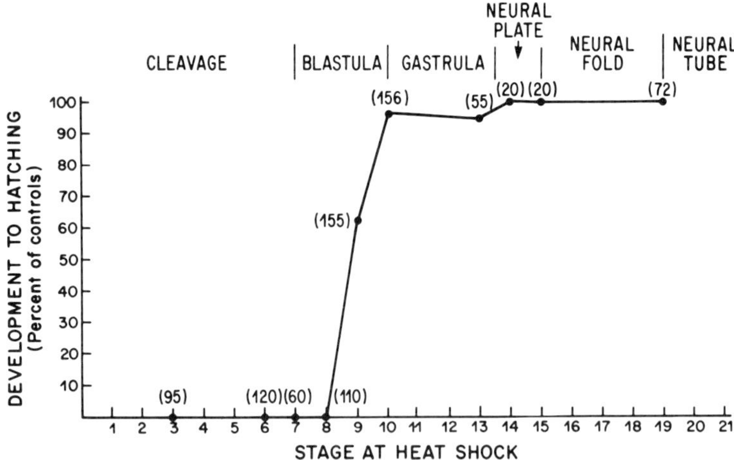

Fig. 6. Thermotolerance of *Xenopus* embryos during development. Embryos were heat shocked at the stages shown on the figure. Heat shock was at 35°C for 20 min. Controls were maintained at 22°C. Embryos were scored for the ability to develop to the hatching stage. This is expressed as a percentage of control embryos developing to hatching. Numbers in parentheses refer to the number of embryos heat shocked at each stage. (From Heikkila *et al.*, 1984.)

presence of hsp70 mRNA in the oocyte and its absence in cleavage-stage embryos suggest that the oocytic heat shock mechanism is somehow erased before cleavage ensues. This erasure may involve the selective loss of hsp70 mRNA.

We have examined the heat shock response during hormone-induced oocyte maturation and have found that a heat shock response continues to be inducible during the early stages of maturation, but this capacity is absent in the unfertilized egg (M. Kloc *et al.*, unpublished observations). We are presently using RNA blot analysis to trace the amounts of hsp70 mRNA during this same interval.

To investigate the developmental significance of the heat shock response, we have examined the developmental potential and gross morphology of embryos that were transiently exposed to heat shock conditions. Embryos were heat shocked for 20 min at 35°C and then returned to 22°C. Development of embryos to the hatching stage (expressed as a percentage of control embryos developing to hatching) is displayed in Fig. 6. Cleavage-stage embryos rapidly become abnormal after heat shock and degenerate. Heat-shocked medium-cell blastulae (stage 8) typically undergo exogastrulation. The endoderm in these embryos remains formless, but the neural axis is more or less normal and is covered dorsally by epidermis. None of these embryos survives to hatching. The earliest stage to exhibit thermotolerance is the fine-cell blastula (stage 9); 62% of these embryos developed to hatching after heat shock. By the gastrula stage more than 90% of heat-shocked embryos developed to hatching, and all embryos at neural plate and later stages were thermotolerant under these conditions. Similar survival rates were found for gastrulae that were heat shocked at 33°C for 90 min—conditions that induce hsp70 synthesis. Thus, acquisition of thermotolerance correlates with the acquisition of the heat shock response. A similar correlation with respect to hsp synthesis and thermotolerance has been reported during both *Drosophila* (Graziosi *et al.*, 1980) and sea urchin development (Roccheri *et al.*, 1981).

IV. MOUSE AND RABBIT PREIMPLANTATION EMBRYOS

Heat shock during mammalian development constitutes another system for the study of gene expression. Additionally, knowledge of the effects of thermal or other stresses on early mammalian development is of obvious importance to livestock management as well as to clinical situations involving maternal hyperthermia during pregnancy. Whereas the detrimental effect of thermal stress on early fetal development has been well documented (Bellvé, 1972; Wildt *et al.*, 1975; Alliston *et al.*, 1965; Ulberg and Sheean, 1973), little information is available on the effect of maternal hyperthermia or other stresses on embryonic gene expression. Therefore, we have initiated studies utilizing rabbit and mouse preimplantation embryos as model systems in which to examine the effect of environmental stress on mammalian embryonic gene expression.

Mammalian embryos, like those of *Drosophila* (Graziosi *et al.*, 1980; Dura, 1981; Zimmerman *et al.*, 1983), sea urchin (Giudice, 1984), and *Xenopus* also appear to respond to heat shock (i.e., synthesis of hsp's) in a stage-dependent manner during early development. Enhanced synthesis of hsp70 was not observed in the spectrum of proteins synthesized by heat-shocked 8-cell-stage mouse embryos (Fig. 7; compare lanes 1 and 2). However, elevation of the incubation temperature of mouse blastocysts produced an increase in the relative synthesis of hsp70 (Fig. 7; compare lanes 3 and 4). In comparable studies, Wittig *et al.* (1983) examined the effect of thermal stress on early mouse development and also were unable to detect enhanced synthesis of hsp70 until the morula/blastocyst stage. Predictably, *in vitro* translation studies identify hsp70 mRNA accumulation only in RNA derived from heat-shocked blastocysts (Fig. 8).

Similarly, heat shock treatment of rabbit 16-cell-stage embryos (43°C for 1 hr) did not affect the pattern of proteins synthesized (Fig. 9). When 6-day-old rabbit blastocysts are subjected to heat shock (43°C for 2 hr), *in vitro* incorporation of labeled amino acid into acid-insoluble protein was reduced 40–50%, relative to the control values. Electrophoretic examination of the newly synthesized proteins reveals an increase in the relative labeling of a 70,000-dalton hsp (Fig. 10;

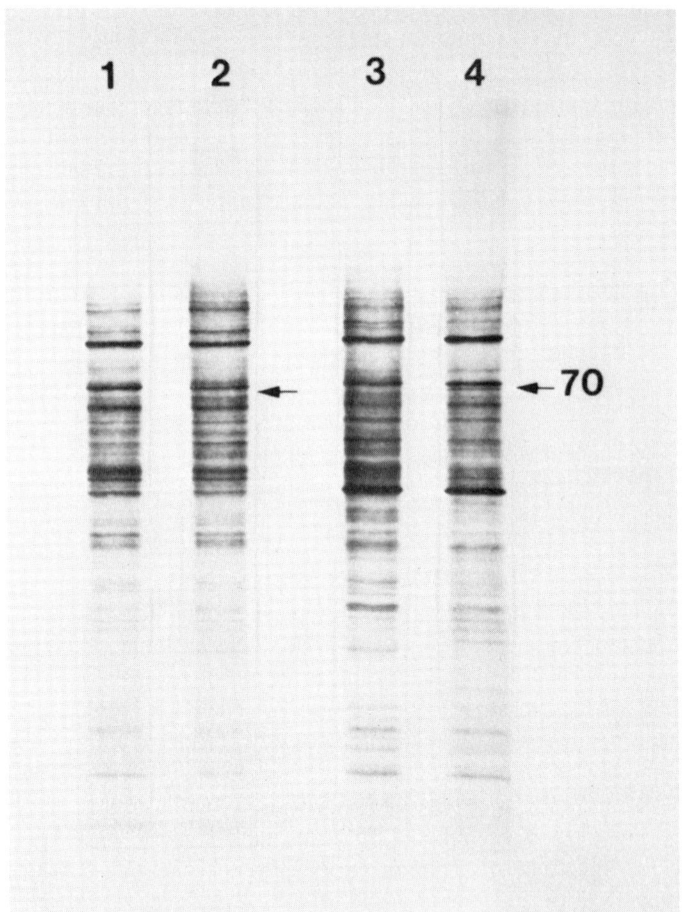

Fig. 7. Stage-dependent induction of hsp70 synthesis during mouse preimplantation embryo development. Isolated mouse 8-cell- or blastocyst-stage embryos (100 embryos per treatment) were incubated at either 37 or 42°C for 1 hr plus a 1-hr incubation period in the presence of [^{35}S]methionine at 37°C. The labeled proteins were analyzed by electrophoresis and fluorography. Lane 1, 8-cell stage, 37°C; lane 2, 8-cell stage, 42°C; lane 3, blastocyst, 37°C; lane 4, blastocyst, 42°C.

compare lanes 1 and 2). The increase in the relative labeling of hsp70 was detectable after only 15-30 min of heat shock treatment and was maximal at 2 hr. Occasionally, an increase in the relative labeling of 95K and 28K protein species was noted.

Examination of the spectrum of proteins in the products of *in vitro* translation reactions programmed with total RNA from heat-shocked rabbit blastocysts (Fig. 10, lanes 3 and 4) also reveals an increase in the relative synthesis of hsp70. The increased labeling of hsp70 reflects an increase in the abundance of its mRNA, suggesting that this 70,000-dalton stress protein is regulated predominantly at the transcriptional level.

Further analysis at the molecular level of this stage-dependent heat shock response during early mammalian development is hampered by the small amounts of embryonic material that can be obtained. Northern hybridization analyses using a mouse hsp70 DNA probe clearly reveal the presence of hsp70

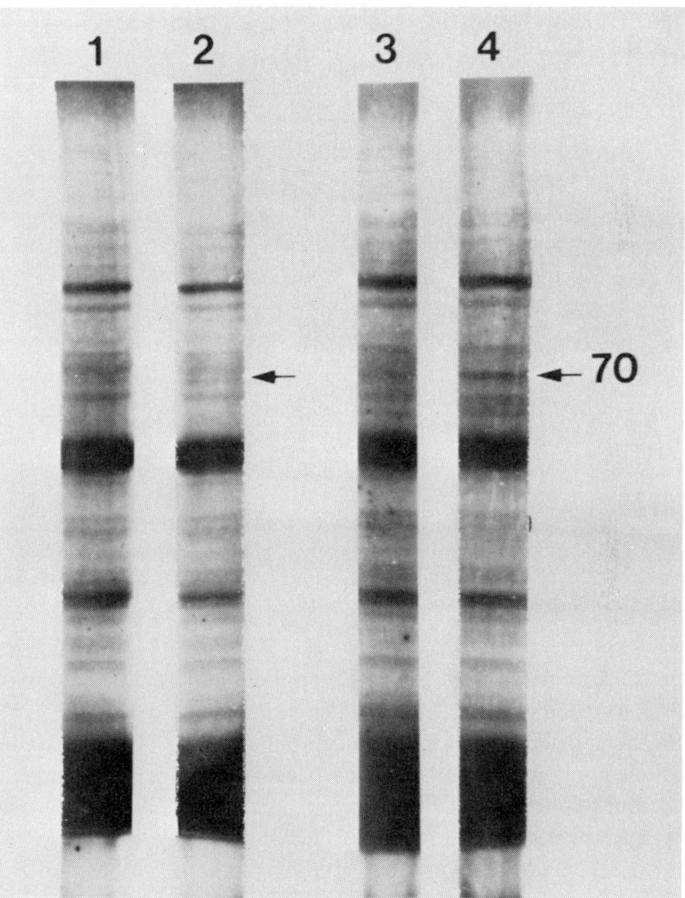

Fig. 8. Comparison of the cell-free products of total RNA isolated from control and heat-shocked mouse 8-cell-and blastocyst-stage embryos. Total RNA was extracted from control (37°C) and heat-shocked (42°C) mouse embryos and translated in a rabbit reticulocyte lysate cell-free system. The [^{35}S]methionine-labeled proteins were analyzed by electrophoresis and fluorography. Lane 1, 8-cell stage, 37°C; lane 2, 8-cell stage, 42°C; lane 3, blastocyst, 37°C; lane 4, blastocyst, 42°C.

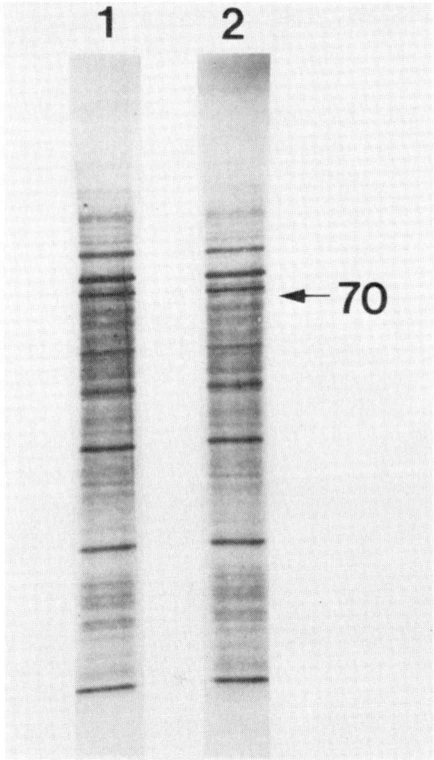

Fig. 9. Effect of heat shock on the pattern of proteins synthesized in rabbit 16-cell-stage embryos. [^{35}S]methionine-labeled proteins from control (37°C for 3 hr) or heat shocked (43°C for 2 hr plus 1 hr at 37°C) 16-cell-stage rabbit embryos (50 embryos per treatment) were analyzed by electrophoresis and fluorography. Lane 1, control; lane 2, heat shock.

mRNA in RNA prepared from heat-shocked 6-day-old rabbit blastocysts but not from the controls (Fig. 11). In the latter experiment, detection of blastocyst hsp70 mRNA is facilitated by its large cell number (80,000). The accumulation of an equivalent number of cells from early cleavage stages is not practical even with techniques of superovulation. Furthermore, the cleavage-stage rabbit embryo has a lower level of protein and RNA synthesis relative to the rabbit blastocyst (Karp *et al.*, 1974; Manes, 1969). We have not been able to detect increased levels of hsp70 mRNA in pools of up to 200 stressed 16-cell-stage rabbit embryos (Fig. 11) by conventional RNA blot analysis. It is possible that trace amounts of hsp70 mRNA are accumulated in stressed cleavage-stage embryos but at levels below the sensitivity of the method. Nonetheless, if a heat shock response does indeed occur at this stage, it is clearly of minimal magnitude compared to that observed at the blastocyst stage. Our current data, however, suggest that cleavage-stage mammalian embryos do not synthesize hsp's in response to stress and follow the pattern observed in early embryos of *Drosophila, Arbacia,* and *Xenopus.*

In many other eukaryotic systems hsp70 can be induced by a variety of stressful agents. For example, in *Drosophila* (Ashburner and Bonner, 1979), fish cells (Kothary and Candido, 1982), and mouse fibroblasts (Bensaude and Morange, 1983) sodium arsenite exposure has been observed to enhance synthesis of hsp70. In rabbit blastocysts exposure to 50 μ*M* sodium arsenite for 3 hr, like heat shock, produced a 50–60% decrease in the incorporation of labeled amino acid into acid-insoluble protein and an enhanced synthesis of hsp70 (Fig.12; compare

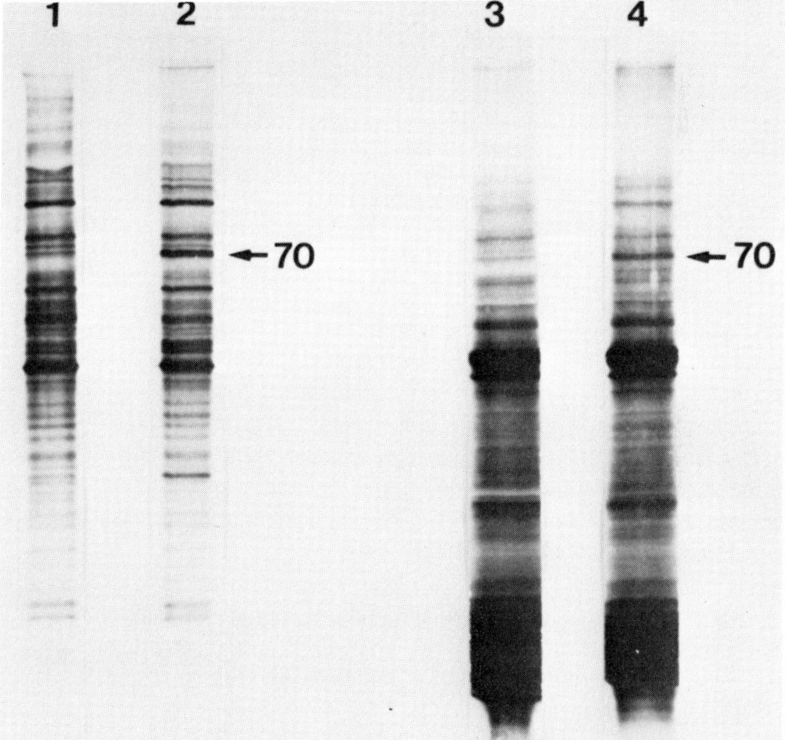

Fig. 10. Induction of hsp70 synthesis and hsp70 mRNA accumulation in rabbit blastocysts by heat shock. Isolated 6-day-old rabbit blastocysts were incubated at either 37°C for 2 hr (lane 1) or at 43°C for 2 hr (lane 2) followed by a 1-hr incubation period in [^{35}S]methionine at 37°C. Alternatively, total RNA (1–2 μg) was extracted from control (lane 3) and heat-shocked (lane 4) blastocysts and translated in a rabbit reticulocyte lysate cell-free system. The labeled proteins were resolved on a 7–17% gradient acrylamide slab gel and detected by fluorography. Equivalent amounts of acid-insoluble radioactivity (40,000 cpm) were loaded for each sample. (From Heikkila and Schultz, 1984.)

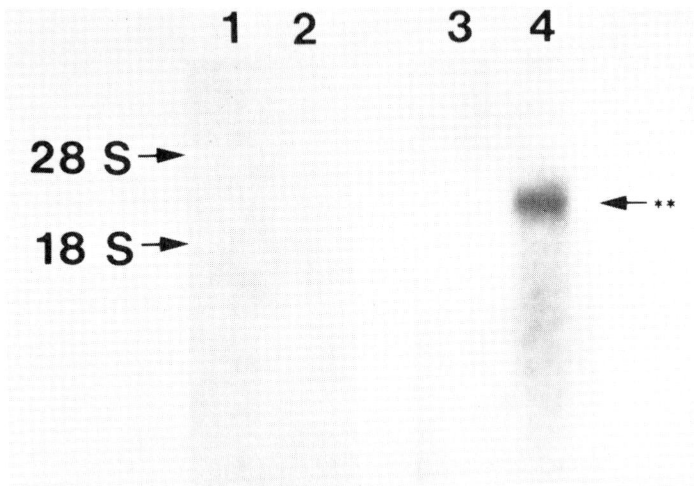

Fig. 11. Stage-dependent accumulation of hsp70 mRNA during rabbit preimplantation embryo development. Total RNA was isolated from control (37°C for 3 hr) and heat-shocked (43°C for 2 hr plus 1 hr at 37°C) 16-cell-stage and 6-day-old blastocyst embryos. The RNA was resolved on a methylmercury hydroxide–agarose gel, transferred to DBM paper, and then hybridized to the ^{32}P-labeled subclone of a mouse hsp70 gene. Lane 1, 16-cell-stage control RNA; lane 2, 16-cell-stage heat shock RNA; lane 3, blastocyst control RNA; lane 4, blastocyst heat shock RNA.

lanes 1 and 4). Even a mechanical stress can elicit a heat shock-like response. When a 6-day-old rabbit blastocyst was punctured with a fine needle, the collapse of the trophectoderm shell was accompanied by an induction of hsp70 while general protein synthesis was reduced (Fig. 12; compare lanes 1 and 3). In a related finding, Currie and White (1981) have found that the act of slicing mammalian tissue results in the preferential synthesis of a 71,000-dalton hsp. Therefore, sodium arsenite exposure, mechanical injury, and heat shock can induce hsp70 synthesis in rabbit blastocysts.

As in the previous systems, the heat shock-induced synthesis of rabbit blastocyst hsp70 appears to be regulated primarily at the transcriptional level. An increase in the abundance of mRNA coding for hsp70 after heat shock was observed through *in vitro* translation experiments (Fig. 10) and verified by Northern hybridization analyses (Fig. 11). To determine whether mechanical injury and sodium arsenite treatment induce hsp70 mRNA accumulation, total nucleic acid was extracted from control and stressed rabbit blastocysts (1–2 embryos per treatment) and analyzed by RNA blot analyses. A typical autoradiogram resulting from this type of experiment is shown in Fig. 13. Hybridization of the mouse hsp70 DNA probe to control RNA was not readily detectable (lane 1). However, the hsp70 DNA probe did hybridize to an RNA species from mechanically injured (lane 2), sodium arsenite-treated (lane 3), and heat-shocked

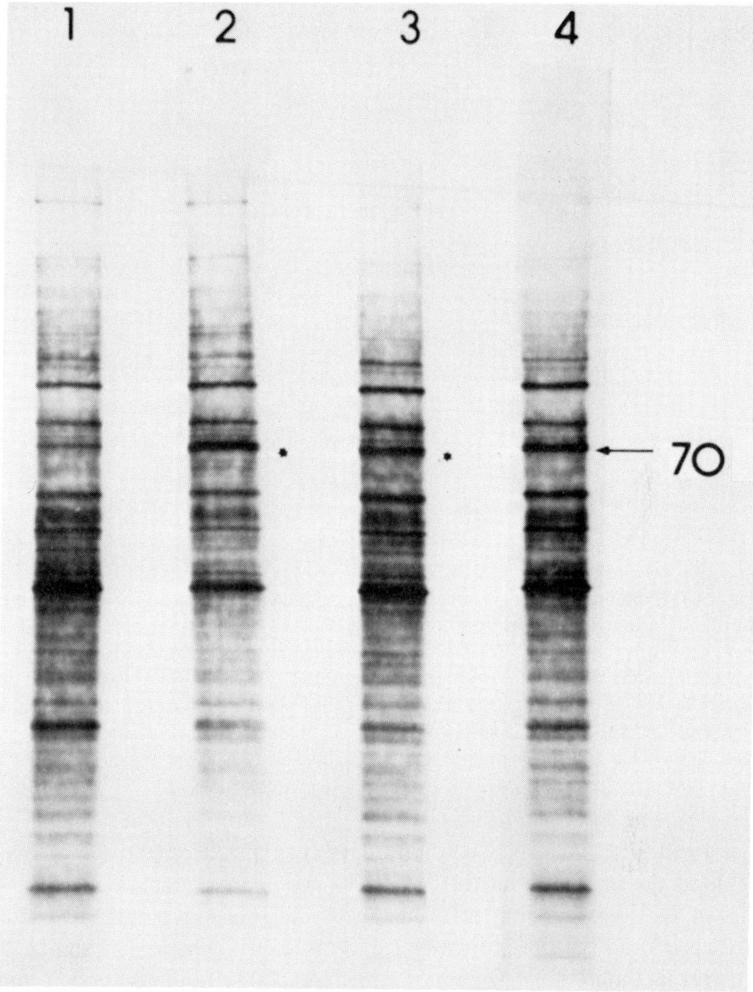

Fig. 12. Induction of hsp70 in rabbit blastocysts by sodium arsenite treatment and mechanical injury. Six-day-old rabbit blastocysts were incubated at either 37°C or heat shocked at 43°C for 2 hr as described in Fig. 10. Other blastocysts were incubated for 3 hr in 50 μM sodium arsenite at 37°C followed by a 1-hr incubation in sodium arsenite-free media with [^{35}S]methionine. Some rabbit blastocysts were punctured with a fine needle, which resulted in the structural collapse of the blastocysts. These embryos were incubated at 37°C for 2 hr followed by a 1-hr labeling period. The labeled blastocyst proteins were analyzed as in Fig. 10. Lane 1, control; lane 2, heat shock; lane 3, mechanical injury; lane 4, sodium arsenite. (From Heikkila and Schultz, 1984.)

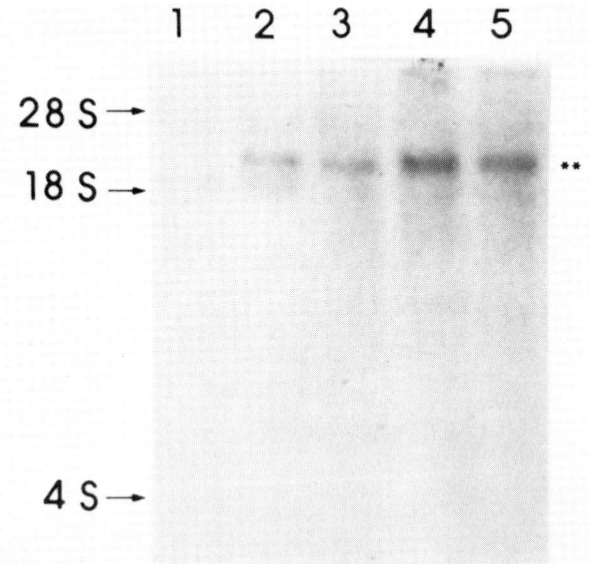

Fig. 13. Hybridization of a genomic clone of a mouse hsp70 gene complementary to the 5′ end of hsp70 mRNA to RNA isolated from control, heat-shocked, sodium arsenite-treated, and mechanically injured rabbit blastocysts. Total RNA was extracted from rabbit blastocysts (two embryos per treatment) following the stressful treatments, as described in Fig. 12. RNA was resolved on methylmercury hydroxide–agarose gels, transferred to DBM paper, and then hybridized with the ^{32}P-labeled hsp70 probe. Lane 1, control, 37°C; lane 2, mechanical injury; lane 3, 50 μM sodium arsenite; lanes 4 and 5, heat shock for 2 hr at 43°C plus 1-hr recovery at 37°C. (From Heikkila and Schultz, 1984.)

(lanes 4 and 5) blastocysts. Therefore, the three types of stress used in this study may affect the transcription of the hsp70 gene, producing an increase in the levels of its corresponding mRNA. The rabbit blastocyst mRNA coding for hsp70 appears to be approximately 2.6 kb in length, which is of similar size to the corresponding fish or *Drosophila* mRNA. These findings support antibody and DNA hybridization studies, suggesting that the gene encoding hsp70 is highly conserved (Kelley and Schlesinger, 1982; Tanguay, 1983; Lowe *et al.*, 1983).

As in the *Xenopus* system outlined in the previous section, we again used a recombinant DNA probe to analyze a consitutive mRNA (actin mRNA). In this experiment we treated the Northern blot shown in Fig. 13 with 90% formamide to remove the labeled mouse hsp70 recombinant DNA probe and hybridized with the ^{32}P-labeled clone of the *Drosophila* actin gene. This actin probe hybridized very strongly to 1.8- to 2.0-kb RNA species in all the rabbit blastocyst samples (Fig. 14). Therefore, neither heat shock nor any of the other stresses used in this study was observed to result in the degradation of blastocyst actin mRNA. This finding supports the *in vitro* translation results (Fig. 10, lanes 3 and 4) showing that a full spectrum of mRNA's of rabbit blastocyst proteins plus hsp70 are synthesized with mRNA from heat-shocked embryos.

It must be noted that all of the above studies were performed on embryonic material heat shocked *in vitro*. It is not known whether a similar stage-dependent synthesis of hsp's can occur in mammalian embryos *in vivo* (i.e., maternal hyperthermia). That this type of stress response may occur in blastocysts *in vivo* is possible since physiologically relevant increases in body temperature induced by drugs or high ambient temperatures have been shown to induce the expression of a heat shock gene in mammalian tissue (Heikkila *et al.*, 1981; Hammond *et al.*, 1982). At a physiological level, periods of transitory infertility are one of the most notable effects of thermal stress on the reproductive capacity of females of mammalian species (Ulberg and Sheean, 1973). Severe maternal hyperthermia has been reported to retard fetal development as well as significantly increase the incidence of abnormal cleavage and pre- and postimplantation mortality (Bellvé, 1972; Bellvé, 1973; Wildt *et al.*, 1975; Alliston *et al.*, 1965; Ulberg and Sheean,

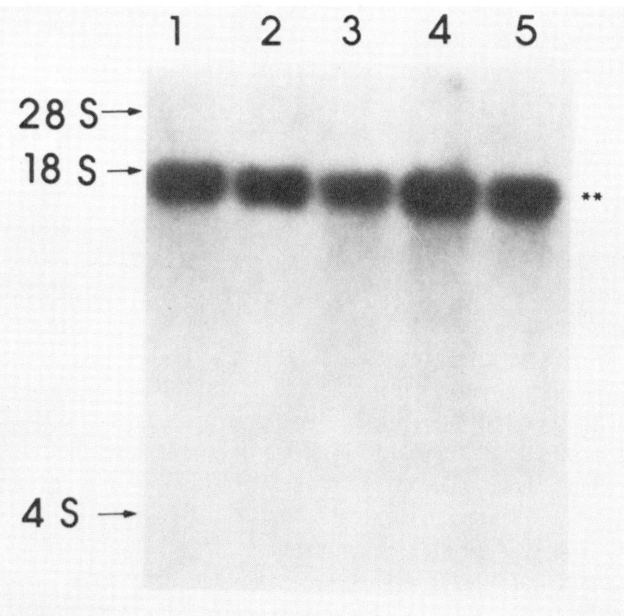

Fig. 14. Hybridization of a *Drosophila* actin genomic clone to RNA isolated from control, heat-shocked, sodium arsenite-treated, and mechanically injured rabbit blastocysts. The DBM paper shown in Fig. 13 was treated with 90% formamide to remove the ^{32}P-labeled mouse 70K hsp probe. The covalently bound RNA on the DBM paper was rehybridized to the ^{32}P-labeled actin probe. Lane 1, control; lane 2, mechanically injured; lane 3, 50 μM sodium arsenite; lanes 4 and 5, heat shock for 2 hr at 43°C plus a 1-hr recovery period at 37°C. (From Heikkila and Schultz, 1984.)

1973). The most sensitive period with respect to heat stress appears to be between early cleavage and implantation (Wildt *et al.*, 1975; Bellvé, 1972, 1973; Ulberg and Sheean, 1973), and thermal resistance has been found to increase with the age and cell number of the embryo (Ulberg and Sheean, 1973). Given these results, it is tempting to suggest that the stage-dependent expression of hsp's in mammals may be involved in the acquisition of thermotolerance during early development. Future studies examining the correlation of embryonic heat shock gene expression during periods of maternal hyperthermia with embryonic thermotolerance may shed some light on this problem.

V. CONCLUSIONS

This study has shown that the ability of animal embryos to synthesize hsp's in response to heat shock is dependent on the stage of development. Since delayed acquisition of the heat shock response has now been found during early insect (Graziosi *et al.*, 1980; Dura, 1981), echinoderm, amphibian, and mammalian development, it is likely that this phenomenon has been conserved through evolution. In each of the organisms we have examined, the inability to undergo a major heat shock response (i.e., synthesis of hsp's) correlates with the inability to detect hsp mRNA accumulation by either *in vitro* translation or Northern hybridization assays. Thus, control over this phenomenon is likely to be exerted at either the transcriptional and/or posttranscriptional level but not at the translational level.

The effect of heat shock on the fate of preexisting mRNA was also investigated. In early cleavage-stage embryos of *Xenopus* (32-cell stage) we detected a decrease in the levels of actin mRNA after heat shock. A similar decrease in actin mRNA was noted in 16-cell-stage sea urchin embryos. Thus, in early cleavage stage heat shock may affect the stability of this and other mRNA species. Since these embryos do not recover from thermal stress, mRNA degradation probably increases with time. In later blastula and gastrula stages of *Xenopus* and sea urchin and in blastocyst stages of rabbit, actin mRNA levels are not noticeably reduced by heat shock. Therefore, during early embryonic development heat shock-mediated reductions in the steady-state level of some constitutive mRNA's may also be a stage-dependent phenomenon.

An important question that has been approached by other investigators is whether hsp's are synthesized during normal development in the absence of environmental stress. The normal accumulation of hsp70 mRNA in *Xenopus* oocytes (Bienz and Gurdon, 1982) has already been discussed. In a detailed analysis of heat shock gene expression during *Drosophila* oogenesis and early embryogenesis, Zimmerman *et al.* (1983), utilizing recombinant DNA methodology, have found that the mRNA's for three of the seven hsp's, namely hsp83, hsp28, and hsp26, accumulate in adult ovaries in the absence of heat shock. *In situ* hybridization experiments revealed that these hsp mRNA's are synthesized initially in the nurse cells of the egg chamber and are then passed into the oocyte. These hsp mRNA's appear to be fully processed and are detectable in the early embryo until the blastoderm stage. The synthesis of hsp's in the absence of environmental stress has been found in later developmental stages of *Drosophila* as well. A number of laboratories have reported that four small hsp's (hsp28, hsp26, hsp23, and hsp22) are transcribed during late larval and early pupal stages (Sirotkin and Davidson, 1982; Cheney and Shearn, 1983). During early mouse development hsp70 has been detected in the spectrum of proteins synthesized normally at the two-cell stage but not at the one-cell stage (Bensaude *et al.*, 1983). In this latter report it was suggested that hsp70 may be one of the first products of zygotic gene activity. The ability of an embryo to express heat shock genes in response to environmental stress as well as during specific developmental periods suggests the presence of multiple regulatory signals that can activate the transcription of a heat shock gene(s).

As mentioned in Section I, the ability of a cell to synthesize a set of hsp's in response to thermal stress has been correlated with an increase in thermotolerance. During early sea urchin and *X. laevis* development the capacity to synthesize hsp's also coincides with the acquisition of thermotolerance. At this time it is not known whether a similar correlation exists with early mammalian embryos. A full understanding of the relationship between hsp's and thermotolerance may require the identification of their cellular function. Whereas specific roles of the hsp's within the cell are unknown, it has been reported that some hsp's (i.e., hsp70) migrate into the nucleus after their synthesis and become associated with the nucleoskeleton (Tanguay and Vincent, 1982; Sinbaldi and Morris, 1981; Duband *et al.*, 1983). Therefore, it is possible that these hsp's may be performing some role at the nuclear level leading to cellular homeostasis.

Note Added in Proof

After this manuscript was written, Bienz (1984) demonstrated by nuclease S, mapping that hsp70 mRNA is undetectable in *Xenopus* embryos after fertilization and becomes heat inducible at gastrulation.

Acknowledgments

This research has been supported by a Medical Research Council (M.R.C.) of Canada postdoctoral fellowship to John J. Heikkila along with a research allowance from the Alberta Heritage Foundation for Medical Research (A.H.F.M.R.), a grant from M.R.C. of Canada to Gilbert A. Schultz, an A.H.F.M.R. research fellowship to Malagorzata Kloc, and a grant from the Natural Sciences and Engineering Research Council to Leon W. Browder. Some of this work was conducted at the Marine Biological Laboratory, Woods Hole, Massachusetts, in the embryology program sponsored by a National Institutes of Health training grant. The authors are grateful to Mrs. Jillian Wilkes for her excellent technical assistance.

REFERENCES

Alliston, C. W., Howarth, B., Jr., and Ulberg, L. C. (1965). Embryonic mortality following culture *in vitro* of one- and two-cell rabbit eggs at elevated temperatures. *J. Reprod. Fertil.* **9,** 337–341.

Alwine, J. C., Kemp, D. J., and Stark, G. R. (1977). Method for detection of specific RNAs in agarose gels by transfer to diazobenzyloxymethyl paper and hybridization with DNA probes. *Proc. Natl. Acad. Sci. U.S.A.* **74,** 5350–5354.

Ashburner, M., and Bonner, J. J. (1979). The induction of gene activity in *Drosophila* by heat shock. *Cell* **17,** 241–254.

Bachvarova, R., and Davidson, E. H. (1966). Nuclear activation at the onset of amphibian gastrulation. *J. Exp. Zool.* **163,** 285–295.

Bailey, J. M., and Davidson, N. (1976). Methylmercury as a reversible denaturing agent for agarose gel electrophoresis. *Anal. Biochem.* **70,** 75–85.

Bellvé, A. R. (1972). Viability and survival of mouse embryos following parental exposure to high temperatures. *J. Reprod. Fertil.* **30,** 71–81.

Bellvé, A. R. (1973). Development of mouse embryos with abnormalities induced by parental heat stress. *J. Reprod. Fertil.* **35,** 393–403.

Bensaude, O., and Morange, M. (1983). Spontaneous high expression of heat-shock proteins in mouse embryonal carcinoma cells and ectoderm from day 8 mouse embryo. *EMBO J.* **2,** 173–177.

Bensaude, O., Babinet, C., Morange, M., and Jacob, F. (1983). Heat shock proteins, first major products of zygotic gene activity in mouse embryos. *Nature (London)* **305,** 331–332.

Bienz, M. (1984). Developmental control of the heat shock response in *Xenopus*. *Proc. Natl. Acad. Sci. U.S.A.* **81,** 3138–3142.

Bienz, M., and Gurdon, J. B. (1982). The heat shock response in *Xenopus* oocytes is controlled at the translational level. *Cell* **27,** 811–819.

Brandhorst, B. P., and Humphreys, T. (1972). Stability of nuclear and messenger RNA molecules in sea urchin embryos. *J. Cell Biol.* **53,** 474–482.

Browder, L. W. (1984). "Developmental Biology," 2nd ed. Saunders, Philadelphia, Pennsylvania.

Bushman, F. D., and Crain, W. R., Jr. (1983). Conserved pattern of embryonic actin gene expression in several sea urchins and a sand dollar. *Dev. Biol.* **98,** 429–436.

Cheney, C. M., and Shearn, A. (1983). Developmental regulation of *Drosophila* imaginal disc proteins: synthesis of a heat shock protein under non-heat shock conditions. *Dev. Biol.* **95,** 325–330.

Currie, R. W., and White, F. P. (1981). Trauma-induced protein in rat tissues: a physiological role for a "heat shock" protein? *Science* **214,** 72–73.

Duband, J. L., Valet, J. P., and Tanguay, R. M. (1983). Cellular localization of heat shock-induced proteins in *Drosophila* cells. *J. Cell Biol.* **97,** 150a.

Dura, J. M. (1981). Stage-dependent synthesis of heat shock induced proteins in early embryos of *Drosophila melanogaster*. *Mol. Gen. Genet.* **184,** 381–385.

Fyrberg, E. A., Kindle, K. L., and Davidson, N. (1980). The actin genes of *Drosophila*: a dispersed multigene family. *Cell* **19,** 365–378.

Galau, G. A., Lipson, E. D., Britten, R. J., and Davidson, E. H. (1977). Synthesis and turnover of polysomal mRNAs in sea urchin embryos. *Cell* **10,** 415–432.

Giebelhaus, D. H., Heikkila, J. J., and Schultz, G. A. (1983). Changes in the quantity of histone and actin mRNA during the development of pre-implantation mouse embryos. *Dev. Biol.* **98,** 148–154.

Graziosi, G., Micali, F., Marzari, R., De Cristini, F., and Savioni, A. (1980). Variability of response of early *Drosophila* embryos to heat shock. *J. Exp. Zool.* **214,** 141–145.

Hammond, G. L., Lai, Y.-K., and Markert, C. L. (1982). Diverse forms of stress lead to new patterns of gene expression through a common and essential metabolic pathway. *Proc. Natl. Acad. Sci. U.S.A.* **79,** 3485–3488.

Heikkila, J. J., and Schultz, G. A. (1984). Different environmental stresses can activate the expression of a heat shock gene in rabbit blastocyst. *Gamete Res.* **10,** 45–56.

Heikkila, J. J., Cosgrove, J. W., and Brown, I. R. (1981). Cell-free translation of free and membrane-bound polysomes and polyadenylated mRNA from rabbit brain following administration of d-lysergic acid diethylamide *in vivo. J. Neurochem.* **35,** 1229–1238.

Heikkila, J. J., Schultz, G. A., Iatrou, K., and Gedamu, L. (1982). Expression of a set of fish genes following heat or metal ion exposure. *J. Biol. Chem.* **257,** 12,000–12,005.

Heikkila, J. J., Kloc, M., Bury, J., Schultz, G. A., and Browder, L. W. (1984). Acquisition of the heat shock response and thermotolerance during early development of *Xenopus laevis. Dev. Biol.* (in press).

Howlett, S., Miller, J., and Schultz, G. A. (1983). Induction of heat shock proteins in early embryos of *Arbacia punctulata. Biol. Bull. (Woods Hole, Mass.)* **165,** 500.

Humphreys, T. (1971). Measurements of messenger RNA entering polysomes upon fertilization of sea urchin eggs. *Dev. Biol.* **26,** 201–208.

Karp, G. C., Manes, C., and Hahn, W. E. (1974). Ribosome production and protein synthesis in the preimplantation rabbit embryo. *Differentiation (Berlin)* **2,** 65–73.

Kelley, P. M., and Schlesinger, M. J. (1982). Antibodies to two major heat shock proteins cross-react with similar proteins in widely divergent species. *Mol. Cell. Biol.* **2,** 267–274.

Kothary, R. K., and Candido, E. P. M. (1982). Induction of a novel set of polypeptides by heat shock or sodium arsenite in cultured cells of rainbow trout, *Salmo gairdnerii. Can. J. Biochem.* **60,** 347–355.

Krüger, C., and Benecke, B.-J. C. (1981). *In vitro* translation of heat shock and non-heat shock mRNAs in heterologous and homologous cell-free systems. *Cell* **23,** 595–603.

Li, G. C., and Werb, Z. (1982). Correlation between synthesis of heat shock proteins and development of thermotolerance in Chinese hamster fibroblasts. *Proc. Natl. Acad. Sci. U.S.A.* **79,** 3268–3272.

Livak, K. J., Freund, R., Schweber, M., Wensink, P. C., and Meselson, M. (1978). Sequence organization and transcription at two heat shock loci in *Drosophila. Proc. Natl. Acad. Sci. U.S.A.* **75,** 5613–5617.

Loomis, W. F., and Wheeler, S. A. (1980). Heat shock response of *Dictyostelium. Dev. Biol.* **79,** 399–408.

Lowe, D. G., Fulford, W. D., and Moran, L. A. (1983). Mouse and *Drosophila* genes encoding the major heat shock protein (hsp 70) are highly conserved. *Mol. Cell. Biol.* **3,** 1540–1543.

McAlister, L., and Finkelstein, D. B. (1980). Heat shock proteins and thermal resistance in yeast. *Biochem. Biophys. Res. Commun.* **93,** 819–824.

Manes, C. (1969). Nucleic acid synthesis in preimplantation rabbit embryos. I. Quantitative aspects, relationship to early morphogenesis and protein synthesis. *J. Exp. Zool.* **172,** 303–310.

Maniatis, T., Fritsch, E. F., and Sambrook, J. (1982). "Molecular Cloning: A Laboratory Manual." Cold Spring Harbor Lab., Cold Spring Harbor, New York.

Newport, J., and Kirschner, M. (1982a). A major developmental transition in early *Xenopus* embryos: 1. Characterization and timing of cellular changes at the midblastula stage. *Cell* **30,** 675–686.

Newport, J., and Kirschner, M. (1982b). A major developmental transition in early *Xenopus* embryos. II. Control of the onset of transcription. *Cell* **30,** 687–696.

Perlman, S., and Rosbash, M. (1978). Analysis of *Xenopus laevis* ovary and somatic cell polyadenylated RNA by molecular hybridization. *Dev. Biol.* **63,** 197–212.

Roccheri, M. C., Di Bernardo, M. G., and Giudice, G. (1981). Synthesis of heat shock proteins in developing sea urchin. *Dev. Biol.* **83,** 173–177.

Schlesinger, M. J., Ashburner, M., and Tissières, A., eds. (1982). "Heat Shock, from Bacteria to Man." Cold Spring Harbor Lab., Cold Spring Harbor, New York.

Shiokawa, K., Tashiro, K., Misumi, Y., and Yamana, K. (1981). Non-coordinated synthesis of RNAs in pre-gastrular embryos of *Xenopus laevis*. *Dev. Growth Differ.* **23,** 589–597.

Sinbaldi, R. M., and Morris, P. W. (1981). Putative function of *Drosophila melanogaster* heat shock proteins in the nucleoskeleton. *J. Biol. Chem.* **256,** 10,735–10,738.

Sirotkin, K., and Davidson, N. (1982). Developmentally regulated transcription from *Drosophila melanogaster* chromosomal site 67B. *Dev. Biol.* **89,** 196–210.

Storti, R. V., Scott, M. P., Rich, A., and Pardue, M. L. (1980). Translational control of protein synthesis in response to heat shock in *D. melanogaster* cells. *Cell* **22,** 825–834.

Tanguay, R. M. (1983). Genetic regulation during heat shock and function of heat shock proteins: a review. *Can. J. Biochem. Cell Biol.* **61,** 387–394.

Tanguay, R. M., and Vincent, M. (1982). Intracellular translocation of cellular and heat shock induced proteins upon heat shock in *Drosophila* Kc cells. *Can. J. Biochem.* **60,** 306–315.

Ulberg, L. C., and Sheean, L. A. (1973). Early development of mammalian embryos in elevated temperatures. *J. Reprod. Fertil. Suppl.* **19,** 155–161.

Wildt, D. E., Riegle, G. D., and Dukelow, W. R. (1975). Physiological temperature response and embryonic mortality in stressed swine. *Am. J. Physiol.* **229,** 1471–1475.

Wittig, S., Hensse, S., Keitel, C., Elsner, C., and Wittig, B. (1983). Heat shock gene expression is regulated during teratocarcinoma cell differentiation and early embryonic development. *Dev. Biol.* **96,** 507–514.

Woodland, H., and Gurdon, J. (1969). RNA synthesis in an amphibian nuclear-transplant hybrid. *Dev. Biol.* **20,** 89–104.

Zimmerman, J. L., Petri, W., and Meselson, M. (1983). Accumulation of a specific subset of *D. melanogaster* heat shock mRNAs in normal development without heat shock. *Cell* **32,** 1161–1170.

8

Effects of Stress on the Gene Expression of Amphibian, Avian, and Mammalian Blood Cells

BURR G. ATKINSON AND ROB L. DEAN

I. INTRODUCTION

The synthesis of heat shock proteins (hsp's) in response to a rapid elevation in temperature (heat shock) is characteristic of the cells from all organisms exam-

Changes in Eukaryotic Gene Expression
in Response to Environmental Stress

ISBN 0-12-066290-6

ined (Schlesinger *et al.*, 1982). To date, the only cells which fail to exhibit a detectable hsp response are those in very early stages of embryogenesis (see Heikkila *et al.*, Chapter 7, this volume). The synthesis of heat shock proteins is dramatic, rapid, and reversible, and, thus, provides a useful experimental tool for studying the regulation of gene expression. Unfortunately, while this response appears to be universal in differentiating or differentiated cells, little is known regarding the function of newly generated hsp's. In our studies of this phenomenon, we felt that it should be possible to select a single cell type ideal for investigating both the function of hsp's and the means by which their expression is regulated, that is, a cell type which (1) responds to stress and is, in its natural surroundings, continuously involved with environmental stressors, and (2) has properties which greatly facilitate its use in regulatory studies. Cells from vertebrate blood appeared to fulfill our conditions and seemed ideal since they are easy to obtain in large quantities, are readily separated from other components of blood tissue, and can be maintained in culture.

In our initial investigations with isolated blood cells, we explored the heat shock response in different types of blood cells from a variety of organisms. As shown in Fig. 1, we found that human white blood cells (WBC's), avian (quail) red blood cells (RBC's) and white blood cells, and amphibian (frog) red blood cells all showed a pronounced heat shock response. In each case, the new and/or enhanced synthesis of a few proteins is evident in blood cells which were heat shocked in culture (compare control and heat-shocked cells in Fig. 1). Although heat-shocked blood cells from mammals (human and murine WBC's exhibit a similar response), avians and amphibians all show a significant degree of synthesis of a protein having a molecular weight (M_r) of approximately 70,000 [the amphibian 70,000 (70K) protein is slightly larger than the mammalian or avian 70K hsp] and of one having an M_r of approximately 90,000, a number of hsp's appear to be class specific; heat-shocked blood cells from mammals synthesize hsp's of 110K and 115K which are not found in heat-shocked avian or amphibian cells, and heat-shocked avian blood cells synthesize a 26K hsp which is not detected in heat-shocked mammalian or amphibian cells.

In reviewing our results with heat-shocked blood cells from these different organisms, we felt that the heat shock response detected in avian RBC's coupled with the limited gene expression normally found in these cells [90% of the protein produced in these cells is globin (Lasky *et al.*, 1978; Williamson and Tobin, 1977)] made them ideal for elucidating the regulatory mechanisms governing the expression of heat shock genes. Moreover, because the nucleated erythroid cells from the peripheral circulation of normal and anemic birds are well characterized and a great deal is known concerning the differential gene expression exhibited by these cells, we chose the avian RBC for further study. Thus, the remainder of this chapter is concerned with our results from studies on the effects of thermal and chemical stress on gene expression in avian RBC's.

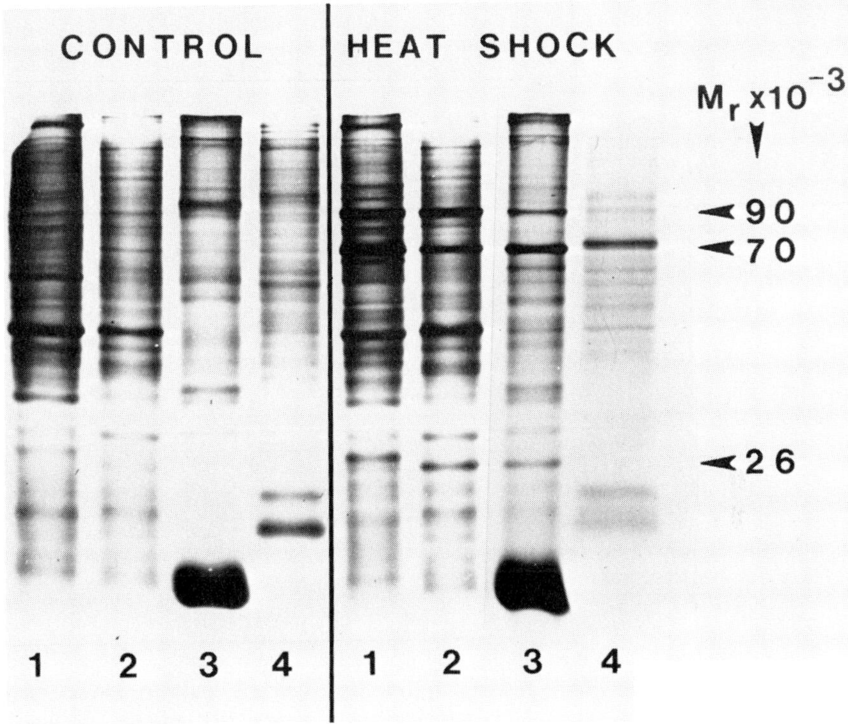

Fig. 1. Fluorograms from one-dimensional SDS–PAGE separations of polypeptides synthesized by (1) human white blood cells; (2) quail white blood cells; (3) quail red blood cells, and (4) frog red blood cells at control and heat shock temperatures (37 and 43, 37 and 45, 37 and 45, and 22 and 37°C, respectively). All species show pronounced synthesis of an hsp with a molecular weight (M_r) of 70,000 (slightly larger in the frog) and one with an M_r of 90,000. In this study, hsp's with M_r values of 110,000 and 115,000 were found only in human WBC's, and synthesis of an hsp with an M_r of 26,000 was detected only in blood cells from the quail.

II. ELABORATION OF A THERMAL STRESS RESPONSE IN CULTURED RED BLOOD CELLS FROM NORMAL (NONANEMIC) AND PHENYLHYDRAZINE-TREATED (ANEMIC) ADULT QUAIL

Three successive daily injections of phenylhydrazine into adult quail reduces their mean hematocrit from ~46 to ~20. Forty-eight hours after the final injection, the hematocrit shows an increase as the birds begin to recover from anemia (Atkinson and Dean, 1983). Smears of red blood cells from whole blood of normal quail and those recovering from anemia show marked differences in cell

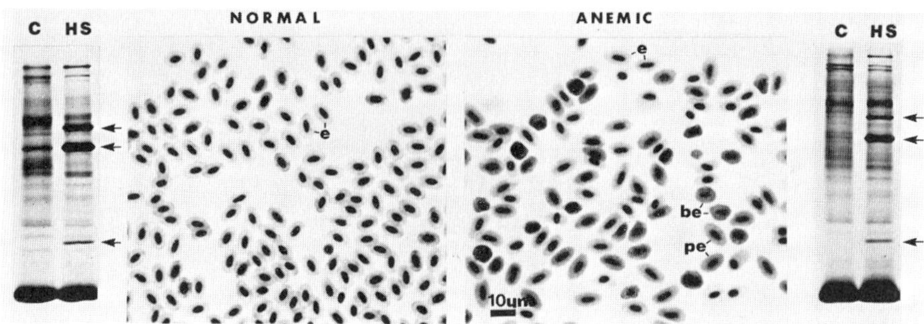

Fig. 2. Light micrographs of blood smears from untreated quail (normal) and from quail 2 days after a third daily phenylhydrazine injection (anemic). Almost all the red blood cells from untreated quail are mature erythrocytes (e) with highly condensed nuclei and cytoplasms lacking basophilia. Red blood cells from quail recovering from anemia are predominantly immature cells, either basophilic erythroblasts (be) or erythrocytes showing varying degrees of basophilia [polychromatic erythrocytes (pe)]. Side panels show fluorograms from one-dimensional PAGE separations of lysates of normal and anemic red blood cells incubated for 3 hr at 37 (C) or 45°C (HS) with new or enhanced synthesis of three polypeptides (M_r values of 90,000, 70,000, and 26,000; see arrows). There are slight differences between the synthetic patterns of control cells, notably a greater rate of synthesis of a 70K polypeptide in normal cells.

composition (Fig. 2). While mature erythrocytes are the primary cell type found in blood from normal quail, few mature erythrocytes are found in blood from anemic quail; blood from anemic quail contains a number of immature erythrocytes showing varying degrees of basophilia and a large number of basophilic erythroblasts. Red blood cells from normal and anemic birds respond to heat stress (45°C) in a similar fashion which is characterized by (1) the new or enhanced synthesis of three polypeptides with M_r values of 90,000, 70,000, and 26,000, and (2) the depressed synthesis of polypeptides previously made (Fig. 2).

A method for obtaining a cell fraction enriched in immature cells from the blood of anemic ducks (Imaizumi-Scherrer *et al.*, 1982) was used to fractionate the RBC's collected from both normal and anemic quail. Cells were separated by centrifugation into three fractions (top, middle, and bottom fractions) depending on their final position in the cell pellet. This method did not produce significant enrichment of immature cells from normal quail blood since the cells were almost entirely mature erythrocytes exhibiting no basophilia (Fig. 3, normal, fractions 1, 2, and 3). However, when used to fractionate RBC's from anemic quail, this method provided an upper fraction enriched in immature cells (90% immature cells) as shown in Fig. 3. This was confirmed by inspection of electron micrographs made from sections of the top, middle, and bottom fractions of RBC's collected from anemic birds (Fig. 4). Approximately 200 cells in sections of each fraction, in adjacent spaces of a copper grid, were characterized as erythroblasts, intermediate RBC's, or mature erythrocytes according to criteria described in Fig. 4. When cultured, RBC's from the top, middle, and bottom cell fractions of normal and anemic quail all responded to heat stress by synthesizing similar, if not identical, heat shock polypeptides (Fig. 3).

Our preliminary results with mature RBC's, isolated from normal or anemic adult quail, suggest that they may be more synthetically active than is generally supposed. Indeed, we find that more than 90% of the RBC's from normal adult

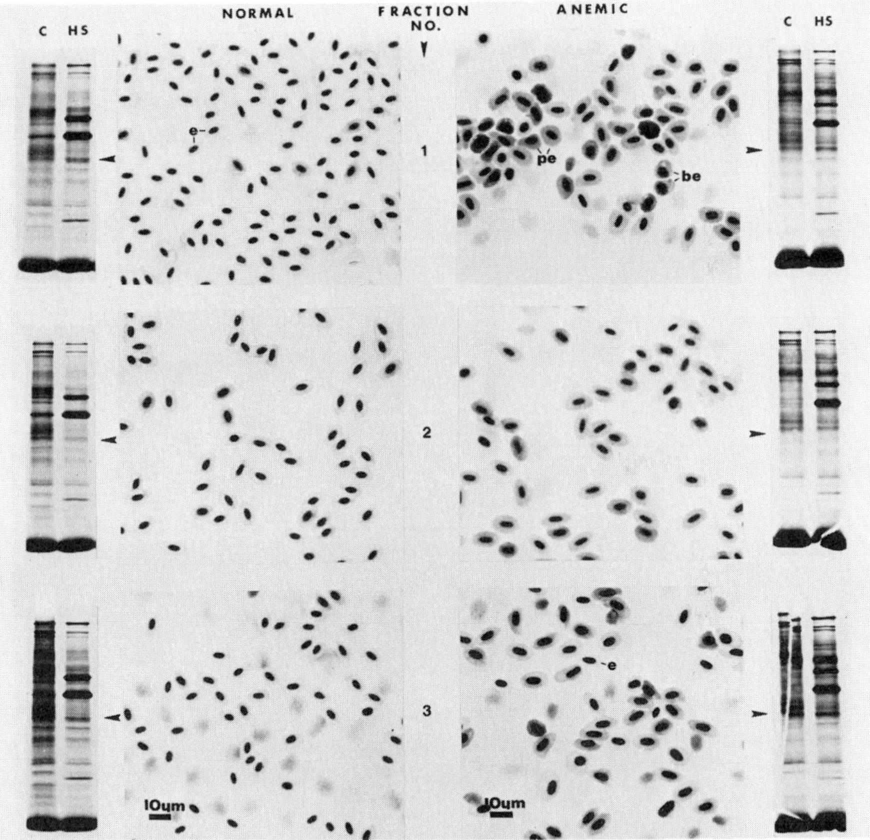

Fig. 3. Representative light micrographs of smears of red blood cells taken from normal and anemic quail and separated into a top (fraction 1), middle (fraction 2), and bottom fraction (fraction 3) by the method of Imaizumi-Scherrer *et al.* (1982). The upper fraction of red blood cells from anemic quail is enriched in immature cell types. Since RBC's from normal quail are almost entirely mature erythrocytes, no obvious enrichment in immature cells is seen. Side panels show fluorograms from one-dimensional SDS–polyacrylamide electrophoretic separation of polypeptides synthesized by RBC's from each fraction in response to a 3-hr incubation at 37° [control (C)] or 45°C [heat shock (HS)]. [^{35}S]Methionine was added to the medium for the last 2 hr of incubation. All cell fractions from heat-shocked cells responded to elevated temperature by synthesizing the three major hsp's. Symbols: erythrocytes (e); basophilic erythrocytes (be); polychromatic erythrocytes (pe).

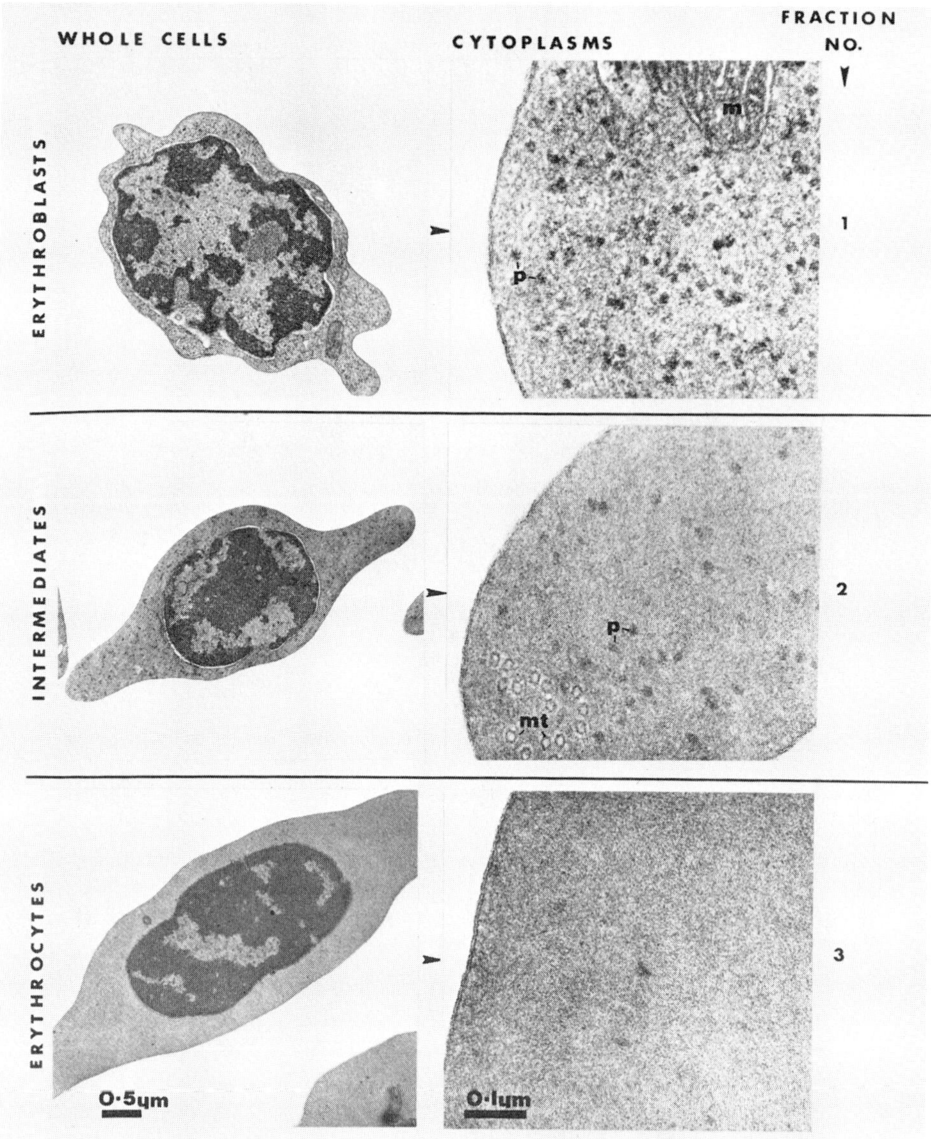

Fig. 4. Electron micrographs of red blood cells at different stages of maturation. Red blood cells from untreated quail are almost exclusively mature erythrocytes. These exhibit an ovoid profile with a regular nucleus, highly condensed heterochromatin, no nucleoli, no cytoplasmic membranes or membrane-bound organelles, and, apparently, few to no ribosomes. Red blood cells from phenylhydrazine-treated quail are only ~10% mature erythrocytes. The remaining cells are erythroblasts or erythrocytes in a variety of stages of maturation. Quail erythroblasts have an irregular outline, a

quail consist of mature RBC's [similar to that found in adult goose blood (Adams and Neelin, 1972)] and that these RBC's are (1) apparently capable of synthesizing hsp's in response to thermal or chemical stress and (2) only 3–5 times less active in protein synthesis than immature RBC's. Studies are currently underway to establish whether the synthesis of hsp's by mature RBC's is due to a small number (<10%) of immature RBC's present in our samples or, in fact, is an intrinsic property of mature RBC's. The currently, unresolved nature of the mature RBC synthetic response to stress, coupled with the greater synthetic activity of immature RBC's and the ease with which these immature cells can be generated, prompted us to use the immature quail RBC as a model system for investigating the effects of stress on gene expression.

III. CHARACTERIZATION OF THE HEAT SHOCK AND STRESS PROTEINS INDUCED IN CULTURED RED BLOOD CELLS FROM ANEMIC ADULT QUAIL

As Lindquist has pointed out (Lindquist and DiDomenico, this volume, Chapter 4), heat shock proteins show a capacity for independent regulation in certain aspects of their expression. Two of the so-called noncoordinate aspects of hsp regulation have been demonstrated in quail RBC's. First, there is variation in the rate and level of hsp synthesis at different temperatures and times after the onset of hyperthermia. We found that the amount of [^{35}S]methionine incorporated into the protein of cultured RBC's is not significantly altered by a 1- to 3-hour shift from 37°C to a temperature between 38 and 45°C (see, for example, Table I; cpm incorporated [^{35}S]methionine/10^6 cells at 37 and 45°C). However, incubation at temperatures at or above 41°C results in the new and/or enhanced synthesis of proteins with molecular weights of 90K and 70K and at temperatures at or above 42°C the new and/or enhanced synthesis of a 26K protein is apparent (Fig. 5A). When RBC's from normal or anemic quail are heat shocked for various times at 45°C and radioactively labeled at 37°C with either [^{14}C]leucine (not shown) or [^{35}S]methionine (Fig. 5B), new and/or enhanced synthesis of the 90K and 70K hsp's is easily observed after 30 min at 45°C; detectable synthesis of the 26K hsp requires at least 45 min at 45°C. Although radioactive labeling of heat-shocked cells at 37°C clearly establishes the new and/or enhanced synthesis of hsp's, little depression in the synthesis of other RBC proteins is evident (Fig. 5C). Radioactive labeling of heat-shocked cells at the heat shock temperature, however,

---

large irregular nucleus, highly dispersed heterochromatin, numerous polysomes, and always some mitochondria. In addition, as cells mature, interstices between the chromatin and the cytoplasmic matrix appear more electron dense as a result of hemoglobin accumulation (Tooze and Davies, 1963; Small and Davies, 1970). Polysomes (p), microtubules (mt), and mitochrondria (m).

TABLE I

Influence of Actinomycin D on the Incorporation of [³H]Uridine into RNA and [³⁵S]Methionine into Protein[a]

Actinomycin D (μg/ml)	Incubation temp. (°C)	[³H]Uridine incorporation		[³⁵S]Methionine incorporation	
		cpm/10⁶ cells	% of control	cpm/10⁶ cells	% of control
0 (Control)	37	10,790	—	25,300	—
0.5	37	3,690	34.0	23,600	93.0
1.0	37	2,350	21.0	24,300	96.0
2.0	37	725	6.7	18,400	73.0
3.0	37	525	4.9		
0 (Control)	45	10,223	—	21,300	—
0.5	45	3,025	29.0	20,500	96.0
1.0	45	1,441	15.0	14,200	67.0
2.0	45	758	7.5	14,200	67.0
3.0	45	616	6.0		

[a] Actinomycin D drastically reduces [³H]uridine incorporation but has a relatively slight effect on [³⁵S]methionine incorporation. This suggests that new messenger RNA synthesis is severely curtailed in the presence of actinomycin D while protein synthesis remains relatively unaffected.

results in the depressed synthesis of other RBC proteins and enhanced synthesis of hsp's (Fig. 5C). When RBC's are maintained at 45°C for 5 hr or more, protein synthesis is severely depressed and a newly synthesized 55K polypeptide is evident; at this time it is not clear whether this polypeptide represents a new hsp or a breakdown product of another hsp.

A second noncoordinate regulatory feature is the variation observed in the basal level of different hsp's synthesized at normal temperatures. Two-dimensional analysis of lysates of heat-shocked and control cells reveals that hsp's with M_r values of approximately 70,000 are constitutively synthesized at normal temperatures but that their rate of synthesis is enhanced during heat stress (Fig. 6A). In addition to the enhanced synthesis of these polypeptides, heat shock also results in the synthesis of two polypeptides (M_r of 90,000, p*I* of 5.0 and M_r of 26,000, p*I* of 5.4) not detected in cells maintained at 37°C (Fig. 6B).

Fig. 5. Assessment of the factors [temperature (A) and time (B and C)] involved in the kinetics of heat shock protein induction in cultured RBC's. (A) Fluorograms from one-dimensional SDS–PAGE separations of lysates from red blood cells incubated for 3 hr at 37, 40, 41, 42, 43, 44, and 45°C. In this case, [³⁵S]methionine was added to the medium after 1 hr. (B) Fluorograms from one-dimensional SDS–PAGE separations of lysates from red blood cells which have been incubated at 45°C, for 0, ¼, ½, ¾, 1, 2, and 3 hr and then transferred to medium containing [³⁵S]methionine and incubated for 2 hr at 37°C. (C) Fluorograms from one-dimensional SDS–PAGE separations of lysates from red blood cells which have been incubated at 45°C for 3, 4, 5, 6, 7, and 8 hr to which [³⁵S]methionine was added for the last 2 hr of incubation.

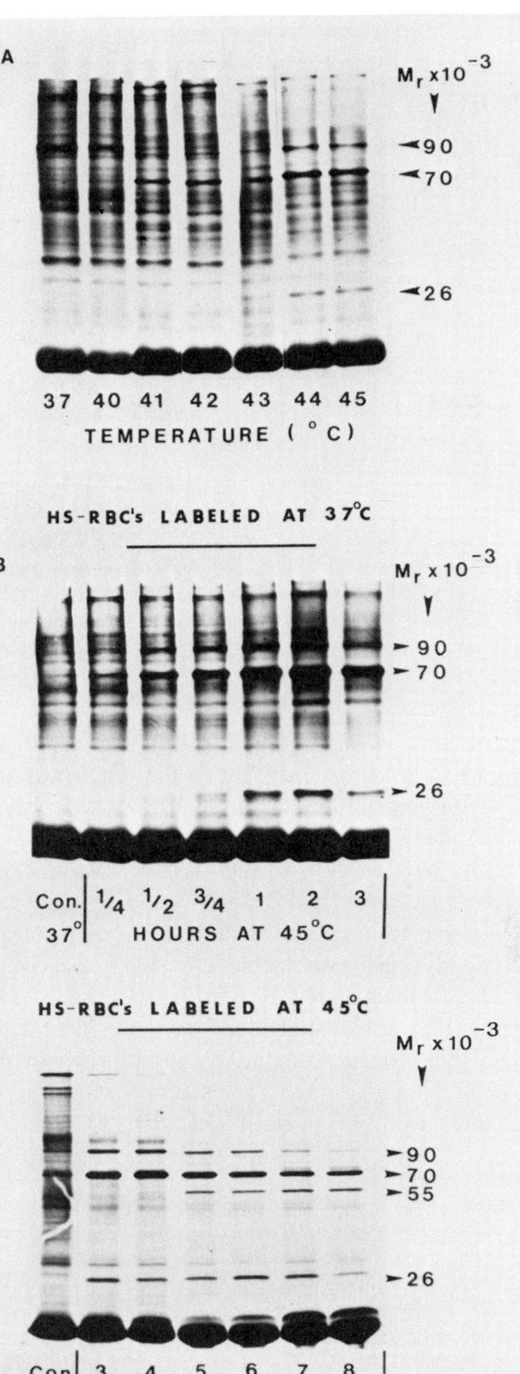

A
$M_r \times 10^{-3}$
90
70
26
37 40 41 42 43 44 45
TEMPERATURE (°C)
HS-RBC's LABELED AT 37°C
B
$M_r \times 10^{-3}$
90
70
26
Con. 37°
1/4 1/2 3/4 1 2 3
HOURS AT 45°C
HS-RBC's LABELED AT 45°C
C
$M_r \times 10^{-3}$
90
70
55
26
Con. 37°
3 4 5 6 7 8
HOURS AT 45°C

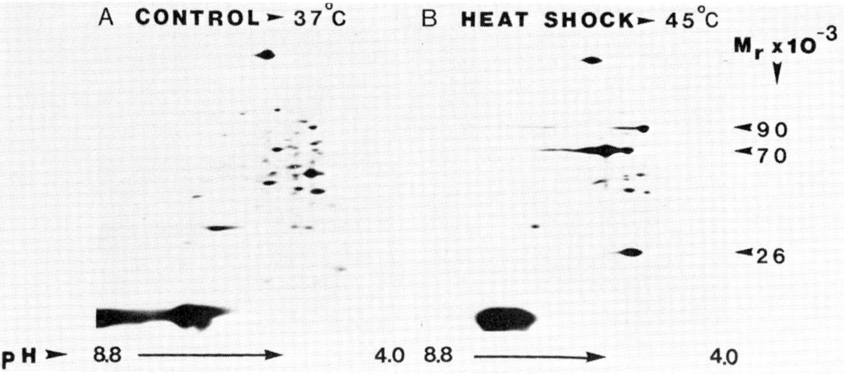

Fig. 6. Fluorograms from two-dimensional electrophoretic separations of lysates from quail RBC's maintained at 37 and 45°C. These gels were prepared as described elsewhere (Atkinson, 1981) except that the second-dimension gels consisted of a 7.5–17.5% polyacrylamide gradient. Equal amounts of acid-precipitable radioactivity ([^{35}S]methionine) were applied to the first dimension of each two-dimensional gel and the fluorograms were developed at −70°C for the same period of time. In order to emphasize the induction of hsp's and the repressed synthesis of those proteins made prior to heat shock, we have deleted the newly synthesized globins from the gels presented in this figure.

In the course of our studies, we found that the isoelectric point of the newly synthesized quail 90K hsp was dependent on the solution used to lyse the quail RBC's. While the prominent 90K hsp from RBC's freshly lysed in triethanolamine buffer (TEA) has a p*I* of approximately 5.0, the most prominent 90K hsp's from RBC's freshly lysed in urea or SDS have p*I* values of approximately 5.5 and 6.0, respectively (Atkinson and Dean, 1983). The solution used to lyse quail RBC's was also found to affect the stability of newly synthesized heat shock proteins. While all of the hsp's detectable in freshly lysed RBC's were present in SDS lysed RBC's stored at −20°C for 5–30 days, none of these hsp's were detectable in RBC–TEA lysates stored for 5 days at −20°C, and the 90K hsp was not found in RBC–urea lysates stored for 5 days at −20°C (Atkinson and Dean, 1983).

A brief incubation (1 or 2 hr) of quail RBC's at 37°C with arsenite, cadmium

---

Fig. 7. Fluorographic analysis of the stress-induced polypeptides synthesized in cells cultured in the presence of arsenite (A), cadmium ions (B), and zinc ions (C). Proteins synthesized in response to arsenite closely resemble the hsp's (shown in lane 5). An exception to this similarity is the new and/or enhanced synthesis of a 35K polypeptide in cells exposed to 2×10^{-5} *M* arsenite (see arrow). This polypeptide is not synthesized by cells exposed to higher arsenite concentrations or to elevated temperatures. Cadmium and zinc ions induced new and/or enhanced synthesis of hsp-like proteins with M_r values of 90,000 and 70,000. Both ions also induced the synthesis of a 78K polypeptide (see arrows) not seen in the heat shock response, but failed to induce the 26K polypeptide that is synthesized in response to arsenite or elevated temperatures.

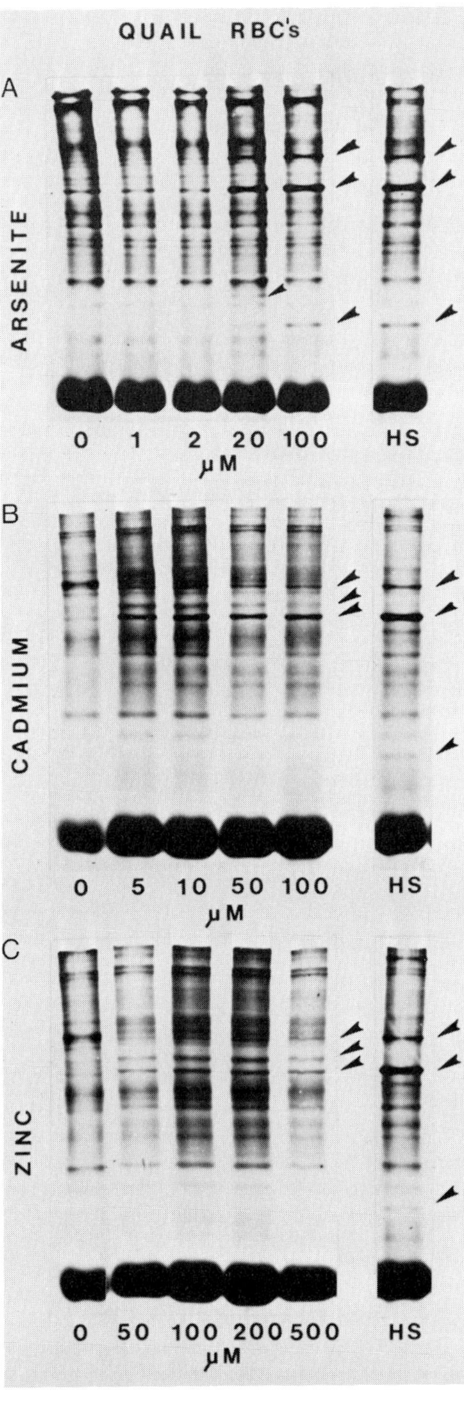
QUAIL RBC's
A
ARSENITE
0 1 2 20 100 HS
µM
B
CADMIUM
0 5 10 50 100 HS
µM
C
ZINC
0 50 100 200 500 HS
µM

ions, or zinc ions can also cause pronounced changes in the types of proteins synthesized by these cells. The major proteins synthesized by RBC's in response to elevated levels of arsenite (at or exceeding 2×10^{-5} M) are similar in relative molecular weight (M_r; Fig. 7A), isoelectric point (p*I*; not shown), and products derived from partial proteolytic digestions (Fig. 8) to those synthesized by heat-shocked RBC's as well as to those synthesized by primary cultures of heat-shocked or arsenite-treated quail myoblasts (Atkinson, 1981; Atkinson and Pol-

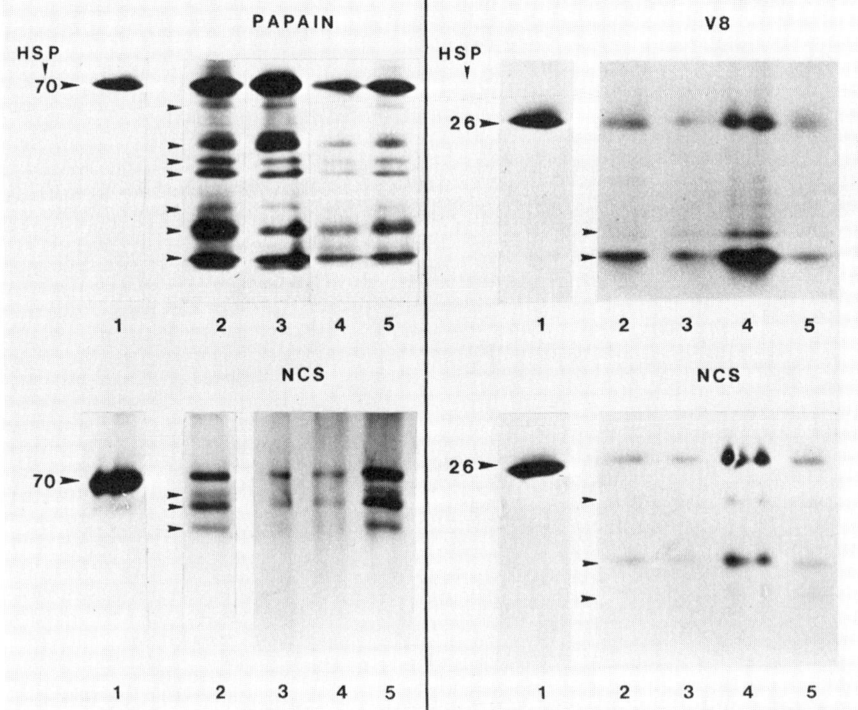

Fig. 8. Fluorogram resulting from partial proteolytic digestions of [^{35}S]methionine-labeled and SDS–PAGE-purified 70K and 26K hsp's. These hsp's were obtained from quail RBC's (lanes 2 and 3) and from 24-hr-old primary cultures of quail myoblasts (lanes 4 and 5; see Atkinson, 1981; Atkinson *et al.*, 1983) which had been heat shocked at 45°C (lanes 2 and 4) or treated for 2 hr with arsenite (75 μ*M*; lanes 3 and 5). Enzyme digests [papain (3.2 μg/ml) and the V8 protease from *Staphylococcus aureus* (12.8 μg/ml)] were performed according to the methods of Cleveland *et al.* (1977), as modified by Tijssen and Kurstak (1983). The *N*′-chlorosuccinimide (NCS) digestions followed the method of Lischwe and Ochs (1982). The results from the electrophoretic separations clearly establish that the products derived from partial proteolytic digestion of the RBC 70K hsp as well as those derived from the RBC 26K hsp are the same as those similarly derived from proteins with similar M_r values found in heat-shocked or arsenite-treated myogenic cells or in arsenite-treated RBC's.

lock, 1982; Atkinson *et al.*, 1983). A noticeable exception to these similarities is the new and/or enhanced synthesis of a 35K polypeptide by RBC's treated with 2×10^{-5} *M* arsenite; synthesis of this polypeptide is not detectable in RBC's treated with higher levels of arsenite or in heat-shocked RBC's. [Johnston *et al.* (1980) and Kim *et al.* (1983) reported a similar arsenite effect in chick embryo cells and avian myoblasts, respectively.]

While low, but elevated above normal, levels ($>1 \times 10^{-6}$ *M*) of either cadmium or zinc induce new and/or enhanced synthesis of hsp-like polypeptides with M_r values of 90,000 and 70,000 and a polypeptide with an M_r of approximately 78,000 (this last protein does not appear to be induced by heat shock or arsenite), neither metal ion (at levels up to and including those inhibiting protein synthesis) causes RBC's to synthesize a 26K polypeptide similar to that found in heat-shocked or arsenite-treated RBC's (compare A, B, and C in Fig. 7). The absence of any detectable synthesis of a 26K heat shocklike protein by RBC's stressed with copper or zinc was surprising since quail myogenic cells (Atkinson, 1981; Atkinson *et al.*, 1983; Somerville, 1984) as well as cells from other avian systems (Levinson *et al.*, 1979, 1980; Schlesinger, this volume, Chapter 9), respond to these metals by synthesizing a 26K protein. We suspect that the lack of response to copper or zinc by RBC's may be related to the fact that they are blood cells and may have specific means, unlike other cell types, of detoxifying some heavy metals and preventing them from accumulating in a free form within the cells (Cousins, 1982).

IV. COMPARISON OF QUAIL RED BLOOD CELL HEAT SHOCK PROTEINS INDUCED IN CULTURE WITH THOSE INDUCED *IN SITU*

Birds may become hyperthermic as a result of fever accompanying bacterial infection (D'Alecy and Kluger, 1975) or when subjected to high environmental temperatures. Such heat-stressed birds characteristically regulate body temperature 2–4°C above that of unstressed birds at rest (Calder and King, 1974). To determine whether RBC's in heat-stressed birds synthesize hsp's like those synthesized by RBC's heat stressed in culture, we induced hyperthermia in adult anemic quail. Leaving the head and neck free, the quail were restrained in a cotton cloth and wrapped in an electric heating pad. Cloacal temperature was monitored constantly, and the heating pad was switched on for experimental birds and left off for the controls. Control birds maintained core temperatures within normal limits (for normal diurnal temperature fluctuations see M. Kavaliers and M. Hirst, unpublished) and none exceeded 41°C for more than 5 min. Core temperatures of experimental quail rose rapidly above 41°C and reached a plateau between 42 and 43.2°C. The quail were able to regulate core tem-

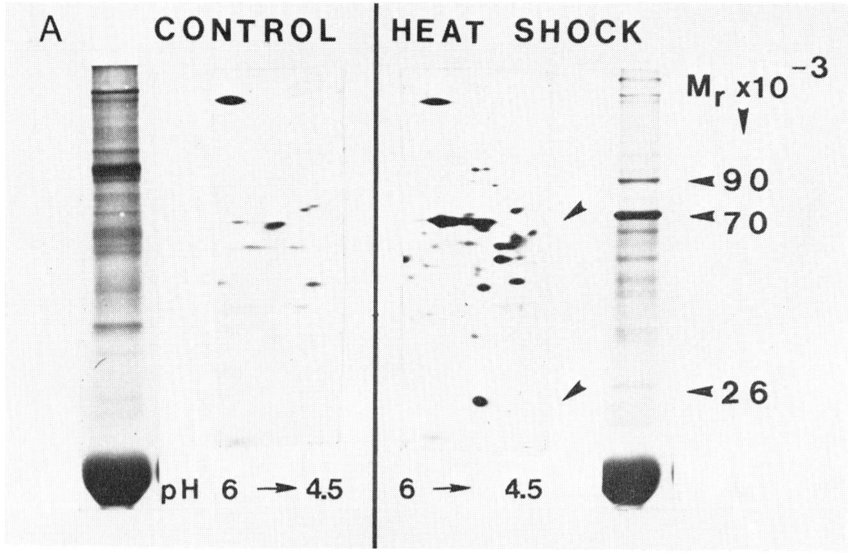

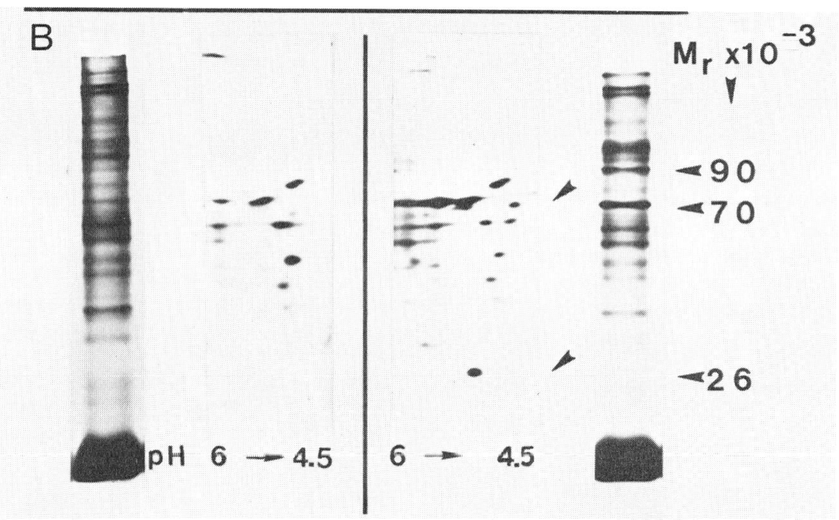

Fig. 9. One- and two-dimensional electrophoretic comparisons of quail red blood cell hsp's induced in culture (A) with those induced *in situ* (B). When adult anemic quail are made hyperthermic (43°C for 80 minutes; see details in text) their RBC's synthesize polypeptides which correspond in M_r and pI values to the hsp's synthesized by RBC's heat-shocked in culture.

peratures within this narrow though elevated range by panting and gular flutter (Calder and King, 1974). After hyperthermic or control episodes of varying durations, blood was collected from the birds and the red blood cells were isolated and labeled in a culture medium at 37°C. Figure 9 reveals that the red blood cells from heat-shocked, intact quail show a pattern of hsp synthesis identical to that seen in RBC's heat shocked in culture (Fig. 9A) (Dean and Atkinson, 1984b). The kinetics of the *in situ* response to hyperthermia (Fig. 9B) also appear similar to the heat shock response observed in cultured RBC's; no hsp synthesis was detected below 41°C in cultured RBC's, and no hsp response was found in RBC's from control birds in which core temperature rarely, and then only briefly, exceeded 41°C. Also, a more pronounced stress response was seen in RBC's from experimental birds in which the core temperature remained above 41°C for longer periods than in those which experienced shorter hyperthermic episodes.

V. CHARACTERIZATION OF THE RESPONSE OF RED BLOOD CELLS FROM ANEMIC QUAIL TO HEAT SHOCK AND CHEMICAL STRESS

An intrinsic and fundamental property of RBC's involves the selective expression of globin genes and a repression in the expression of most other genes [90% of protein synthesis is normally devoted to globin (Williamson and Tobin, 1977)]. The highly restricted gene expression found in these cells involves regulation at all levels of information transfer, from the genome to the phenotype (Imaizumi-Scherrer *et al.*, 1982). The fact that heat shock or chemical stress rapidly and selectively modulates further gene expression in these highly specialized cells prompted us to examine whether the RBC response to stress (hsp synthesis and depressed synthesis of globin and other proteins) involves changes in transcription as well as translation.

A. Cell-Free Translation of Polysomal and Nonpolysomal Cytoplasmic mRNA's from Control and Heat-Shocked Red Blood Cells

The work of Spirin (1969) first revealed the compartmental organization of mRNA's outside the polyribosomes. Since then, nonpolyribosomal cytoplasmic mRNA's complexed with proteins (i.e., free mRNP's) have been reported in a variety of eukaryotic cells, including avian RBC's (Spohr *et al.*, 1970, 1972; Gander *et al.*, 1973). The mRNP complexes in avian RBC's have been shown to consist predominantly of mRNA's coding for globin chains [this includes about 1400 different mRNA species of which globin mRNA's represent 44% of the total mass (Imaizumi-Scherrer *et al.*, 1982)]. Since it had been suggested (Vincent *et al.*, 1980) that the mRNA's in these mRNP's may represent a biochemical basis for regulation at the cytoplasmic level, we extracted and translated the mRNA's from polyribosomes and from free mRNP's isolated from the same control and heat-shocked RBC's (Dean and Atkinson, 1984a). Figure 10 shows the comparative fluorographic results obtained from two-dimensional gel separations of the products synthesized in culture by control and heat-shocked cells (A), in a cell-free system by RNA's extracted from polyribosomes (B), and by RNA's extracted from free mRNP's (C) of similarly treated cells.

In these studies, similar amounts of RNA were extracted from polyribosomes [isolated according to the method of Imaizumi-Scherrer *et al.* (1982)] of control and heat-shocked RBC's and translated in a rabbit reticulocyte lysate system. Since 90% of the polyribosomal mRNA in erythroid cells is normally globin mRNA [there are about 1500 molecules of each globin mRNA species and about 7 molecules of each nonglobin mRNA species—100–200 different species—in the polyribosomes of each cell (Lasky *et al.*, 1978; Humphries *et al.*, 1976)], the synthesis of nonglobin proteins by RBC polyribosomal mRNA in cell-free preparations is normally low and, as shown in B (control) of Fig. 10 (the newly synthesized globin has been deleted from these two-dimensional pictures), not easy to detect. Although other investigators have circumvented this difficulty by modifying their synthetic techniques to increase the translated amount of these less abundant mRNA's (Lodish, 1974; Imaizumi-Scherrer *et al.*, 1982), we merely required a comparative detection of the most abundant nonglobin products synthesized by these mRNA's and, therefore, simply applied the same amount of acid-precipitable radioactivity to each two-dimensional gel and exposed the PPO-impregnated gels to X-ray film for similar periods of time. The two-dimensional electrophoretic comparison (Fig. 10 A,B) of nonglobin products synthesized in this cell-free system, by polyribosomal mRNA's from control and heat-shocked RBC's, to those synthesized in culture by control and heat-shocked RBC's, discloses (1) that polyribosomal mRNA's from heat-shocked RBC's synthesize products which have molecular weights and isoelectric points the same as hsp's synthesized by RBC's in culture, (2) that these mRNA's are, in addition to globin mRNA's (newly synthesized globins have been deleted from these photographs), major components found on the polyribosomes of heat-shocked RBC's, (3) that the 70K hsp consists of (at least) two polypeptides with different isoelectric points which are translated from different mRNA's, and (4) that these particular hsp mRNA's are absent, are not translated by, or are rare components of polyribosomes from control RBC's.

Quail RBC mRNP's were isolated according to the method of Spohr *et al.* (1972), and the mRNA's were extracted from them and translated in a rabbit reticulocyte lysate system. A comparison of the nonglobin products synthesized in this cell-free system by mRNA's from free mRNP's of control and heat-shocked RBC's (Fig. 10C) discloses that the most abundant translatable non-

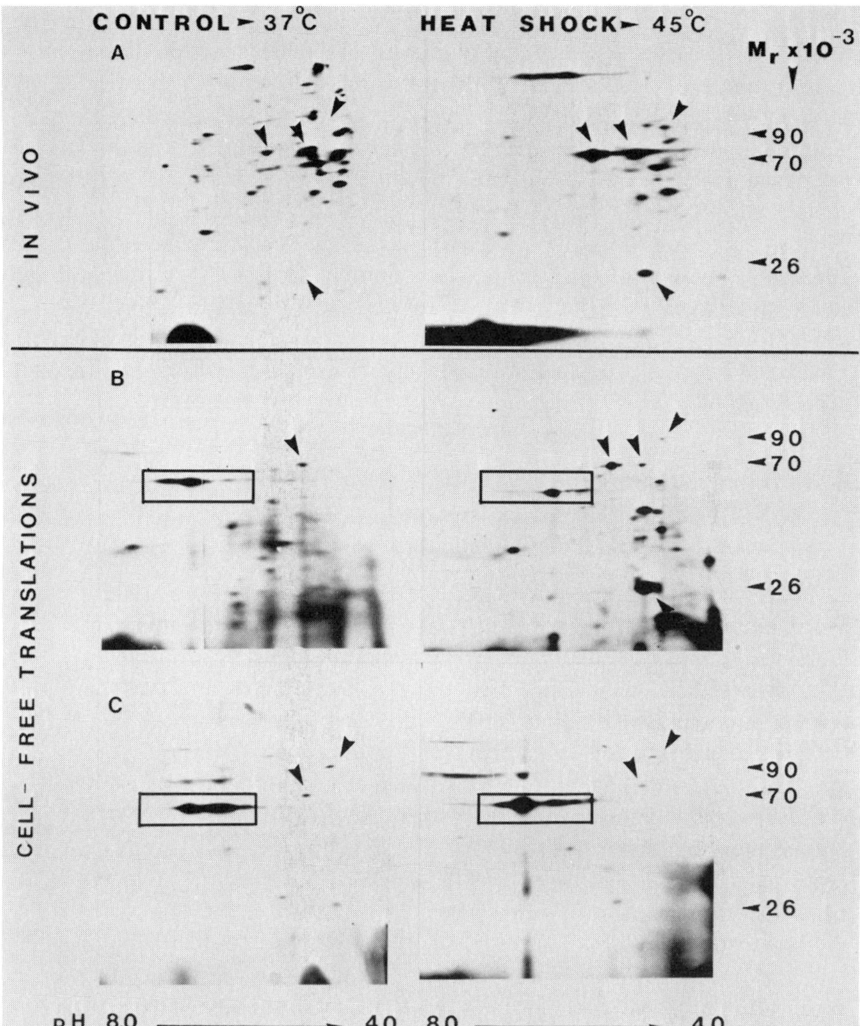

Fig. 10. Fluorographic comparison of the proteins synthesized in culture by control and heat-shocked RBC's (A) to those synthesized in a cell-free system by polyribosomal mRNA's (B) and by free RNP mRNA's (C) obtained from control and heat-shocked RBC's. Equal amounts of acid-precipitable radioactivity and equal developing times were used for each pair of gels shown in this figure. In each case newly synthesized globin chains have been deleted from the figures. The areas of radioactivity enclosed in boxes in the fluorograms shown in B and C result from contaminants in the reticulocyte lysate used in these studies.

globin mRNA's found in control-cell mRNP's are similar to those found in the mRNP's from heat-shocked cells. Comparison of the nonglobin products synthesized by the mRNA's from free mRNP's with those synthesized by mRNA's from polyribosomes of control and heat-shocked RBC's (Figs. 10B,C) reveals that, although the mRNP's from control and heat-shocked RBC's contain no detectable mRNA for the 26K hsp (as measured by *in vitro* translation), they do contain mRNA's for a 90K polypeptide, which has a p*I* value the same as the 90K hsp, and for a 70K polypeptide, which has a p*I* (5.4) the same as one of the 70K hsp's. These results suggest but do not prove that mRNA's for the 90K hsp and for at least one form of the 70K hsp may be under some translational control, similar to that reported for the 70K hsp in *Xenopus* oocytes (Bienz and Gurdon, 1982), and not wholly dependent on new and/or enhanced transcription for their expression.

B. Effect of Actinomycin D on RNA Synthesis and on the Synthesis of Heat Shock Proteins in Red Blood Cells from Anemic Adult Quail

Control or heat-shocked quail RBC's incubated in the presence of actinomycin D are transcriptionally inactive when the concentration of this inhibitor is equal to or greater than 2 μg/ml of culture medium (1×10^8 cells/ml) (Table I). This effect is dependent on the concentration of the inhibitor present; lower concentrations of actinomycin D inhibit RNA synthesis to a lesser extent (Table I). Although the concentrations of actinomycin D (up to and including 2 μg/ml) did not greatly affect the rate of protein synthesis in control or heat-shocked RBC's (Table I) nor the types of proteins synthesized by the control RBC's, they did depress the synthesis of hsp's by heat-shocked cells (Fig. 11). In this case, the depressed synthesis of hsp's by heat-shocked RBC's appears to be directly related to the concentration of actinomycin D used and/or the amount of RNA synthesis inhibited (Table I and Fig. 11); the higher the concentration of inhibitor, the more depressed RNA and hsp synthesis becomes. The fluorogram depicted in Fig. 11 also demonstrates that RBC's heat shocked in the presence of actinomycin D synthesize proteins (other than hsp's) which correspond in molecular weight to those synthesized by RBC's incubated at control temperatures in the presence or absence of actinomycin D. These data suggest that actinomycin D is blocking transcription of hsp mRNA's (and perhaps other mRNA's) but not affecting, even under heat shock conditions, the translation of previously synthesized cytoplasmic mRNA's. Indeed, RBC's heat shocked in the presence of 2 μg/ml of actinomycin D do not synthesize heat shock proteins, nor, unlike *Drosophila* cells (see Lindquist and DiDomenico, Chapter 4, this volume), do they exhibit a repression (block) in the synthesis of preexisting mRNA's. This latter observation suggests that new mRNA synthesis, whether it

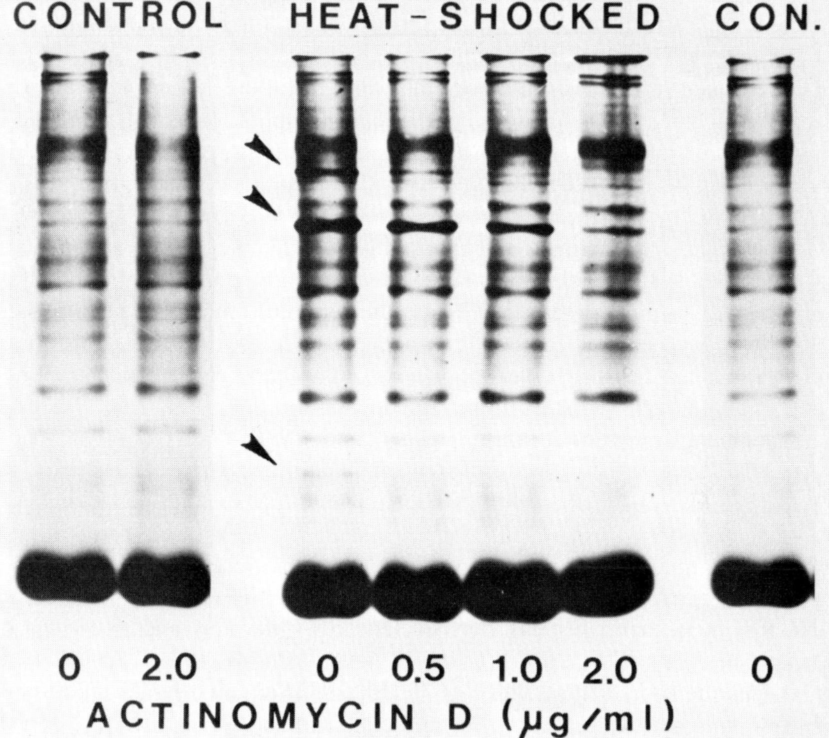

Fig. 11. Actinomycin D inhibits the synthesis of RBC hsp's in a dose-dependent fashion. Shown are fluorograms from one-dimensional SDS–PAGE separations of proteins synthesized by red blood cells during incubation at 37 (control) or 45°C (heat-shocked) without actinomycin D or in the presence of 0.5, 1.0, or 2.0 μg/ml of actinomycin D. Actinomycin D does not affect the type of proteins synthesized at 37°C but blocks, in a dose-dependent fashion, the synthesis of hsp's at 45°C. When cells are heat shocked in the presence of 2.0 μg/ml of actinomycin D, they synthesize the same proteins as cells incubated at 37°C with or without the inhibitor.

be mRNA's for hsp's or other mRNA's, is a prerequisite for the normally observed, repressed synthesis of preexisting mRNA's in heat-shocked RBC's.

VI. CONCLUSION

The data presented in this review were primarily derived from studies on the effects of thermal or chemical stress on the gene expression of erythroid cells from the peripheral circulation of birds. We have undertaken similar studies with mammalian blood cells (human and mouse leukocytes; Rodenhiser *et al.*, 1983, 1984), and investigations on amphibian blood cells are in progress. While our choice of nucleated blood cells for the study of stress on the regulation of gene expression involved a number of practical considerations, the ease with which we can induce hsp synthesis in blood cells, both in culture and *in situ,* suggests that this phenomenon has important physiological implications and is more than a useful way of studying gene regulation. Indeed, a number of studies with other systems using heat-shocked, cultured cells (Loomis and Wheeler, 1980; Ashburner and Bonner, 1979; Mitchell *et al.*, 1979; Li and Werb, 1982; Li *et al.*, 1983) or tissues from animals heat shocked *in situ* (Mitchell *et al.*, 1979; Dean and Atkinson, 1983) have associated the heat shock response (i.e., the production of hsp's) with an increased capacity for thermotolerance. Our results with RBC's from heat-shocked quail strongly support the contention that hsp or stress protein synthesis may be part of a homeostatic mechanism commonly used by blood cells for coping with environmental stress.

The temperatures used for inducing hsp synthesis in cultured quail RBC's are higher than those generally used for mammalian cells but were empirically derived (see Fig. 2) and are well within the range of temperatures to which quail are often exposed (Brush, 1965; Hudson and Brush, 1964; Calder and King, 1974); the normal core temperatures of quail are also much higher than those found in mammals (M. Kavaliers and M. Hirst, unpublished). The response of avian RBC's to thermal stress is characterized by the new and/or enhanced noncoordinate synthesis of polypeptides with M_r values of 90,000, 70,000, and 26,000 and the depressed synthesis of globin and other polypeptides previously made. Although hsp's with M_r values of 90,000 and 26,000 and p*I* values of 5.0 and 5.4, respectively, are not detectably synthesized by control cells, low amounts of hsp's with M_r values of 70,000 and p*I* values of 5.4 and 5.8, respectively, are synthesized in control RBC's.

Cell-free translational studies with mRNA's isolated from the polyribosomes and from the free RNP cytoplasmic particles of control and heat-shocked RBC's suggest that new synthesis of hsp mRNA's is required for synthesis of the 26K hsp and for synthesis of the more basic 70K hsp. While mRNA's for the 26K, 70K, and 90K hsp's were detected on polyribosomes from the heat-shocked cells, only mRNA for the more acidic 70K hsp was detected on polyribosomes from control cells. In addition, mRNA's for the 90K and the more acidic (p*I* of 5.4) 70K hsp's were found in the free RNP cytoplasmic particles from control and heat-shocked cells. These results suggest that the RBC mRNA's for the 90K hsp and for at least one form of the 70K hsp may be under some translational control, similar to that reported for the 70K hsp in *Xenopus* oocytes (Beinz and Gurdon, 1982), and may not be wholly dependent on new or enhanced transcription for their expression.

The rapid but selective modulation of gene expression which results from stressing these highly specialized cells appears to reflect major changes in transcription. The syntheses of RNA and hsp are both repressed in heat-shocked cells by levels of actinomycin D which do not greatly affect the rate of protein synthesis in control or heat-shocked cells nor the types of proteins synthesized by control cells. The proteins that are synthesized by RBC's heat shocked in the presence of actinomycin D correspond in molecular mass to those synthesized by RBC's incubated at control temperatures in the presence or absence of actinomycin D. These results indicate that new mRNA synthesis (either hsp mRNA's and/or mRNA's necessary for the synthesis of hsp's) is required for the new and/or enhanced synthesis of hsp's and may be necessary during heat shock, for depressing the translation of mRNA's made prior to heat shock.

The fact that RBC's—cells in which gene expression is so highly regulated and committed to the synthesis of globin [90% of protein synthesized is globin and 90% of mRNA found on RBC polyribosomes is globin mRNA (Williamson and Tobin, 1977; Imaizumi-Scherrer *et al.*, 1982)]—respond to stressful conditions by rapidly changing their intrinsic "program" of gene expression and concentrate their resources on synthesizing hsp's implies a functional need for these proteins and suggests that a hierarchy of regulation must exist. Moreover, the selective and rapid synthesis of hsp's in such highly specialized cells as RBC's suggests that their synthesis must be of a high order and involve regulatory mechanisms operating in concert at all levels of gene expression.

Acknowledgments

This research has been supported by a grant from the Natural Sciences and Engineering Research Council of Canada and by a grant from the Academic Development Fund of the University of Western Ontario, London, Ontario. The authors are grateful to Mr. Tim Blaker for his excellent technical assistance and to Mr. Ian Craig for his photographic expertise.

REFERENCES

Adams, G. H. M., and Neelin, J. M. (1972). Large scale separation of avian erythroblasts in density-gradients of Ficoll. *Cytobios* **6,** 133–142.

Ashburner, M., and Bonner, J. J. (1979). The induction of gene activity in *Drosophila* by heat shock. *Cell* **17,** 241–254.

Atkinson, B. G. (1981). Synthesis of heat shock proteins by cells undergoing myogenesis. *J. Cell Biol.* **89,** 666–673.

Atkinson, B. G., and Dean, R. L. (1983). Heat shock- and chemical stress-induction of new and/or enhanced gene expression in terminally differentiating or differentiated avian red blood cells. *J. Cell Biol.* **97,** 151a (abstr.).

Atkinson, B. G., and Pollock, M. (1982). Effect of heat shock on gene expression in human epidermoid carcinoma cells (strain KB) and in primary cultures of mammalian and avian cells. *Can. J. Biochem.* **60,** 316–327.

Atkinson, B. G., Cunningham, T., Dean, R. L., and Somerville, M. (1983). Comparison of the effects of heat shock and metal-ion stress on gene expression in cells undergoing myogenesis. *Can. J. Biochem. Cell Biol.* **61,** 404–413.

Bienz, M., and Gurdon, J. B. (1982). The heat shock response in *Xenopus* oocytes is controlled at the translational level. *Cell* **29,** 811–819.

Brush, A. H. (1965). Energetics, temperature regulation and circulation in resting, active and defeathered California quail, *Lophortyx californicus. Comp. Biochem. Physiol.* **15,** 399–421.

Calder, W. A., and King, J. R. (1974). Thermal and caloric relations of birds. *In* "Avian Biology" (D. S. Farner, and J. R. King, eds.), Vol. 4, pp. 259–413. Academic Press, New York.

Cleveland, D. W., Fischer, S. G., Kirschner, M. W., and Laemmli, U. K. (1977). Peptide mapping by limited proteolysis in sodium dodecyl sulfate and analysis by gel electrophoresis. *J. Biol. Chem.* **252,** 1102–1106.

Cousins, R. J. (1982). Relationship of metallothionein to zinc metabolism. *Fed. Proc. Fed. Am. Soc. Exp. Biol.* **41,** 624.

D'Alecy, L. G., and Kluger, M. J. (1975). Avian febrile response. *J. Physiol.* **253,** 223–232.

Dean, R. L., and Atkinson, B. G. (1983). The acquisition of thermal tolerance in larvae of *Calpodes ethlius* (Lepidoptera) and the *in situ* and *in vitro* synthesis of heat shock proteins. *Can. J. Biochem. Cell Biol.* **61,** 472–479.

Dean, R. L., and Atkinson, B. G. (1984a). Heat shock- and chemical stress-induction of new and/or enhanced gene expression in terminally differentiating avian red blood cells. Submitted for publication.

Dean, R. L., and Atkinson, B. G. (1984b). Synthesis of heat shock proteins in quail red blood cells following brief, physiologically relevant increases in whole body temperature. *Comp. Biochem. Physiol.* (in press).

Gander, E. S., Stewart, A. G., Morel, C. M., and Scherrer, K. (1973). Isolation and characterization of ribosome-free cytoplasmic messenger-ribonucleoprotein complexes from avian erythroblasts. *Eur. J. Biochem.* **38,** 443–452.

Hudson, J. W., and Brush, A. H. (1964). A comparative study of the cardiac and metabolic performance of the dove *Zenaidura macroura,* and the quail, *Lophortyx californicus. Comp. Biochem. Physiol.* **12,** 157–170.

Humphries, S., Windass, J., and Williamson, R. (1976). Mouse globin gene expression in erythroid and non-erythroid tissues. *Cell* **7,** 267–277.

Imaizumi-Scherrer, M. T., Maundrell, K., Civelli, O., and Scherrer, K. (1982). Transcriptional and post-transcriptional regulation in duck erythroblasts. *Dev. Biol.* **93,** 126–138.

Johnston, D., Oppermann, H., Jackson, J., and Levinson, W. (1980). Induction of four proteins in chick embryo cells by sodium arsenite. *J. Biol. Chem.* **255,** 3230–3233.

Kim, Y-J., Schuman, J., Sette, M., and Przybyla, A. (1983). Arsenite induces stress proteins in cultured rat myoblasts. *J. Cell Biol.* **96,** 393–400.

Laskey, L., Nozick, N. D., and Tobin, A. J. (1978). Few transcribed RNAs are translated in avian erythroid cells. *Dev. Biol.* **67,** 23–39.

Levinson, W., Idris, J., and Jackson, J. (1979). Metal binding drugs induce synthesis of four proteins in normal cells. *Biol. Trace Elem. Res.* **1,** 15–23.

Levinson, W., Oppermann, H., and Jackson, J. (1980). Transition series metals and sulfhydryl reagents induce the synthesis of four proteins in eukaryotic cells. *Biochim. Biophys. Acta* **606,** 170–180.

Li, G. C., and Werb, Z. (1982). Correlation between synthesis of heat shock proteins and development of thermotolerance in Chinese hamster fibroblasts. *Proc. Natl. Acad. Sci. U.S.A.* **79,** 3218–3272.

Li, G. C., Meyer, J., Mak, J. Y., and Hahn, G. M. (1983). Heat induced protection of mice against thermal death. *Cancer Res.* **43,** 5758–5760.

Lischwe, M. A., and Ochs, D. (1982). A new method for partial peptide mapping using N'-chlorosuccinimide/urea and peptide silver staining in sodium dodecyl sulfate polyacrylamide gels. *Anal. Biochem.* **127,** 453–457.

Lodish, H. F. (1974). Model for the regulation of mRNA translation applied to haemoglobin synthesis. *Nature (London)* **251,** 385–388.

Loomis, W. F., and Wheeler, S. A. (1980). Heat shock response of *Dictyostrium*. *Dev. Biol.* **79,** 399–408.

Mitchell, H. K., Moller, G., Petersen, N. S., and Lipps-Sarmiento, L. (1979). Specific protection from phenocopy induction by heat shock. *Dev. Genet.* **1,** 181–192.

Rodenhiser, D. I., Jung, J. H., and Atkinson, B. G. (1983). Human lymphocytes respond to mild hyperthermia *in vitro* by a rapid and dramatic change in gene expression. *Bull. Genet. Soc. Can.* **14,** A11 (abstr.).

Rodenhiser, D. I., Jung, J. H., and Atkinson, B. G. (1984). The effect of thermal and chemical stress on the gene expression of human lymphocytes. Submitted for publication.

Schlesinger, M. J., Ashburner, M., and Tissières, A., eds. (1982). "Heat Shock, from Bacteria to Man". Cold Spring Harbor Lab., Cold Spring Harbor, New York.

Small, J. V., and Davies, H. A. (1970). The haemoglobin in the condensed chromatin of mature amphibian erythrocytes: a further study. *J. Cell Sci.* **7,** 15–33.

Somerville, M. J. (1984). Stress-induced alterations of gene expression in myogenic and erythroid cells from quail. Master's thesis, Univ. Western Ontario, London.

Spirin, A. S. (1969). Informosomes. The second Sir Hans Krebs lecture. *Eur. J. Biochem.* **10,** 20–35.

Spohr, G., Granboulan, N., Morel, C., and Scherrer, K. (1970). mRNA in HeLa cells: an investigation of free and polyribosome-bound cytoplasmic mRNP by kinetic labelling and electron microscopy. *Eur. J. Biochem.* **17,** 296–318.

Spohr, G., Kayibanda, B., and Scherrer, K. (1972). Polyribosome-bound and free-cytoplasmic-hemoglobin-messenger RNA in differentiating avian erythroblasts. *Eur. J. Biochem.* **31,** 194–208.

Tijssen, P., and Kurstak, E. (1983). An efficient two-dimensional sodium dodecyl sulfate-polyacrylamide gel electrophoresis method for simultaneous peptide mapping of protein contained in a mixture. *Anal. Biochem.* **128,** 26–35.

Tooze, J., and Davies, H. G. (1963). The occurrence and possible significance of haemoglobin in the chromosomal regions of mature erythrocyte nuclei of the newt *Triturus cristatus cristatus*. *J. Cell Biol.* **16,** 501–511.

Vincent, A., Civelli, O., Maundrell, K., and Scherrer, K. (1980). Identification and characterization of the translationally repressed cytoplasmic globin messenger-ribonucleoprotein particles from duck erythroblasts. *Eur. J. Biochem.* **112,** 617–633.

Williamson, F. L., and Tobin, A. J. (1977). The heterogenous nuclear RNA of chicken erythroblasts. *Biochim. Biophys. Acta* **475,** 366–382.

9

Stress Response in Avian Cells

MILTON J. SCHLESINGER

I. INTRODUCTION

Chicken embryos and tissue culture cells derived from them have proved to be valuable biological systems for a wide variety of studies. In particular, they have been employed as efficient hosts for replicating animal viruses, and experiments utilizing virus-infected chicken embryo fibroblasts have provided an enormous amount of data about the molecular biology of virus gene expression as well as other events in virus formation. It was in the course of studying the replication in chicken embryo fibroblasts of Sindbis virus, a small enveloped RNA animal virus (Schlesinger and Kaariainen, 1980), that we first observed the dramatic response of the host chicken cells to stress (Kelley and Schlesinger, 1978). This chapter reviews these initial studies as well as relates information about the response of chicken cells in the intact embryo and in tissue culture to various kinds of environmental stress.

Changes in Eukaryotic Gene Expression
in Response to Environmental Stress

ISBN 0-12-066290-6

II. STRESSORS OF AVIAN CELLS

The initial observations which led to the discovery of heat shock in animal cells were made during experiments in which amino acid analogs were added to chicken embryo fibroblasts in order to interfere with posttranslational proteolytic processing events. What was found after a long (12-hr) incubation with certain analogs was a dramatic change in the kinds of proteins synthesized by the normal cell, as revealed by SDS–PAGE. Most of the normal cell protein synthesis ceased and was quantitatively replaced by unusually high levels of synthesis of a small number of polypeptides. The synthesis of the three major new polypeptides with subunits M_r = 24,000, 70,000, and 89,000 was subsequently found to be induced by a large variety of chemical, physical, and biological agents—all of which are now categorized as "stressors." Table I lists the various kinds of inducing agents; it would be surprising if this list did not continue to expand.

All these agents *can* elicit a common response, but one should hasten to add that complex variations emerge on more detailed study with the different agents in different avian biological systems; it is unwise to overgeneralize the commonality of the stress response. In fact, it has become evident that both the quantity and quality of stress proteins (sp's) is very much dependent on cell type, state of cell differentiation, and kind of stress.

For example, sp24 was not induced in heat-shocked, differentiated, fused myotubes of quail, although treatment of these cells with arsenite or heavy metals (Cu or Zn) induced this protein (Atkinson *et al.*, 1983). Undifferentiated

TABLE I

Stressors of Chicken Embryo Fibroblasts

Temperature	Cuprous chloride
Amino acid analogs	Zinc chloride
Canavanine	Cadmium chloride
o-Methylthreonine	Iodoacetamide
Hydroxynorvaline	*p*-Chlormercuribenzoate
p-Fluorophenylalanine	8-Hydroxyquinoline
Puromycin	*o*-Phenanthroline
Diamide	Thiosemicarbazide
Kethoxal bisthiosemicarbazone	Glucose starvation
Disulfiram	Virus infection
Copper sulfate	Paramyxovirus
Sodium arsenite	Rous sarcoma virus
Sodium arsenate	Sindbis virus
Mercuric chloride	Vesicular stomatitis virus

myoblasts or those that had differentiated but had not fused were capable of inducing sp24 after temperature stress. However, as cells became more differentiated in culture, stronger stress levels, that is, increased times at higher temperature, were required to obtain levels of stress protein induction comparable to those in undifferentiated cells (Atkinson, 1981). Other variations have been noted, such as changes in a particular isoelectric variant. In contrast, arsenite failed to induce synthesis of proteins with M_r values of 82,000 and 88,000 in quail myogenic cultures even though these proteins were induced by heat shocking these cells (Atkinson *et al.*, 1983). In fibroblasts arsenite concentration and time of treatment affected the pattern of induced protein; at low drug levels an sp35 appeared but was replaced by sp27 (noted here as sp24) at higher drug levels (Johnston *et al.*, 1980). Changing the stress from heat (1 hr, 45°C) to arsenate (4 hr, 150 μM) led to induction of a new protein with M_r = 30,000 in minced breast tissue from a 13-day-old chicken embryo (Kim *et al.*, 1983). No sp24 was found in brain tissue from 9-day-old quail embryos heated at 45°C for 1–2 hr although high levels of this protein were induced in breast and limb bud tissue. Brains from heated 11-day-old chick embryos also lacked a 24K protein, but a protein with M_r of ~50,000 appeared which reacted with antibodies against the chicken 24K protein (C. Malfer and M. J. Schlesinger, unpublished observations). This protein may be the same as that reported in heat-shocked myotube cultures in which a new protein with M_r of ~50,000 was found (Atkinson *et al.*, 1983). Another variation is presented in Fig. 1. The protein synthesis pattern from secondary 11-day-old chicken embryo fibroblasts is shown after 1–4 hr of stress imposed by different temperatures. Synthesis of sp70 is clearly seen as a strong sensitive response to stress, yet after 3 hr at the intermediate temperature of 43°C synthesis of this stress protein has shut down. The explanation for this phenomenon is described in detail by Lindquist and DiDomenico (Chapter 4, this volume). Briefly, sp70 regulates its *own* biosynthesis, and the levels required for this feedback effect are strongly stress dependent. Thus, at the "intermediate" stress of 2 hr at 43°C, enough sp70 has accumulated to suppress further synthesis.

Infection of chicken fibroblasts by several kinds of lytic viruses [i.e., Newcastle disease virus (Collins and Hightower, 1982), Sendai virus (Peluso *et al.*, 1978), Sindbis and vesicular stomatitis virus (Garry *et al.*, 1983)] induces the two larger stress proteins but not the smaller one. In the infection by Sindbis virus, mobilities of these proteins in SDS–PAGE was shifted (Garry *et al.*, 1983). Virus induction is somewhat unusual since the viruses shut down host cell mRNA synthesis and translation, but stress protein synthesis during virus infection is actinomycin D sensitive. The sp mRNA's tend to resemble virus mRNA's in their resistance to changes in intracellular Na^+ and K^+ concentrations (Garry *et al.*, 1983). With less virulent infections by Newcastle disease virus, enhanced

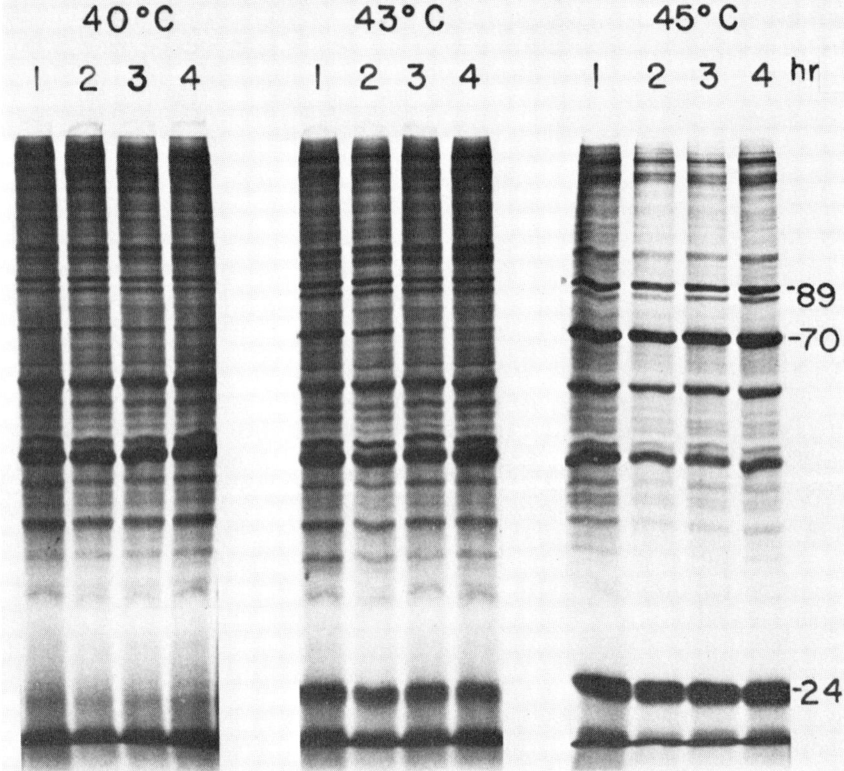

Fig. 1. The heat shock response in chicken embryo fibroblasts exposed for varying lengths of time to different temperatures. After the temperature treatment, cells were incubated for 1 hr with [^{35}S]methionine at 37°C and equivalent amounts of extracts examined by SDS–PAGE. Note the decrease in sp70 synthesis after 3 hr at 43°C.

levels of stress proteins appeared even though mRNA levels were not significantly greater than in the more virulent case (Collins and Hightower, 1982).

There is the expectation that the cornucopia of stressors is activating some common cellular metabolic signal since virtually all turn on the same or very similar genes and it is on the DNA where this response to stress initiates. Based on similar responses in bacteria to heat and oxidants, oxidation has been proposed as the common agent that triggers synthesis of unusual dinucleotides (AppppA) which act as effectors (Lee *et al.*, 1983). For those "exceptions" noted above, one may postulate that different cells have distinct thresholds for this putative signal and, further, that changes may occur in the structure of the heat shock proteins themselves.

III. INDUCTION AND DEINDUCTION

None of the stress genes in the avian system have been identified, although cDNA clones have been isolated which selectively respond to mRNA's from heat-shocked cells (U. Bond and M. J. Schlesinger, unpublished observations). *De novo* synthesis of stress mRNA's is required for expression of stress proteins since actinomycin D blocks induction of protein synthesis. Induction proceeds in the presence of cycloheximide, thus, no activator protein need be made for turning on the appropriate genes. Only in the case of amino acid analogs is protein synthesis needed for stress gene activation (Kelly and Schlesinger, 1978). For stress incurred by amino acid analogs, it has been postulated that the accumulation of improperly folded, denatured proteins which might lead to increased proteolytic or secretory activity generates the stress signal (Hightower, 1980). The stress mRNA's dominate the avian cells' protein synthetic machinery, but there are no clear data as to the fate of normal mRNA's. Voellmy and Bromley (1982) reported no reduction in the *in vitro* translation of normal mRNA's after heat shock; however, Kelly *et al.*, (1980) showed that in several

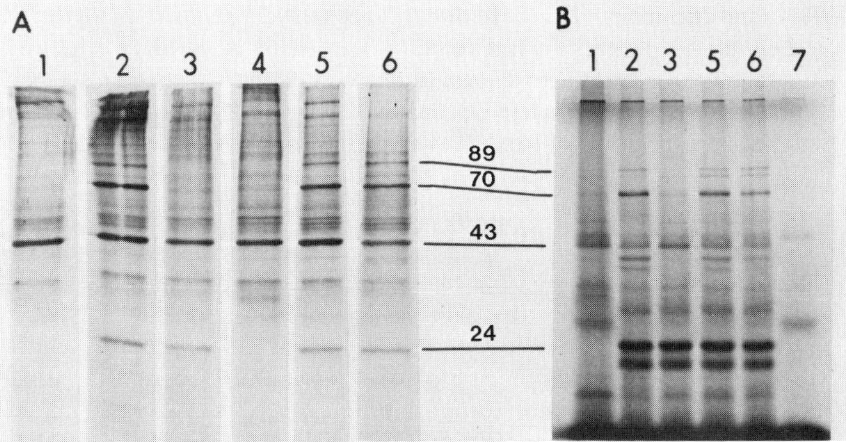

Fig. 2. Recovery of normal protein synthesis after heat shock is inhibited by actinomycin D and cycloheximide. (A) Chicken embryo fibroblasts were heat shocked for 1 hr at 45°C and labeled with [^{35}S]methionine (1 hr incorporation) immediately (lane 2) or after 8 (lane 3) and 24 (lane 4) hr at 37°C. Actinomycin D (1 μg/ml) was added to cells at the start of recovery, and cells were analyzed 8 hr later (lane 5). Cycloheximide (25 μg/ml) was present during an 8-hr recovery but removed prior to labeling (lane 6). (B) Poly(A)$^+$ RNA was isolated from cells [larger amounts than for (A)] that were treated according to the numbered lanes noted for (A) with lane 7 containing no added RNA and translated *in vitro* in a reticulocyte extract system. Samples in lane 1 in both panels are from cells incubated at 37°C. For reference, several stress proteins are indicated. (From Schlesinger *et al.*, 1982.)

heterologous *in vitro* cell-free translation systems primed with mRNA from stressed and nonstressed cells the protein products precisely mimicked the *in vivo* stress (heat shock) pattern (see also Fig. 2). The mRNA's from heat-shocked fibroblasts for sp 89, 70, and 24 had sedimentation coefficients of 20 S, 18 S, and 13 S, respectively. Johnston, *et al.* (1980) concluded that there were increased levels of mRNA's for stress proteins after arsenite treatment of fibroblasts on the basis of *in vitro* translation experiments. New bands of mRNA from stressed cells were detected in methylmercury–agarose gels, and those with minimum molecular weights of 0.9×10^6 and 1.3×10^6 were able to direct synthesis *in vitro* of sp73 and sp89, respectively.

In avian as well as in other systems, cells can recover to their normal states provided the stress has not been too severe or been applied for too long. Functional half-lives of the stress mRNA's are 2–4 hr, considerably less than those of most eukaryotic cell mRNA's. In contrast, the newly made stress proteins remain quite stable for very long times after recovery; thus, they apparently do not interfere significantly with normal cell metabolism. But there is apparently a need for *new* protein synthesis in order for recovery to proceed since cycloheximide can block the cells' return to their prestressed state (Fig. 2, and Hightower, 1980). The data in Fig. 2 show further that actinomycin D blocked recovery of prestress protein synthesis when the drug was added at the time cells were returned to normal growth conditions. Hightower (1980) also showed that recovery was blocked by actinomycin D. Clearly, the normal RNA's present prior to heat shock are incapable of simply resuming their original synthetic roles.

IV. MAJOR AVIAN STRESS PROTEINS

There are three major proteins whose synthesis is most noticeably altered during the stress response; they have subunit molecular weights of 24,000–27,000, 68,000–70,000, and 85,000–89,000. In their native states they may be present in higher-molecular-weight complexes, and there is substantial evidence that they are heterogeneous in isoelectric points, often with multiple forms detectable in isoelectric focusing gels. Atkinson and Pollock have reported that the newly synthesized stress proteins from quail myoblasts with M_r values of 94,000, 64,000, and 25,000 had p*I* values of 5.2, 5.8, and 5.4, respectively, but they also found enhanced formation after stress of proteins with M_r values of 94,000, 88,400, 82,000, and 64,000 and with p*I* values of 5.1, 5.2, 5.2, and 5.4, respectively (Atkinson, 1981; Atkinson and Pollock, 1982).

A. Stress Protein 89

Stress protein 89 is primarily distinguishable by four properties: (1) it is a highly phosphorylated polypeptide with phosphate groups esterified to serine residues (Kelley and Schlesinger, 1982; Oppermann *et al.*, 1981a,b); (2) it is synthesized under normal cell culture conditions, and after stress its formation continues at a relatively elevated level; (3) it is exclusively localized to the cytosol compartment of the cell; (4) the protein forms a complex with the Rous sarcoma virus (RSV) phosphotyrosine kinase shortly after formation of this enzyme in virus-transformed chicken cells.

During purification of the protein from heat-shocked, chicken embryo fibroblasts there were substantial amounts of sp89 material in higher-molecular-weight complexes, but the precise composition of these complexes has yet to be determined (Kelley and Schlesinger, 1982). Although homogeneous preparations have been obtained and antibodies raised against sp89, very little primary structure information is available. Proteolytic digests carried out with trypsin or *Staphylococcus aureus* V8 gave patterns of the avian sp89 which were similar to those obtained from stress proteins of similar molecular weights (84,000) from human and *Drosophila* cells (Voellmy *et al.*, 1983). There is one report showing methylation of sp89 on lysine (Wang *et al.*, 1982), and a size-resolved doublet was detected by Johnston *et al.* (1980). Polyclonal antibodies raised in rabbits to an SDS-denatured form of sp89 cross-reacted with proteins of similar molecular weight in virtually all tissues of the chicken as well as in other species ranging in diversity from the fly to the human. No proteins cross-reacting to anti-sp89 antibodies were in membrane ghosts prepared from human erythrocytes (Kelley and Schlesinger, 1982). After heat shocking intact 18-day-old chicken embryos (2.5 hr at 43–44°C), sp84 synthesis was notably increased in brain, heart, liver, and lung (Voellmy and Bromley, 1982). On the basis of immunoblotting, which measures the levels of proteins, sp89-like proteins are a normal constituent of embryonic and adult chicken cells from virtually every tissue (Kelley and Schlesinger, 1982).

One of the most intriguing properties of sp89 is its presence in complex with newly synthesized pp60^{src} together with a 50,000-dalton protein (Brugge *et al.*, 1981; Adkins *et al.*, 1982; Oppermann *et al.*, 1981a,b). In the steady state this complex turns over rapidly (half-life of 9–15 min), and only a fraction of sp89 is in the complex (Brugge *et al.*, 1983). The pp60^{src} in this complex has not yet become active as a phosphotyrosine-specific protein kinase and appears to be in transit to the cell's plasma membrane. Other strains of retroviruses that transform chicken cells and contain phosphotyrosine-specific protein kinases show complexes between sp89 and the kinase (Lipsich *et al.*, 1982). Interestingly, mutants of RSV temperature sensitive (*ts*) in pp60^{src} produce proteins which at the nonpermissive temperatures accumulate in the sp89 complex and fail to move, or do so much more slowly, on to the plasma membrane. If cells infected with the *ts* mutants are permitted to accumulate mature active kinase at the cell's membrane but then are subsequently shifted to high temperatures under conditions that shut down newly made kinase, the preexisting pp60^{src} leaves the membrane and is found in the sp89 complex (Courtneidge and Bishop, 1982). It is as if there may be some recycling of this protein between the membrane and the complex. Precisely what role sp89 plays in the maturation and/or cycling of this kinase is not at all clear. The $pp60^{src}$ protein in complex with sp89 is not phosphorylated on tyrosines, nor is the state of sp89 phosphorylation affected by complexing with this kinase. The $pp60^{v\text{-}src}$ has long-chain fatty acids acylated to it (Sefton *et al.*, 1982; Garber *et al.*, 1983), and the incorporation of this moiety may occur in the complex.

Stress protein 89 is found complexed to the nucleocapsids of Sindbis and vesicular stomatitis viruses in avian cells infected by these viruses (Garry *et al.*, 1983). It is also one of the proteins controlled by the availability of glucose; its synthesis is inhibited when cells are deprived of glucose and stimulated on refeeding of the sugar (Lanks *et al.*, 1982). There is evidence that $pp60^{src}$ has properties indistinguishable from those of a phosphoglycerol kinase (Graziani *et al.*, 1983). This information and the relationship to glucose metabolism lead one to speculate that sp89 may be part of the cell's glycolytic machinery.

B. Stress Protein 70

The protein most characteristic of the stress response in many different kinds of cells has a subunit molecular weight of about 70,000. In avian cells this protein is the first whose synthesis is markedly altered after a stress (for example, see Fig. 1), yet there are closely related (if not identical) forms of the protein present in cells of most tissues of embryonic and adult birds. In embryonic chicken myotubes or fibroblasts grown in culture, synthesis of at least two isoelectric forms (p*I* values of 5.7 and 5.95) are dramatically increased after stress by temperature (45°C, 1 hr) or arsenite (25 μM, 4 hr) (Wang *et al.*, 1981). These two proteins contain methyl groups on their lysines and arginines (Wang *et al.*, 1981, 1982) with some subtle but definitive differences between the proteins from normal cells and those from stressed cells.

The sp70 associates with cytoskeleton structures and is detected in nuclear as well as cytoplasmic components. Wang *et al.* (1981) reported that one form of sp70 copurified with brain microtubules and intermediate filaments, and both forms remained associated with intermediate filament-rich cytoskeletons from tissue culture cells prepared by treating cells with 0.5% Triton X-100 and 0.6 *M* KCl. From immunofluorescence studies, the proteins appeared to be components of chicken myofibrils or were bound to stress fibers. In heat-shocked chicken fibroblasts most of the protein is in the cytoplasmic compartment of the cell and little is strongly associated with membranes (Kelley and Schlesinger, 1982). Purification of the protein from fibroblasts revealed that some forms of sp70 are in higher-molecular-weight aggregates, but a major fraction appeared to be present as dimers (Kelley and Schlesinger, 1982).

Polyclonal antibodies prepared in rabbits against a homogeneous preparation of SDS-denatured sp70 cross-reacted with proteins of similar size from widely divergent sources including yeast, flies, amphibians, plants, rodents, and humans. In yeast and *Drosophila* these antibodies detected increased amounts of sp70 after high-temperature stress. The antibodies to chicken sp70 cross-reacted also in an immunoblot assay with a highly purified preparation of a bovine brain enzyme which binds to and uncoats clathrin-coated vesicles (M. J. Schlesinger and J. R. Rothman, unpublished observations). Further analyses have shown that some forms of chicken and *Drosophila* sp70 have an ATP binding site and show ATP-dependent binding to coated vesicles. Activities of the uncoating enzyme increased severalfold after heat shocking *Drosophila* or chicken cells (J. R. Rothman, unpublished observations). These data may provide an important clue to one role for sp70 in the stress response. The interaction of sp70 with coated vesicles should lead to a block in most types of intramembranal transport and turnover of surface membranes during stress.

C. Stress Protein 24

The smallest avian stress protein appears in large amounts after relatively severe stress conditions; its gene appears to be the last to be activated after stress. Furthermore, as described earlier, the kind of stress and the type of cell affect induction of this protein. The protein itself has some rather unusual properties and can appear in very high-molecular-weight aggregates, some of which are covalently linked. At least two isoelectric variants can be detected in myotubes and fibroblasts (Wang *et al.*, 1981; P. M. Kelley and M. J. Schlesinger, unpublished observations). Antibodies prepared in rabbits against an SDS-denatured form of sp24 cross-react with an avian muscle protein of subunit molecular weight 22,000. The latter protein is unusually abundant in chicken embryonic heart as well as in skeleton muscle. Perhaps of greater interest is the fact that the anti-sp24 antibodies cross-react with a β-crystallin prepared from 11-day-old embryonic chicken lens (Fig. 3). The immunological relatedness between sp24 and lens crystallin provides strong evidence for a structural role of this protein in the stress response. Furthermore, it should be recalled that there is homology of polypeptide sequences between the low-molecular-weight stress proteins of *Drosophila* and mammalian lens crystallins (Ingolia and Craig, 1982; Southgate *et al.*, 1983). Thus, the avian low-molecular-weight stress proteins should be considered analogous to the low-molecular-weight *Drosophila* proteins.

It is not surprising that attempts to localize these proteins have led to considerable ambiguity. In chick fibroblasts the sp24 aligns along stress fibers in cytoplasmic and nuclear compartments. There was strong localization to nucleolar regions of heat-shocked quail cells and fractionation of cells revealed

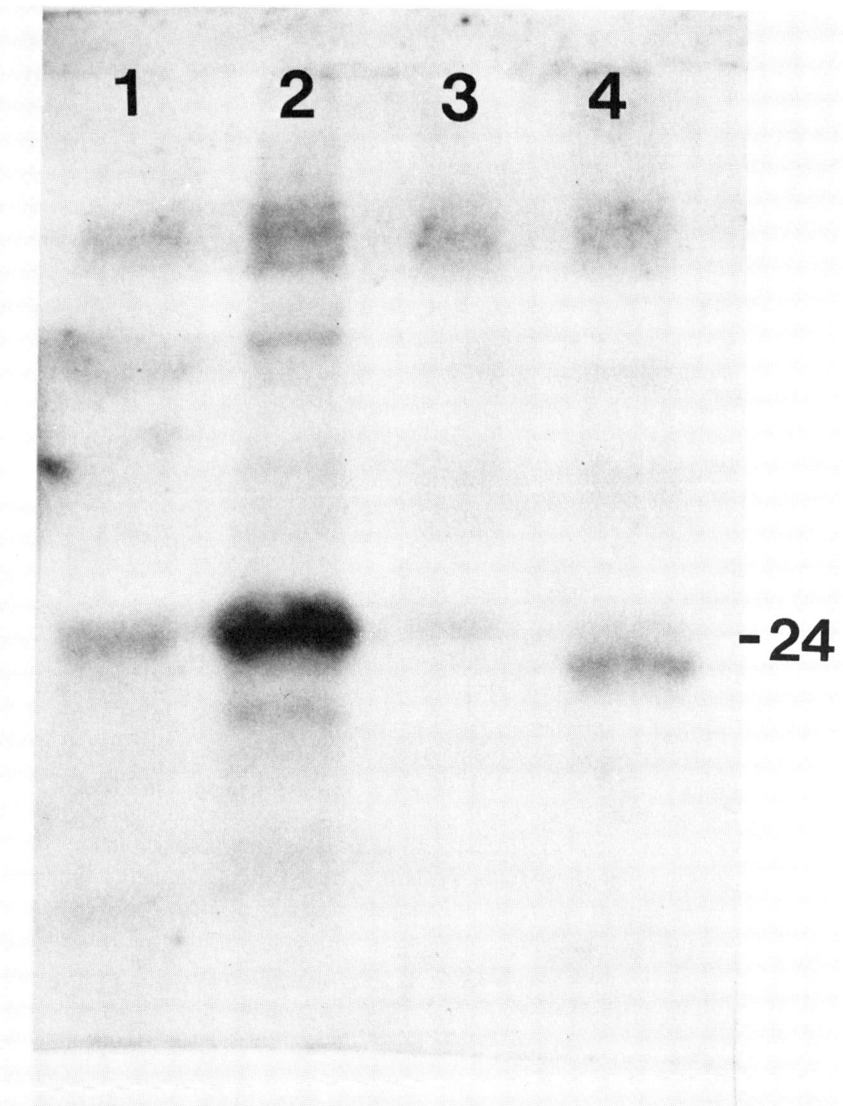

Fig. 3. Immunological cross-reactivity of proteins in normal chicken lens and heart to the sp24. Extracts (100 μg protein) from 11-day-old chicken embryo tissues or fibroblast cultures were subjected to SDS–PAGE and proteins electrophoretically transferred to APT paper (Kelley and Schlesinger, 1982). Rabbit anti-sp24 was used as a probe, followed by ^{125}I-labeled protein A. The lens protein reacting positively was identified as a β-crystallin based on a positive response to anti-β-crystallin antibody, kindly provided by Dr. J. Piatigorsky (NIH). Lanes: (1) lens, (2) heat-shocked (4 hr) fibroblast, (3) normal fibroblast, and (4) embryo heart tissue.

enrichment in the nucleus. Most sp24 is extractable by buffer from stressed cells but readily aggregates after concentration of the extract. After denaturation by SDS, higher-molecular-weight covalent aggregates were detectable in denaturing gels (M. J. Schlesinger, unpublished observations).

V. CONCLUSIONS

The stress response in avian cells has been examined primarily at the level of specific gene activation and subsequent expression of proteins. Very few studies have been reported on other effects, many of which are noted for other species elsewhere in this volume. Little is known about general DNA or RNA synthesis, membrane biogenesis and turnover, changes in ionic state of the cell, and so forth.

We used the anti-chicken heat shock protein antibodies in attempts to test for changes in glycolytic enzyme activities after heat shock. No meaningful differences were observed in a number of glycolytic enzymes, but levels of lactic acid increased about twofold, and increases of ~50% were measured in hexose transport after heat shocking fibroblasts. These data could be important since a recent study of heat-shocked rats revealed that temperature stress in contrast to certain other types of stress had a profound effect on lactic acid accumulation in heart tissue (Hammond *et al.*, 1982). The interpretation was that cellular metabolic activities were shifting to anaerobic glycolysis for cell energy demands, and cellular acidosis could be an important factor in the heat shock response.

After stress by amino acid analogs, P. M. Kelley and M. J. Schlesinger (unpublished observations) found that ribosomal RNA (rRNA) synthesis continued, but there were defects in processing and intermediate forms (30 S and 42 S) of rRNA accumulated. This effect may result from a severe limitation of ribosomal protein synthesis. Johnston *et al.* (1980) also noted a reduction in 18 S and 28 S RNA formation after arsenite stress. What happens to small RNA's and tRNA's could have important consequences on the stress response, particularly in light of the possible involvement of tRNA synthetases in the bacteria stress response. Not only is one of the lysyl-tRNA synthetases a heat shock protein in *Escherichia coli*, but this enzyme can also produce ApppppA—a metabolite whose level rises dramatically under stress conditions (Lee *et al.*, 1983).

Many baffling questions about the stress response in the avian system remain unanswered. There are clear analogies and conservation of protein structure between avian stress proteins and those identified in other species. Furthermore, the stressors for avian cells are common to most other species. It remains to be determined how closely avian stress protein genes will resemble those in *Drosophila*, but one suspects there will be strong similarities there also.

REFERENCES

Adkins, B., Hunter, T., and Sefton, B. M. (1982). The transforming proteins of PRC11 virus and Rous sarcoma virus form a complex with the same two cellular phosphoproteins. *J. Virol.* **43,** 448–455.

Atkinson, B. G. (1981). Synthesis of heat-shock proteins by cells undergoing myogenesis. *J. Cell Biol.* **89,** 666–673.

Atkinson, B. G., and Pollock, M. (1982). Effect of heat shock on gene expression in human epidermoid carcinoma cells (strain KB) and in primary cultures of mammalian and avian cells. *Can. J. Biochem.* **60,** 316–327.

Atkinson, B. G., Cunningham, T., Dean, R. L., and Somerville, M. (1983). Comparison of the effects of heat shock and metal ion stress on gene expression in cells undergoing myogenesis. *Can. J. Biochem. Cell Biol.* **61,** 404–413.

Brugge, J., Erikson, E., and Erikson, R. L. (1981). The specific interaction of the Rous sarcoma virus transforming protein, $pp60^{src}$, with two cellular proteins. *Cell* **25,** 363–372.

Brugge, J., Yonemoto, W., and Darrow, D. (1983). Interaction between the Rous sarcoma virus transforming protein and two cellular phosphoproteins: analysis of the turnover and distribution of this complex. *Mol. Cell. Biol.* **3,** 9–19.

Collins, P. L., and Hightower, L. E. (1982). Newcastle disease virus stimulates the cellular accumulation of stress (heat shock) mRNAs and protein. *J. Virol.* **44,** 703–707.

Courtneidge, S. A., and Bishop, J. M. (1982). Transit of $pp60^{v}$src to the plasmid membrane. *Proc. Natl. Acad. Sci. U.S.A.* **79,** 7117–7121.

Garber, E. A., Krueger, J. G., Hanafusa, H., and Goldberg, A. R. (1983). Only membrane-associated RSV *src* proteins have amino-terminally bound lipid. *Nature, (London)* **302,** 161–163.

Garry, R. F., Ulug, E. T., and Bose, J. R., Jr. (1983). Induction of stress proteins in Sindbis virus and vesicular stomatitis virus-infected cells. *Virology* **129,** 319–332.

Graziani, Y., Erikson, E., and Erikson, P. L. (1983). Evidence that the Rous sarcoma virus transforming gene product is associated with glycerol kinase activity. *J. Biol. Chem.* **258,** 2126–2129.

Hammond, G. L., Lai, Y.-K., and Markert, C. L. (1982). Diverse forms of stress lead to new patterns of gene expression through a common and essential metabolic pathway. *Proc. Natl. Acad. Sci. U.S.A.* **79,** 3484–3488.

Hightower, L. E. (1980). Cultured animal cells exposed to amino acid analogues or puromycin rapidly synthesize several polypeptides. *J. Cell Physiol.* **102,** 407–427.

Ingolia, T. D., and Craig, E. A. (1982). Four small *Drosophila* heat shock proteins are related to each other and to mammalian α-crystallin. *Proc. Natl. Acad. Sci. U.S.A.* **79,** 2360–2364.

Johnston, D., Oppermann, H., Jackson, J., and Levinson, W. (1980). Induction of four proteins in chick embryo cells by sodium arsenite. *J. Biol. Chem.* **255,** 6975–6980.

Kelley, P. M., and Schlesinger, M. J. (1978). The effect of amino acid analogues and heat-shock on gene expression in chicken embryo fibroblasts. *Cell* **15,** 1277–1286.

Kelley, P. M., and Schlesinger, M. J. (1982). Antibodies to two major chicken heat shock proteins cross react with similar proteins in widely divergent species. *Mol. Cell. Biol.* **2,** 267–274.

Kelley, P. M., Aliperti, G., and Schlesinger, M. J. (1980). *In vitro* synthesis of heat shock proteins by mRNAs from chicken embryo fibroblasts. *J. Biol. Chem.* **255,** 3230–3233.

Kim, Y.-J., Shuman, J., Sette, M., and Przybyla, A. (1983). Arsenate induces stress proteins in cultured rat myoblasts. *J. Cell Biol.* **96,** 393–400.

Lanks, K. W., Kasambalides, E. J., Chinkers, M., and Brugge, J. S. (1982). A major cytoplasmic glucose-regulated protein is associated with the Rous sarcoma virus $pp60^{src}$ protein. *J. Biol. Chem.* **257,** 8604–8607.

Lee, P. C., Bochner, B. R., and Ames, B. N. (1983). AppppA, heat shock stress, and cell oxidation. *Proc. Natl. Acad. Sci. U.S.A.* **80,** 7496–7500.

Levinson, W., Mikelens, P., Oppermann, H., and Jackson, J. (1978a). Effect of antabuse (disulfiram) on Rous sarcoma virus and on eukaryotic cells. *Biochim. Biophys. Acta* **519,** 65–75.

Levinson, W., Oppermann, H., and Jackson, J. (1978b). Induction of four proteins in eukaryotic cells by kethoxal bis(thiosemicarbazone). *Biochim. Biophys. Acta* **518,** 401–412.

Levinson, W., Oppermann, H., and Jackson, J. (1980). Transition series metals and sulfhydryl reagents induce the synthesis of four proteins in eukaryotic cells. *Biochim. Biophys. Acta* **606,** 170–180.

Lipsich, L. A., Cutt, J., and Brugge, J. S. (1982). Association of the transforming proteins of Rous, Fujinami, and Y73 avian sarcoma viruses with the same two cellular proteins. *Mol. Cell. Biol.* **2,** 875–880.

Oppermann, H., Levinson, W., and Bishop, J. M. (1981a). A cellular protein that associates with the transforming protein of Rous sarcoma virus is also a heat shock protein. *Proc. Natl. Acad. Sci. U.S.A.* **78,** 1067–1071.

Oppermann, H., Levinson, A. D., Levintow, L., Varmus, H. E., Bishop, M., and Kawai, S. (1981b). Two cellular proteins that immunoprecipitate with the transforming protein of Rous sarcoma virus. *Virology* **113,** 736–751.

Peluso, R. W., Lamb, R. A., and Choppin, R. W. (1978). Infection with paramyxoviruses stimulates synthesis of cellular polypeptides that are also stimulated in cells transformed by Rous sarcoma virus and deprived of glucose. *Proc. Natl. Acad. Sci. U.S.A.* **75,** 6120–6124.

Schlesinger, M. J., and Kaariainen, L. (1980). Translation and processing of alphavirus proteins. *In* "The Togaviruses" (R. W. Schlesinger, ed.), pp. 371–392. Academic Press, New York.

Schlesinger, M. J., Aliperti, G., and Kelley, P. M. (1982). The response of cells to heat shock. *Trends Biochem. Sci.* (*Pers. Ed.*) **7,** 222–225.

Sefton, B. M., Trowbridge, I. S., Cooper, J. A., and Scolnick, E. M. (1982). The transforming proteins of Rous sarcoma virus, Harvey sarcoma virus and Abelson virus contain tightly bound lipid. *Cell* **31,** 465–474.

Southgate, R., Ayme, A., and Voellmy, R. (1983). Nucleotide sequence analysis of the *Drosophila* small heat shock gene cluster at locus 67B. *J. Mol. Biol.* **165,** 35–37.

Voellmy, R., and Bromley, P. (1982). Massive heat-shock polypeptide synthesis in late chicken embryos: convenient system for study of protein synthesis in highly differentiated organisms. *Mol. Cell. Biol.* **2,** 479–483.

Voellmy, R., Bromley, P., and Kocker, H. P. (1983). Structural similarities between corresponding heat-shock proteins from different eucaryotic cells. *J. Biol. Chem.* **258,** 3516–3522.

Wang, C., Gomer, R. H., and Lazarides, E. (1981). Heat shock proteins are methylated in avian and mammalian cells. *Proc. Natl. Acad. Sci. U.S.A.* **78,** 3531–3535.

Wang, C., Lazarides, E., O'Connor, C. M., and Clark, S. (1982). Methylation of chicken fibroblast heat shock proteins at lysyl and arginyl residues. *J. Biol. Chem.* **257,** 8356–8362.

10

Stress Responses in Avian and Mammalian Cells

L. E. HIGHTOWER, P. T. GUIDON, JR., S. A. WHELAN, AND C. N. WHITE

I. INTRODUCTION

A broad variety of noxious chemical, physical, and biological agents stimulate an apparently common response in eukaryotic and possibly prokaryotic cells as well (reviewed in Schlesinger *et al.*, 1982). These stressors stimulate the rapid and copious synthesis of a relatively small set of proteins, originally known as heat shock proteins and now also known as stress proteins, some of which are highly conserved (Kelley and Schlesinger, 1982). Currently, little is known about the interactions between stressors and cells which stimulate the response, how stress genes are regulated, or what the functions of stress proteins are. Based on our own work and that of many others, we have constructed a hypothetical model of this generalized cellular response. We assume that stressors damage target molecules on cell surfaces and/or within cells, that cells possess systems

Changes in Eukaryotic Gene Expression in Response to Environmental Stress

ISBN 0-12-066290-6

for sensing such damage (stress) and signaling gene activation, and that the induced proteins protect cells from further, possibly lethal damage to and/or initiate repair of essential cellular processes.

We are currently testing major features of this model, and the following is a progress report from our laboratory. A purification scheme for three major rat stress proteins (with masses of 88,000; 73,000; 71,000 daltons) is described. Evidence that several stress proteins are released from cultured rat embryo (RE) cells is presented. The kinetics of stress mRNA synthesis in cultured chicken embryo (CE) cells treated with either heat or the arginine analog canavanine are compared. And finally, we show that glycerol and deuterium oxide (D_2O) block the induction of stress proteins by several different stressors, and we suggest that these molecules protect sensitive cellular targets.

II. PURIFICATION OF THREE MAJOR RAT STRESS PROTEINS

Our goal is to purify milligram amounts of the native forms of the major mammalian stress proteins for physical and chemical analyses and to provide material for monoclonal antibody production. The chicken 88-, 71-/72-, and 23-kilodalton stress proteins have been purified under denaturing conditions, and monospecific antisera have been produced from them (Kelley and Schlesinger, 1982). Partial purification of several mammalian stress proteins from HeLa cells has been reported (Welch and Feramisco, 1982); however, the inducible and constitutively synthesized forms of the 72- to 73-kilodalton stress proteins (Hightower and White, 1981) were not separated, and the 88-kilodalton stress protein was contaminated with another major stress protein. In addition, HeLa cells synthesize as many as eight isoelectric variants of the 72- to 73-kilodalton stress proteins (Thomas *et al.*, 1982), which is a potential complication for detailed comparisons of the inducible and constitutive forms.

Rat livers are an excellent source of mammalian stress proteins, and only single isoelectric forms of the major inducible 71-kilodalton protein and its 73-kilodalton cognate protein are produced. A protocol which has been developed to produce electrophoretically pure 88-, 73-, and 71-kilodalton rat stress proteins is outlined in Fig. 1. To heat shock rats, 8- to 10-week-old male CD rats are lightly anesthetized with Nembutol, placed between two heating pads, and heated to a core temperature (monitored by rectal probe) of 41.5°C for 15 min. Typically, 95% of the animals recover and survive for 24 hr. The animals are then sacrificed; their livers are perfused, removed, and homogenized in 0.32 *M* sucrose, 10 m*M* Tris (pH 7.4), 10 m*M* NaCl, 0.5 m*M* EDTA, and aprotinin (10 μg/ml) using a Polytron. The homogenate is centrifuged first at 16,500 *g* for 30 min. at 4°C and then at 100,000 *g* for 4 hr at 4°C. The resulting supernatant is then

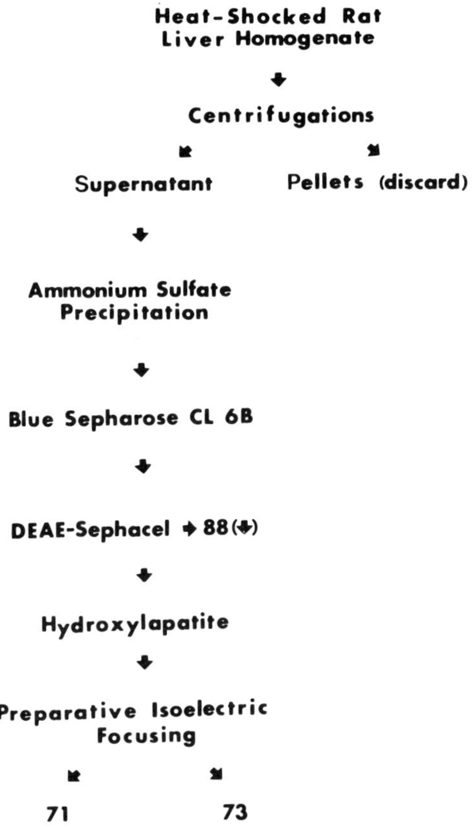

Fig. 1. Purification scheme for the 88-, 73-, and 71-kilodalton rat stress proteins.

brought to 40% of saturation with ammonium sulfate to precipitate the bulk of the major stress proteins. After dialysis and clarification by centrifugation, the sample is loaded on a Blue Sepharose CL-6B column and eluted with a 10 m*M* to 0.4 *M* NaCl gradient. This and all subsequent columns are monitored by analyzing eluted fractions with sodium dodecyl sulfate–polyacrylamide gel electrophoresis (SDS–PAGE). Fractions enriched with stress proteins are pooled, concentrated, dialyzed, and loaded on a DEAE–Sephacel column. This column effectively separates the 88-kilodalton stress protein from the 73- to 71-kilodalton mixture. The 88-kilodalton protein spontaneously precipitates after elution. The precipitate consists of a single major Coomassie blue-stainable band on SDS–PAGE (Fig. 2). Proteins are eluted from the DEAE–Sephacel column, processed as described for the first column, and then loaded on a hydroxylapatite

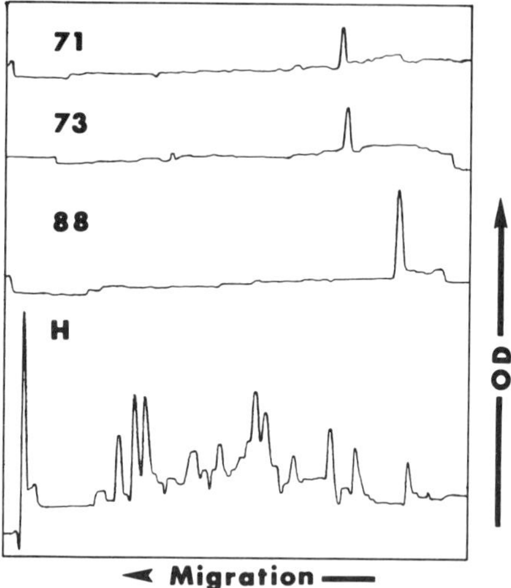

Fig. 2. Densitometer scans of gels containing rat stress proteins purified as outlined in Fig. 1 and rat liver homogenate (H). Samples of the 88-, 73-, and 71-kilodalton proteins along with homogenate were analyzed by SDS–PAGE (Laemmli, 1970) using 9% polyacrylamide gels. The gels were fixed, stained with Coomassie blue, and dried onto cellulose sheets. Individual channels were scanned using a Joyce–Loebl densitometer to determine optical density (OD).

column. This column is developed with a 0–0.3 *M* phosphate gradient, and the 73- to 71-kilodalton mixture elutes in a highly purified form. To separate the 73- and 71-kilodalton stress proteins, we took advantage of their different isoelectric points by using preparative isoelectric focusing.

In the presence of the dissociating agents urea and Nonidet P-40 detergent, the 73- and 71-kilodalton proteins focused with isoelectric points of 5.6 and 5.8, respectively. In the absence of these agents, the 73-kilodalton protein had an isoelectric point (p*I*) of 5.0, and the 71-kilodalton protein had a p*I* of 5.2. Albumin exhibits similar behavior due to the loss of fatty acids when exposed to dissociating agents (Basu *et al.*, 1978). Chloroform/methanol extraction of the purified 73-kilodalton protein from normal rat liver followed by thin layer chromatography of the organic fraction yielded material which comigrated with a free fatty acid standard. Subsequently, the purified 73-kilodalton protein was subjected to acid methanolysis and methyl esterification. Fatty acid methyl esters of palmitic acid, stearic acid, and trace amounts of myristic acid were detected by gas chromatography–mass spectrometry. Since the major 71-kilodalton heat shock protein has a more acidic p*I* (5.2) in the absence of dissociating agents than in their presence (5.8), the 71-kilodalton protein like its cognate protein may be a fatty acid binding protein. Acidic isoelectric points for these proteins in the absence of dissociating agents have also been obtained starting with homogenates of control and heat-shocked rat brain, rat heart, and cultured rat embryo cells. We suspect that partial removal of lipid moieties from the 73-kilodalton protein, which occurs if the protein is exposed to mercaptoethanol or high concentrations of ammonium sulfate, causes this protein and the 71-kilodalton stress protein to behave as extremely heterogeneous proteins on a variety of chromatography columns and during native isoelectric focusing.

III. EXTRACELLULAR APPEARANCE OF RAT STRESS PROTEINS

Searches for the sites where stress proteins function have concentrated on intracellular locations in cultured cells and tissues, primarily in *Drosophila.* In *Drosophila* salivary gland tissues, heat shock proteins are concentrated in nuclei and also at cell boundaries, especially over the lumen of the gland, suggesting membrane associations or possibly secretion (Velazquez *et al.*, 1980). White has obtained evidence of the synthesis and possible transport of the 71-kilodalton rat stress protein by cells associated with brain capillaries (White, 1980). These observations prompted us to search for extracellular stress proteins released by cultured RE cells following heat shock.

Cultured RE cells were either maintained continuously at 37°C or heat shocked for 10 min at 45°C and allowed to recover for 2 hr. The cultures were then labeled with [^{35}S]methionine for 1 hr in serum-free medium. The medium was collected, the proteins were acetone precipitated, and the pellet was dissolved in O'Farrell lysis buffer for two-dimensional gel electrophoresis (Fig. 3). The radioactive proteins that accumulated in the medium over control cultures are shown in Fig. 3A. Most of these proteins accumulated in the medium over heat-shocked cells as well (Fig. 3B). In addition, two inducible proteins (with masses of 110,000 and 71,000 daltons) which were identified previously as rat stress proteins (Hightower and White, 1981) and the 73-kilodalton protein were present.

The time course of appearance of stress proteins in extracellular medium was determined by using a radioisotopic pulse–chase protocol. The 73-kilodalton protein was detectable in the medium from control cells within the first few minutes of chase, and the 73-kilodalton protein along with the 71-kilodalton stress protein were released from heat-shocked cultures within minutes as well. Treatment of heat-shocked RE cells with either the antimicrotubule drug colchicine (5×10^{-7} *M*) or the sodium ionophore monensin (10^{-6} *M*), both of which block release of a variety of proteins from cells, did not block the release of the 71- and 73-kilodalton proteins. We next compared the patterns of [^{35}S]methionine-labeled proteins released from cells within the first 2–3 min of chase with the patterns of radiolabeled proteins released from heat-shocked cells

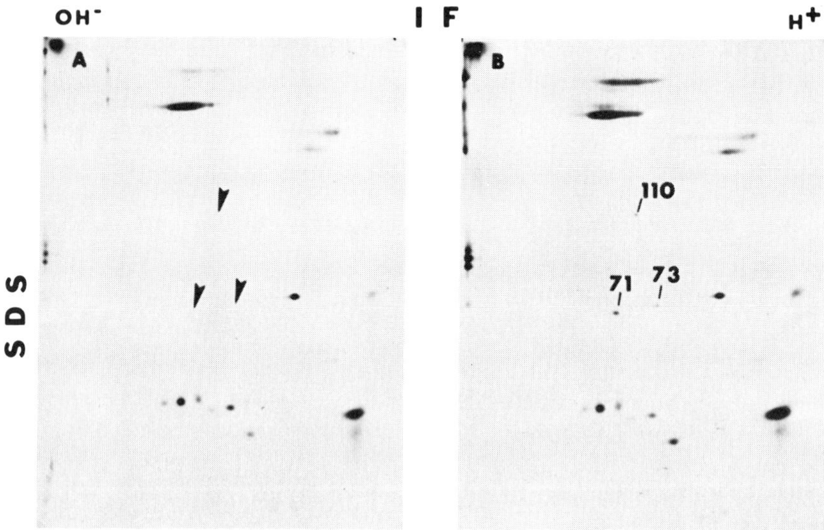

Fig. 3. Fluorograms of [^{35}S]methionine-labeled proteins from medium samples of (A) control and (B) heat-shocked RE cells. Following heat shock and recovery, cultures were incubated at 37°C for 1 hr in the presence of 25 μCi [^{35}S]methionine per milliliter of label medium. The medium was collected and clarified by low-speed centrifugation, ovalbumin carrier was added, and proteins were precipitated with three volumes of cold acetone. Precipitates were dissolved in lysis buffer and processed for two-dimensional polyacrylamide gel electrophoresis by the methods of O'Farrell (1975). Proteins from equal volumes of culture medium were loaded on gels. Isoelectric focusing (IF) was performed in the horizontal dimension and SDS–PAGE in the vertical dimension. Major stress proteins were marked using their apparent sizes in kilodaltons. Darts mark the expected positions of these proteins in control patterns.

by nonionic detergent treatment and dounce homogenization in low-salt buffer. The latter treatments released 71- and 73-kilodalton proteins as well as large amounts of actin, tubulin, and other cytosolic proteins. The pattern of proteins quickly released from heat-shocked and control RE cells during the chase period was different. Tubulins and most of the other proteins released by detergents or homogenization were absent. However, the 73- and 71-kilodalton proteins along with some actin were present. These released proteins were apparently not in blebs or vesicles because they were not sedimentable in a sucrose gradient appropriate for lipid vesicles. While it is difficult to rule out the possibility that the 71- and 73-kilodalton proteins were released from a small population of damaged cells, the pattern of proteins released into the medium did not simply reflect the pattern of cytosolic proteins. We favor the possibility that some of the 71- and 73-kilodalton proteins can be easily released from cells under certain conditions, perhaps stimulated by medium changes in the pulse–chase protocol. Release may be facilitated by the localization of a portion of the intracellular complement of these proteins at the cell membrane and possibly because they are lipoproteins.

IV. STIMULATION OF STRESS mRNA SYNTHESIS IN CHICKEN EMBRYO CELLS EXPOSED TO CANAVANINE OR HEAT

Actinomycin D blocks induction of stress proteins in CE cells treated with canavanine (Kelley and Schlesinger, 1978; Hightower, 1980), heat (Kelley and Schlesinger, 1978), or arsenite (Johnston *et al.*, 1980), indicating that their induction requires RNA synthesis. Also, the levels of functional stress mRNA are elevated in CE cells exposed to heat (Kelley *et al.*, 1980), arsenite (Johnston *et al.*, 1980), and the avian paramyxovirus Newcastle disease virus (Collins and Hightower, 1982). Since the rates of synthesis of the major stress proteins approach that of actin in CE cells (Hightower, 1980), stress mRNA's should likewise constitute a large part of the cellular mRNA synthesized after induction. We expected then that some of these mRNA's should be detectable as [^{3}H]uridine-labeled bands on gels of mRNA extracted from stressed cells.

Figure 4A shows the patterns of mRNA labeled in CE cells during a 1 hr [^{3}H]uridine pulse in the absence (lane 1) or presence (lanes 2–5) of canavanine for increasing lengths of time. Two abundant species appeared and were synthesized continuously throughout the period of canavanine treatment. These species were identified by elution from the gels followed by cell-free translation and by Northern blotting followed by hybridization with cloned *Drosophila* heat shock genes. By these criteria, the higher-molecular mass band was identified as the mRNA encoding the 88-kilodalton chicken stress protein and the more rapidly migrating band as the mRNA for the 71- to 72-kilodalton stress proteins (White and Hightower, 1984).

To compare stress mRNA synthesis induced by heat shock to that induced by canavanine, CE cells were shifted from 40.5 to 44°C and labeled with [^{3}H]uridine for 1 hr at the times shown in Fig. 4B. The first hour at the elevated temperature was marked by the appearance of two bands (lane 2), which were absent from control RNA patterns (lane 1) and which comigrated with the bands induced by canavanine. Unlike the response to canavanine, the heat-induced synthesis of these stress mRNA's was inhibited during the second hour of stress (lane 3). The same result was obtained when cultures were returned to 40.5°C following a 1-hr heat shock.

In *Drosophila* cells, synthesis of functional 70-kilodalton heat shock proteins has been implicated in the shutoff of stress mRNA synthesis and in the degradation of this mRNA (DiDomenico *et al.*, 1982). Because *Drosophila* cells respond very slowly to canavanine as a stressor (DiDomenico *et al.*, 1982; L. E. High-

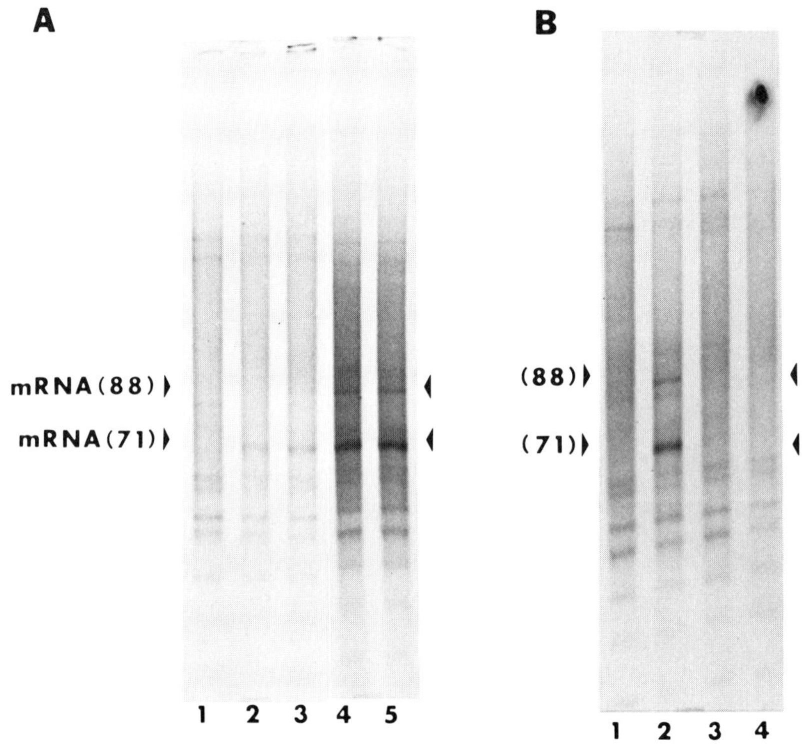

Fig. 4. Fluorograms of [^{3}H]uridine-labeled mRNA induced in CE cells by (A) canavanine or (B) heat shock. Canavanine treatment was done by replacing medium arginine with canavanine, as described before (Hightower, 1980), and heat shock was done by placing culture plates directly on the racks of a 44°C incubator. Cells were labeled either in the presence of canavanine or at 44°C for 1-hr periods with 5–50 μCi [^{3}H]uridine per milliliter of medium. Radioactive mRNA was isolated by oligo(dT)cellulose chromatography and ethanol precipitation, as described previously (Collins *et al.*, 1982), and separated in gels of 1.5% agarose/25 m*M* citrate (pH 3.5)/6 *M* urea (Wertz *et al.*, 1980). The gels were fixed, treated with Enhance, dried, and exposed to preflashed X-ray film at −70°C. Labeling periods were as follows (initiation of canavanine or heat treatment = time 0): (A) lane 1, control in normal medium; lane 2, 0.5–1.5 hr; lane 3, 2.5–3.5 hr; lane 4, 4.25–5.25 hr; lane 5, 6.25–7.25 hr. (B) lane 1, control at 40.5°C; lane 2, 0–1 hr; lane 3, 1.5–2.5 hr; lane 4, 3–4 hr. The induced RNA's were sized relative to vesicular stomatitis virus mRNA's and chicken rRNA markers [mRNA (88) – 1.27×10^6 and mRNA (71) = 0.95×10^6].

tower, unpublished observations), canavanine can be used in this system to render newly synthesized proteins nonfunctional during recovery from heat shock without causing stress protein induction. Under these conditions, stress mRNA synthesis continues and stress mRNA degradation is blocked. In contrast, CE cells synthesize stress mRNA (Fig. 4A) and stress proteins (Hightower, 1980) within 2 hr of canavanine treatment. Therefore, in the experiment shown in Fig. 4A we cannot distinguish between continued induction of stress mRNA synthesis due to the activity of canavanine as a stressor and the failure to inhibit stress mRNA synthesis because putative protein(s) necessary for inhibition are canavanine substituted and probably inactive. However, in the presence of a heat stress, stress mRNA synthesis is inhibited when presumably functional proteins are allowed to accumulate (Fig. 4B). These data are consistent with the *Drosophila* self-regulation model proposed by DiDomenico *et al.* (1982).

At least one aspect of stress mRNA metabolism in CE cells is different from the *Drosophila* system. Chicken stress mRNA's have short half-lives in canavanine-treated as well as in recovering CE cells. The half-lives of the CE stress mRNA's based on an approach-to-steady-state analysis were 89 min for the mRNA encoding the 88-kilodalten protein and 46 min for the mRNA encoding the 71-kilodalton protein (White and Hightower, 1984). Assuming that canavanine substitution renders chicken stress proteins nonfunctional, we concluded that functional stress proteins were not required for rapid degradation of chicken stress mRNA's.

V. INHIBITORS OF THE STRESS RESPONSE

The mechanism(s) of induction of the stress response is unknown. One strategy which may provide insights is the identification and use of agents that block different steps in the induction pathway, beginning with generation of the induction signal and ending with gene activation. Since we hypothesize that stressors damage macromolecules, we sought agents that stabilize macromolecules against thermal denaturation and that protect cells from thermal killing. Both deuterium oxide (D_2O) and polyhydroxyl alcohols like glycerol are known to stabilize macromolecules (Gerlsma and Stuur, 1972; Tuena de Gomez-Puyou *et al.*, 1978; Back *et al.*, 1979) and to protect cells from thermal killing (Henle, 1981; Mas-sicotte-Nolan *et al.*, 1981; Azzam *et al.*, 1982; Fisher *et al.*, 1982). Here we show that D_2O and glycerol block the induction of stress proteins in CE cells by several different stressors.

The effects of glycerol and D_2O on heat-shocked CE cells are shown in Fig. 5. Lane 1 shows a typical gel pattern of [^{35}S]methionine-labeled proteins from control CE cells maintained at 37°C, and lane 2 shows the radioactive proteins from heat-shocked CE cells. The major CE stress proteins (with masses of 88,000; 72,000; 71,000; 23,000 daltons) have been identified previously (Kelley and Schlesinger, 1978; Levinson *et al.*, 1978; Hightower and Smith, 1978). When the cultures were heated briefly in the presence of either 1 *M* glycerol (lane 3) or medium containing 85–99% D_2O (lane 4), the induction of stress proteins was almost completely blocked. Erythritol (1 *M*) was also effective in blocking the induction by heat shock (data not shown).

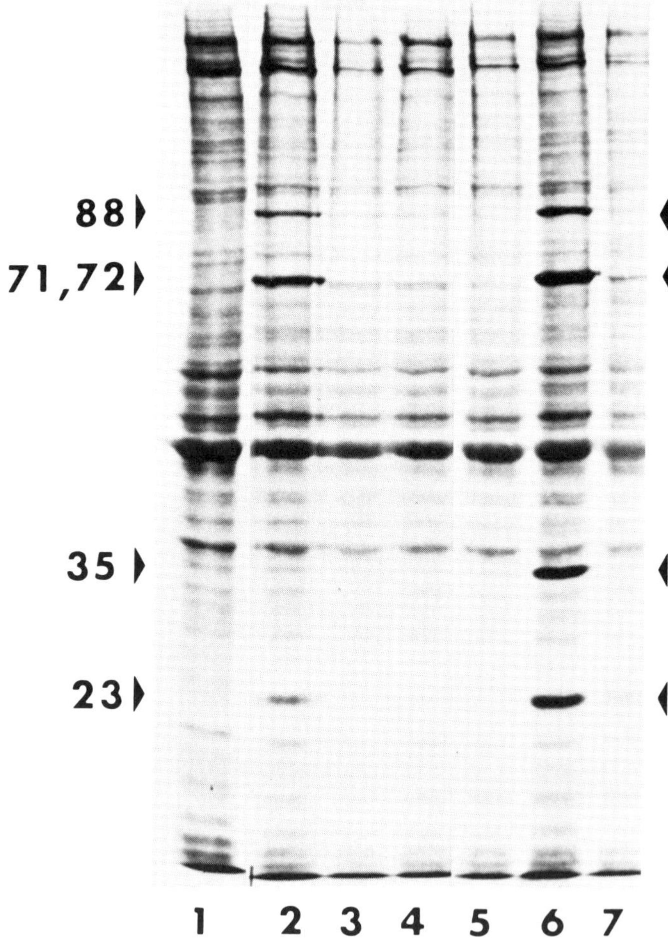

Fig. 5. Effects of glycerol and D_2O on the induction of stress proteins in CE cells by heat or $NaAsO_2$. To induce stress proteins, cultures were either heated at 44°C for 30 min or exposed to 2×10^{-5} *M* arsenite for 3 hr. All incubations were done in serum-free Eagle's minimum essential medium (MEM). To protect cultures, cells were pretreated with 1 *M* glycerol for 20 min prior to addition of stressors, and 1 *M* glycerol was present during the stress. MEM containing 99% D_2O was added at the same time as the stressors. After the appropriate incubation period, cultures were washed and labeled with [^{35}S]methionine (10 μCi/ml) for 30 min at 37°C in medium without stressors and inhibitors. The cultures were then solubilized in gel sample buffer and analyzed by SDS–PAGE on 9% polyacrylamide gels. Fluorograms were prepared as described previously (Hightower, 1980), and protein patterns from the following cultures are shown: lane 1, control cells at 37°C; lane 2, heat-shocked cells; lane 3, cells heat shocked in MEM containing 1 *M* glycerol; lane 4, cells heat shocked in MEM containing 99% D_2O; lane 5, cells exposed to MEM containing 99% D_2O at 37°C; lane 6, cells exposed to 2×10^{-5} *M* arsenite; lane 7, cells exposed to 2×10^{-5} *M* arsenite in MEM containing 99% D_2O.

Interestingly, the induction of stress proteins by agents other than heat is blocked by glycerol and D_2O as well. The protein patterns from cultures treated only with D_2O (lane 5), only with aresenite (lane 6), or with both D_2O and arsenite (lane 7) are shown in Fig. 5. Similar patterns were obtained for copper ions (Cu^{2+}) as a stressor in the presence and absence of D_2O, and glycerol was also quite effective in blocking inductions by either arsenite or copper (data not shown). An additional major CE stress protein (mass of 35 kilodaltons), which was observed by others (Johnston *et al.*, 1980), was strongly induced in CE cells treated with arsenite (lane 6) or copper but not with heat (lane 2).

The block of stress protein induction by glycerol can be overcome by increasing the amount of stressor and, presumably, the level of stress. When the concentration of arsenite was increased 20-fold to 4×10^{-4} *M* or the heat shock temperature was elevated by several degrees, a maximum induction of stress proteins was obtained even in the presence of 1 *M* glycerol (S. A. Whelan, unpublished observations). Similar experiments have not been done for D_2O. Thus, glycerol did not block gene activation or expression. The most likely modes of action for glycerol are to protect stressor-sensitive targets or to inactivate a stress-sensing system within the cell. An experimental result which is consistent with the former possibility is that CE cells heated in the presence of glycerol did not synthesize stress proteins when glycerol was washed out and the cells were returned to 37°C for 2 hr. The cells could be induced to produce stress proteins by heat shock during the recovery period, showing that the mechanism(s) for sensing heat stress and activating stress genes were operational (S. A. Whelan, unpublished observations). If target molecules were altered or damaged by heat in the presence of glycerol, the cell would be expected to respond by synthesizing stress proteins during recovery. Our results suggest that glycerol protected heat-sensitive cellular targets and no stress signals were generated. We suggest that the general term protectors be used for molecules such as glycerol and D_2O which block this cellular response to stress.

VI. SUMMARY

The 88- and 71-kilodalton rat stress proteins and a 73-kilodalton protein closely related to the latter can be obtained in pure form from heat-shocked rat livers using the protocol outlined in Fig. 1, however, overall yields are poor. Currently, this purification protocol is being modified to take advantage of the hydrophobic nature of these proteins, which should result in high yields of the 71- and 73-kilodalton proteins. Biochemical studies of these proteins and their associated lipids may provide clues to their function(s).

Although it is clear that the vast majority of major stress proteins remain within the cells that produce them, some of the 71- and 73-kilodalton proteins are rather easily released from cultured RE cells, and there is suggestive evidence of externalization from experiments on *Drosophila* salivary tissue and incubated slices of rat brain. Because stress proteins account for a substantial fraction of the protein synthetic capacity of stressed cells, the release of only 1% from cells can be substantial. Lindquist (1980) estimated that 1.7×10^7 molecules of the 70-kilodalton heat shock protein are synthesized per heat-shocked *Drosophila* cell in 120 min. If this estimate holds for rat cells, then 1.4×10^6 molecules of the 71-kilodalton stress protein will be synthesized per cell during a 10-min labeling period. Release of 1% of this protein would result in the extracellular accumulation of 1.4×10^4 molecules per cell.

Our stress mRNA patterns (Fig. 4B) show that the heat shock response in CE cells is transient, attaining a maximum level of stress mRNA synthesis within the first hour and undergoing inhibition of synthesis during the second hour. Other experiments (C. N. White, unpublished observations) indicate that stress mRNA produced during heat shock is at least functionally inactivated and probably degraded during the second and third hours of heat shock. It is appealing to think of the response in CE cells to canavanine and heat according to the model from Lindquist's laboratory involving self-regulation of *Drosophila* heat shock gene expression. However, our data can accommodate other interpretations, and the extension of this model to avian and mammalian cells now is probably premature.

The identification of protectors such as glycerol and D_2O, which protect cells from thermal killing, stabilize macromolecules in solution, and block stress protein synthesis, encourages speculation that similar types of molecular damage may be involved in stress states. Massicotte-Noland and co-workers (1981) have noted that a good correlation exists between the effects of mono-, di-, and trihydric alcohols on protein denaturation and thermal killing of cells. Westra and Dewey (1971) suggest that protein denaturation may be the primary cause of heat-induced cell killing. Previously, we suggested (Hightower, 1980) that a common feature of the stress states may be the intracellular accumulation of abnormal proteins. This hypothesis rests on the observation that known stressors either generate abnormal proteins biosynthetically (canavanine) or are known protein denaturants (heat, arsenite). A simple initial model, suggested by these observations, is that stressors damage proteins, resulting in the induction of stress proteins and, ultimately, in cell killing in our example of heat stress. Glycerol and D_2O may protect stress-sensitive proteins from irreversible denaturation possibly by direct interaction but more likely by generalized solvent effects. Minton and co-workers (1982) have proposed that a biological role for heat shock proteins may be to stabilize, in nonspecific ways, stress-susceptible proteins in cells. It would be interesting if small molecules like glycerol or D_2O and stress proteins had somewhat similar stabilizing effects on stress-sensitive cellular components.

Acknowledgments

This work was supported by a grant from the National Science Foundation (PCM 8118285) and benefited from the use of a cell culture facility supported by the National Cancer Institute (CA 14733).

REFERENCES

Azzam, E. I., George, I., and Raaphorst, P. (1982). Alterations in thermal sensitivity of Chinese hamster cells by D_2O treatment. *Radiato Res.* **90,** 644–648.

Back, J. F., Oakenfull, D., and Smith, M. B. (1979). Increased thermal stability of proteins in the presence of sugars and polyols. *Biochemistry* **18,** 5191–5196.

Basu, S. P., Narasinga Rao, S., and Hartsuck, J. A. (1978). Influence of fatty acid and time of focusing on the isoelectric focusing of human plasma albumin. *Biochim. Biophys. Acta* **533,** 66–73.

Collins, P. L., and Hightower, L. E. (1982). Newcastle disease virus stimulates the cellular accumulation of stress (heat shock) mRNAs and proteins. *J. Virol.* **44,** 703–707.

Collins, P. L., Wertz, G. W., Ball, L. A., and Hightower, L. E. (1982). Coding assignments of the five smaller mRNAs of Newcastle disease virus. *J. Virol.* **43,** 1024–1031.

DiDomenico, B. J., Bugaisky, G. E., and Lindquist, S. (1982). The heat shock response is self-regulated at both the transcriptional and post-transcriptional levels. *Cell* **31,** 593–603.

Fisher, G. A., Li, G. C., and Hahn, G. M. (1982). Modification of the thermal response by D_2O. I. Cell survival and the temperature shift. *Radiat. Res.* **92,** 530–540.

Gerlsma, S. Y., and Stuur, E. R. (1972). The effects of polyhydric and monohydric alcohols on the heat-induced reversible denaturation of lysozyme and ribonuclease. *Int. J. Pept. Protein Res.* **4,** 377–383.

Henle, K. J. (1981). Interaction of mono- and polyhydroxy alcohols with hyperthermia in CHO cells. *Radiat. Res.* **88,** 392–402.

Hightower, L. E. (1980). Cultured animal cells exposed to amino acid analogues or puromycin rapidly synthesize several polypeptides. *J. Cell. Physiol.* **102,** 407–427.

Hightower, L. E., and Smith, M. D. (1978). Effects of canavanine on protein metabolism in Newcastle disease virus-infected and uninfected chicken embryo cells. *In* "Negative Strand Viruses and the Host Cell" (B. W. J. Mahy and R. D. Barry, eds.), pp. 395–405. Academic Press, London.

Hightower, L. E., and White, F. P. (1981). Cellular responses to stress: Comparison of a family of 71–73 kilodalton proteins rapidly synthesized in rat tissue slices and canavanine treated cells in culture. *J. Cell. Physiol.* **108,** 261–275.

Johnston, D., Oppermann, H., Jackson, J., and Levinson, W. (1980). Induction of four proteins in chick embryo cells by sodium arsenite. *J. Biol. Chem.* **255,** 6975–6980.

Kelley, P. M., and Schlesinger, M. J. (1978). The effect of amino acid analogs and heat shock on gene expression in chicken embryo fibroblasts. *Cell* **15,** 1277–1286.

Kelley, P. M., and Schlesinger, M. J. (1982). Antibodies to two major chicken heat shock proteins cross react with similar proteins in widely divergent species. *Mol. Cell. Biol.* **2,** 267–274.

Kelley, P. M., Aliperti, G., and Schlesinger, M. J. (1980). *In vitro* synthesis of heat-shock proteins by mRNAs from chicken embryo fibroblasts. *J. Biol. Chem.* **255,** 3230–3233.

Laemmli, U. K. (1970). Cleavage of structural proteins during the assembly of the head of bacteriophage T4. *Nature (London)* **277,** 680–685.

Levinson, W., Oppermann, H., and Jackson, J. (1978). Induction of four proteins in eukaryotic cells by kethoxal bis (thiosemicarbazone). *Biochim. Biophys. Acta* **518,** 401–412.

Lindquist, S. (1980). Translational efficiency of heat-induced messages in *Drosophila* melanogaster cells. *J. Mol. Biol.* **137,** 151–518.

Massicotte-Nolan, P., Glofcheski, D. J., Kruuv, J., and Lepock, J. R. (1981). Relationship between hyperthermic cell killing and protein denaturation by alcohols. *Radiat. Res.* **87,** 284–299.

Minton, K. W., Karmin, P., Hahn, G. M., and Minton, A. P. (1982). Nonspecific stabilization of stress-susceptible proteins by stress-resistant proteins: A model for the biological role of heat shock proteins. *Proc. Natl. Acad. Sci. U.S.A.* **79,** 7107–7111.

O'Farrell, P. H. (1975). High resolution two-dimensional electrophoresis of proteins. *J. Biol. Chem.* **250,** 4007–4021.

Schlesinger, M. J., Ashburner, M., and Tissières, A., eds. (1982). "Heat Shock, from Bacteria to Man." Cold Spring Harbor Lab., Cold Spring Harbor, New York.

Thomas, G. P., Welch, W. J., Mathews, M. B., and Feramisco, J. R. (1982). Molecular and cellular effects of heat-shock and related treatments of mammalian tissue culture cells. *Cold Spring Harbor Symp. Quant. Biol.* **46,** 985–996.

Tuena de Gomez-Puyou, M., Gomez-Puyou, A., and Cerbon, J. (1978). Increased conformational stability of mitochondrial soluable ATPase (F_1) by substitution of H_2O for D_2O. *Arch. Biochem. Biophys.* **187,** 72–77.

Velazquez, J. M., DiDomenico, B. J., and Lindquist, S. (1980). Intracellular localization of heat shock proteins in *Drosophila. Cell* **20,** 679–689.

Welch, W. J., and Feramisco, J. R. (1982). Purification of the major mammalian heat shock proteins. *J. Biol. Chem.* **257,** 14,949–14,959.

Wertz, G. W., Davis, N. L., and Edgell, M. H. (1980). High resolution preparative gel electrophoresis: separation and recovery of functional messenger RNA species. *Anal. Biochem.* **106,** 148–155.

Westra, A., and Dewey, W. C. (1971). Variation in sensitivity to heat shock during the cell-cycle of Chinese hamster cells *in vitro. Int. J. Radiat. Biol. Relat. Stud. Phys. Chem. Med.* **19,** 467–477.

White, C. N., and Hightower, L. E. (1984). Stress mRNA metabolism in canavanine-treated chicken embryo cells. *Mol. Cell. Biol.* **4,** 1534–1541.

White, F. P. (1980). The synthesis and possible transport of specific proteins by cells associated with brain capillaries. *J. Neurochem.* **35,** 88–94.

11

Effect of Hyperthermia and LSD on Gene Expression in the Mammalian Brain and Other Organs

IAN R. BROWN

I. INTRODUCTION

One approach to increasing our knowledge of the molecular biology of the mammalian brain is to determine how biochemical processes in the nervous system are affected by the introduction of physiologically relevant perturbations. We have utilized elevation of body temperature (i.e., hyperthermia) and the potent psychotropic drug LSD (lysergic acid diethylamide) as experimental tools to manipulate molecular processes in the mammalian brain.

These perturbations rapidly induce specific changes in gene expression in the brain at both the transcriptional and translational levels. Elevation of body tem-

Changes in Eukaryotic Gene Expression
in Response to Environmental Stress

ISBN 0-12-066290-6

perature, to levels similar to that attained during fever reactions, induces the activation of a translational control mechanism which results in a rapid global inhibition of protein synthesis in the brain. Hyperthermia also induces the selective expression of a gene coding for a brain protein which is similar in molecular weight to one of the major heat shock proteins reported in unicellular organisms and tissue culture systems following elevation of ambient temperature (Ashburner and Bonner, 1979; Schlesinger *et al.*, 1982).

Fever in susceptible children can provoke convulsions, hallucinations, and a variety of neurological disorders; however, little is known of the consequences of elevated body temperature on underlying neurochemical events. Seizures induced by increases in body temperature (i.e., febrile convulsions) are an important clinical problem in about 3–5% of children between the ages of 6 months and 5 years. Seizures induced by fever occur in normal children with a genetic predisposition for this response and in children with brain damage or epilepsy. In our studies we have sought to investigate the effects of physiologically relevant increases in body temperature on molecular events in both the developing brain and specific neural systems such as the retina. Given the possibility that heat shock proteins may be involved in the response of the body to fever reactions, it is important to increase our knowledge of these proteins in intact mammalian organs.

II. INHIBITORY EFFECT OF LSD ON BRAIN PROTEIN SYNTHESIS

Several years ago, we became interested in analyzing whether the powerful psychotropic drug LSD influenced gene expression in the mammalian nervous system. LSD was known to bind to neurotransmitter receptors in the brain and alter patterns of nerve firing. At the psychological level, the drug induces profound changes in sensory perception. Virtually nothing was known, however, concerning whether LSD influenced gene expression in the brain (see Brown *et al.*, 1982a, for a recent review). As an experimental animal, the rabbit was selected because the large size of the brain offered sufficient tissue from one animal for biochemical analysis, and rabbits, like monkeys and humans, are very sensitive to the drug LSD.

The first question which was addressed was whether the intravenous injection of physiologically relevant doses of LSD affects protein synthesis in the mammalian brain. As an index of protein synthesis, our initial experiments involved an analysis of the integrity of brain polysomes following their fractionation into size classes on sucrose gradients. A marked disaggregation of polysomes to monosomes was observed in all brain regions which were examined (Holbrook and Brown, 1976; Heikkila and Brown, 1981). Maximal polysome disaggregation was observed 30–60 min after drug injection, with a gradual return to normal polysome levels within 4 hr (Holbrook and Brown, 1976). The effect was dose dependent and required the binding of LSD to neurotransmitter receptors in the brain (Holbrook and Brown, 1977a; Heikkila and Brown, 1979a).

As shown in Fig. 1, administration of LSD to a pregnant female rabbit at 50 μg/kg body weight induced a massive disaggregation of polysomes after 1 hr in both the maternal brain and fetal organs such as kidney, liver, and brain. Mild stress (i.e., physical restraint) accentuated the degree of drug-induced polysome disaggregation such that doses of LSD as low as 1 μg/kg body weight induced a marked disaggregation of brain polysomes when combined with the restraining procedure (Heikkila *et al.*, 1978). This drug dose is well within levels utilized by man (Kalant and Khanna, 1975; Rossi, 1971).

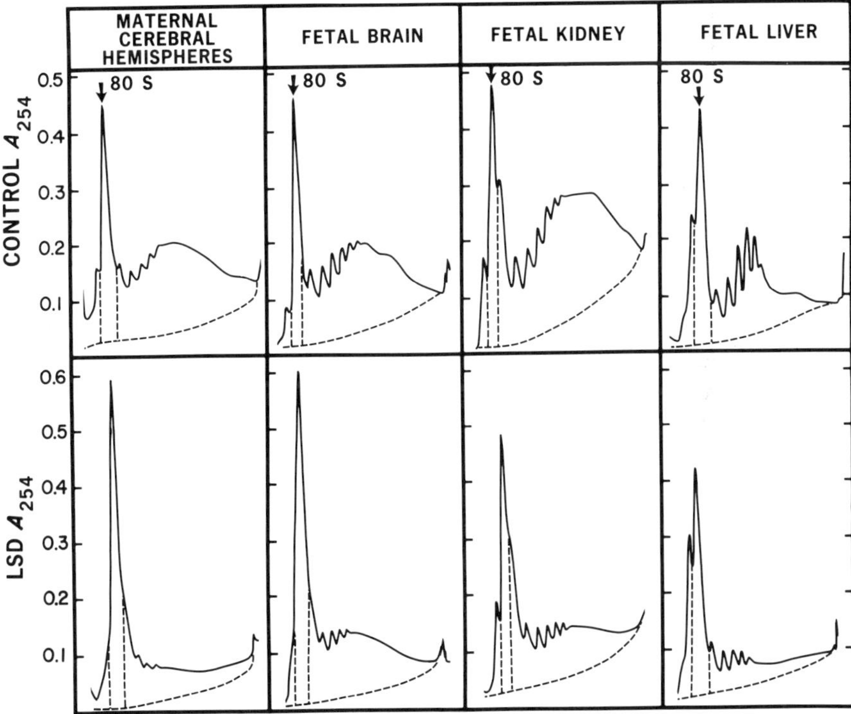

Fig. 1. Disaggregation of polysomes from maternal brain and fetal organs after administration of LSD *in vivo*. Pregnant female rabbits were injected intravenously with LSD at 50 μg/kg body mass (control animals received saline). After 1 hr animals were sacrificed and polysomes from maternal brain and fetal organs were fractionated into size classes on 15–45% sucrose density gradients following centrifugation for 75 min. at 40,000 rpm in a Beckman SW-41 rotor. (From Heikkila *et al.*, 1979.)

In a series of studies, the mechanism of LSD-induced polysome disaggregation was analyzed. Polysomes did not shift to monosomes due to RNase nicking of mRNA (Holbrook and Brown, 1976; Heikkila *et al.*, 1978; Mahony and Brown, 1979), nor did premature termination of translation occur (Holbrook and Brown, 1977b). Our evidence suggests that LSD affects brain protein synthesis in the mammalian brain via a lesion in the process of reinitiation of translation (Brown *et al.*, 1982a).

The inhibitory effect of LSD on protein synthesis in the rabbit brain was confirmed by two other experimental methods: the decreased incorporation of labeled amino acids into brain proteins *in vivo* (Freedman *et al.*, 1981) and the transient inhibition of translational capacity in an initiating cell-free system (Cosgrove *et al.*, 1981). The initiating cell-free system was developed to facilitate an analysis of the steps in protein synthesis which are affected by LSD. Utilizing this cell-free system (i) a translational inhibitory factor was found to be transiently present in the postribosomal supernatant fraction (Cosgrove *et al.*, 1981), and (ii) inhibition of the formation of 40 S and 80 S initiation complexes was apparent at the time of maximal polysome disaggregation (Cosgrove and Brown, 1984).

III. EFFECT OF HYPERTHERMIA ON BRAIN PROTEIN SYNTHESIS

When rectal temperature measurements were taken in conjunction with the experiments on drug-induced disaggregation of brain polysomes, it was apparent that LSD induced a rapid elevation in body temperature in rabbits, and that the kinetics of polysome disaggregation were very similar to the rise and fall in body temperature (Heikkila and Brown 1979a). As shown in Table I, an increase in

TABLE I

Effect of LSD Dosage on Brain Polysome Disaggregation and Hyperthermia[a]

Drug dosage (μg/kg)	% Monosomes/ total ribosomes	Rectal temp. (°C)
Saline	23.6	39.7
1	23.8	40.3
10	32.3	41.2
50	41.9	41.4

[a] Rabbits were injected with LSD or an appropriate volume of saline and sacrificed after 1 hr. Cerebral hemisphere polysomes were isolated and the percentage of monosomes per total ribosomes was determined by analysis on sucrose gradients as in Fig. 1. (Adapted from Heikkila and Brown, 1979a.)

both the degree of polysome disaggregation and the increment of elevation of body temperature was observed as the dose of the drug was increased.

An obvious question was whether protein synthesis in the mammalian brain was sensitive to hyperthermia induced by other means. Elevation of body temperature by 2°C, which was produced either by placement of animals at high ambient temperature (37°C) or by induction of fever by injection of a bacterial pyrogen, resulted in a massive disaggregation of brain polysomes to monosomes (Heikkila and Brown, 1979a,b). A linear relationship was apparent between increase in body temperature and degree of brain polysome disaggregation. These results indicate that physiologically relevant increases in body temperature influence macromolecular synthesis in the intact mammalian brain (Brown *et al.*, 1982a,b).

IV. INDUCTION OF HEAT SHOCK PROTEIN IN INTACT MAMMALIAN ORGANS

Hyperthermia induced by LSD, bacterial pyrogen, or elevation of ambient temperature induced an overall inhibition of protein synthesis in the rabbit brain. To address the question of whether all brain proteins were similarly affected, *in vivo*-labeled brain proteins were resolved by two-dimensional gel fluorography. Elevation of body temperature by 2–3°C following intravenous injection of LSD induced a selective increase in synthesis of a brain protein of molecular weight 74,000 (74K) at a time when the synthesis of other brain proteins was reduced (Freedman *et al.*, 1981; Cosgrove and Brown, 1983). Hyperthermia also induced the synthesis of the 74K protein in the kidney. This protein is similar in molecular weight to one of the major heat shock proteins previously reported to be induced in a wide range of tissue culture systems and unicellular organisms after an increase of ambient temperature (Ashburner and Bonner, 1979; Schlesinger *et al.*, 1982). The 74K brain protein appeared to be similar to a heat shock protein induced in other systems since it was precipitated by antibodies prepared against the chicken 70K heat shock protein (Cosgrove and Brown, 1983). Induction of the 74K brain protein was correlated to drug-induced hyperthermia since prevention of hyperthermia by injection of the drug into rabbits maintained in a cold room at 4°C eliminated the phenomenon; however, behavioral effects of the drug and overall inhibition of brain protein synthesis were still apparent (Cosgrove *et al.*, 1981; Cosgrove and Brown, 1983). Synthesis of the 74K protein in brain and kidney was also induced by elevation of body temperature by other means such as placement of animals at increased ambient temperature (Cosgrove and Brown, 1983).

The induction of synthesis of the 74K protein in intact mammalian organs was also demonstrated by the analysis of the translation products of polysomes isolated from animals subjected to hyperthermia. As shown in Fig. 2, a selective increase in the labeling of the 74K protein was apparent in the cell-free transla-

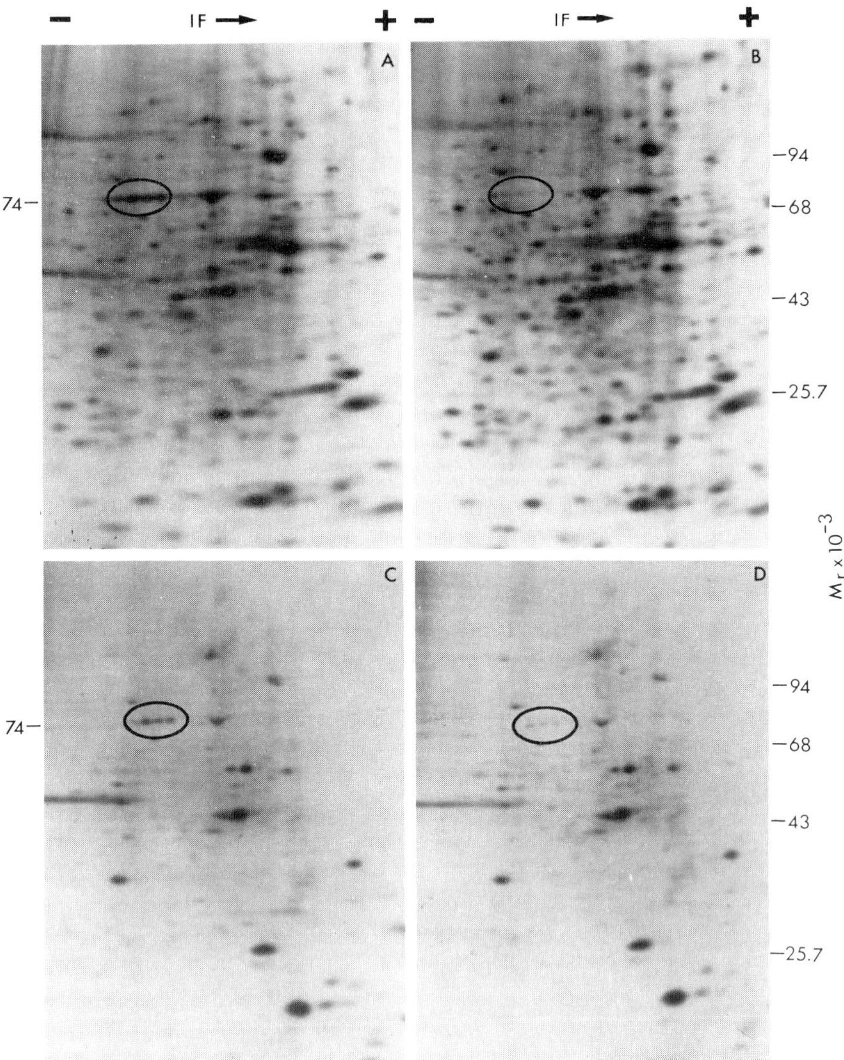

Fig. 2. Induction of synthesis of the 74K heat shock protein by polysomes isolated from brain and heart. Polysomes were isolated from rabbit brain and heart 1 hr after the intravenous injection of LSD at 100 μg/kg body mass, which increased body temperature by 3°C. The purified polysomes were translated in a reticulocyte cell-free system with [^{35}S]methionine, and the labeled translation products were analyzed by gel electrophoresis and fluorography. Equal amounts of acid-precipitable radioactivity (200,000 cpm) from drug-treated and saline-injected control animals were analyzed. IF, isoelectric focusing. The position of the 74K heat shock protein is encircled. Brain polysomes: (A) drug-induced hyperthermia, (B) saline control. Heart polysomes: (C) drug-induced hyperthermia, (D) saline control. (From Cosgrove and Brown, 1983.)

tion products of brain and heart polysomes isolated 1 hr following elevation of body temperature by 3°C.

The demonstration of an increased labeling of the 74K protein in the translation products of a heterologous cell-free translation system suggests that the phenomenon is due to an increase in the relative abundance in polysomes of mRNA coding for this protein and not to a change in posttranslational modification. Analysis of the cell-free translation products of brain polysomes isolated at various time points following drug injection indicated that the appearance of heat shock mRNA in brain polysomes was transient and paralleled the rise and fall of body temperature, which was induced by the drug (Cosgrove and Brown, 1983).

V. DEVELOPMENTAL CHANGES IN THE INDUCIBILITY OF HEAT SHOCK PROTEIN

A developmental increase in the temperature required to induce synthesis of the 74K heat shock protein in intact organs of the rat was apparent during early postnatal growth. In the 2-day-old rat, elevation of body temperature from 32 to 38°C induced synthesis of the 74K protein, as determined by the analysis of cell-free translation products of isolated brain polysomes (Figs. 3A, B). At 7 days of

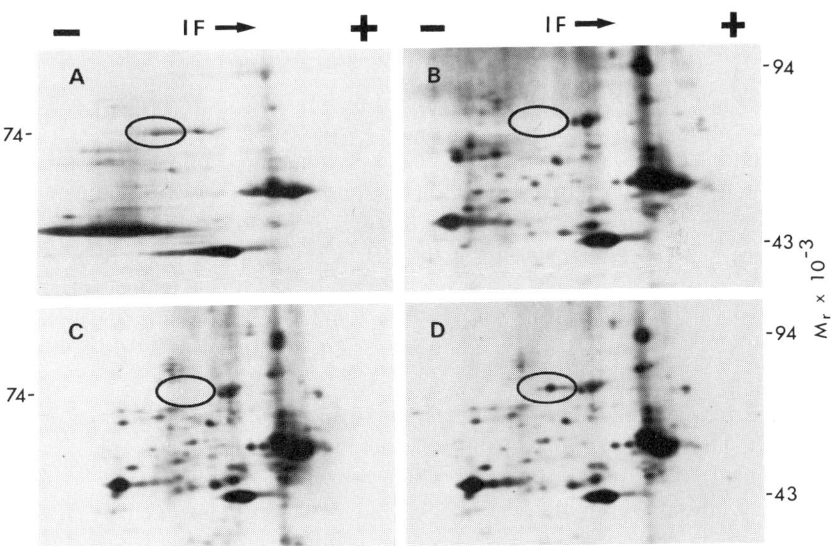

Fig. 3. Developmental change in the inducibility of the 74K brain heat shock protein. Following elevation of body temperature from 32 to 38 or 43°C, polysomes were isolated from brains of 2- and 7-day-old rat pups and translated in an mRNA-dependent reticulocyte cell-free system with [^{35}S]methionine. Labeled translation products were then analyzed as in Fig. 2. The position of the 74K heat shock protein is encircled. Two-day-old rat pups: (A) 38°C hyperthermia, (B) control, Seven-day-old rat pups: (C) 38°C hyperthermia, (D) 43°C hyperthermia. (From Brown, 1983.)

postnatal growth, the 74K protein was not apparent in the translation products of brain polysomes isolated from animals subjected to a 38°C heat shock (Fig. 3C); however, it was apparent when body temperature was increased to 43°C (Fig. 3D). Similar results were observed for other body organs such as heart and kidney. These results suggest that organs of the newborn mammal are particularly sensitive to induction of the 74K heat shock protein. As early postnatal development proceeds, higher body temperatures are required to induce synthesis of this protein.

An interesting parallel is apparent between the sensitivity of the 2-day-old rat pup to induction of the heat shock protein and the sensitivity of the neonatal rat to hyperthermia-induced seizures. In the 2-day-old rat, elevation of body temperature from 32 to 37°C has been reported to induce seizure activity, whereas in 7-day-old pups, seizures do not occur until body temperature is increased to 43°C (Holtzman *et al.*, 1981).

Elevation of maternal body temperature has a marked effect on protein synthesis in fetal organs. As shown in Fig. 4, elevation of body temperature of a pregnant female rabbit induced the synthesis of the 74K heat shock protein by polysomes isolated from the maternal brain and from fetal organs such as brain, heart, and liver.

VI. HEAT SHOCK PROTEIN IN SPECIFIC CELLULAR SYSTEMS IN BRAIN

Our initial studies on the effect of hyperthermia on protein synthesis in the rabbit brain were carried out on total cerebral hemispheres, a tissue in which functional and topographical heterogeneity are great. Recently we have examined more specific cellular systems in the nervous system such as the retina of the eye. This has permitted the examination of the question of whether newly synthesized heat shock protein in retinal ganglion cells is transported down the optic nerve by axonal transport to synaptic termini in the visual projection areas of the brain.

In the retina a transient disaggregation of polysomes was observed which parallels the rise and fall of drug-induced hyperthermia (Clark and Brown, 1982a). During this phase of overall reduction in retinal protein synthesis, induction of synthesis of a 74K heat shock protein was demonstrated by *in vivo*-labeling studies and analysis of cell-free translation products of isolated retinal polysomes (Fig. 5). Induction of synthesis of a 95K protein was also observed. This protein did not appear to be induced in the cerebral hemispheres or the cerebellum.

We have utilized the rabbit visual system to investigate the fate of heat shock proteins in neurons following their synthesis. Neurons are very active in syn-

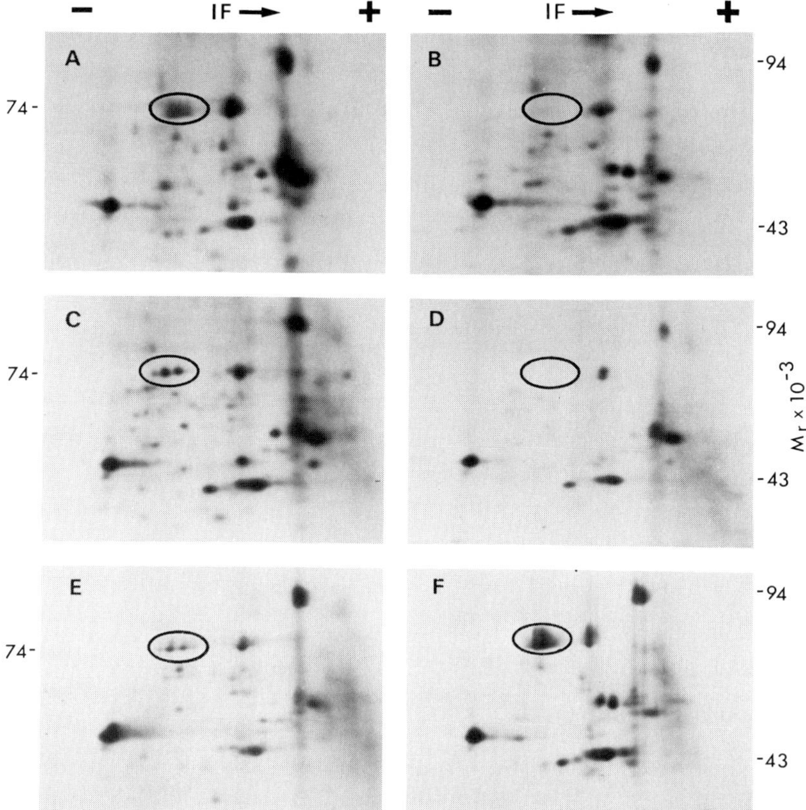

Fig. 4. Induction of a 74K heat shock protein in fetal brain and other organs of the rabbit following elevation of maternal body temperature. An mRNA-dependent reticulocyte cell-free system was programmed with polysomes isolated from fetal organs and maternal brain following elevation of maternal body temperature by 3°C. Labeled translation products were analyzed as in Fig. 2. The position of the 74K heat shock protein is encircled. Hyperthermic treatment: (A) fetal brain, (C) maternal brain, (E) fetal liver, (F) fetal heart. Control: (B) fetal brain, (D) maternal brain. (From Brown, 1983.)

thesizing classes of proteins which are transported down the axon to the synaptic terminus. Specific classes of proteins move down the axon in waves at characteristic transport rates. For example, glycoproteins move by fast axonal transport (400 mm/day), whereas cytoskeletal proteins such as microtubular and neurofilament proteins move by slow axonal transport (0.2 mm/day) (Willard *et al.*, 1974). Each transport group is composed of proteins which associate to form specific cellular structures or perform similar functions. This has been termed the structural hypothesis of axonal transport (Tytell *et al.*, 1981).

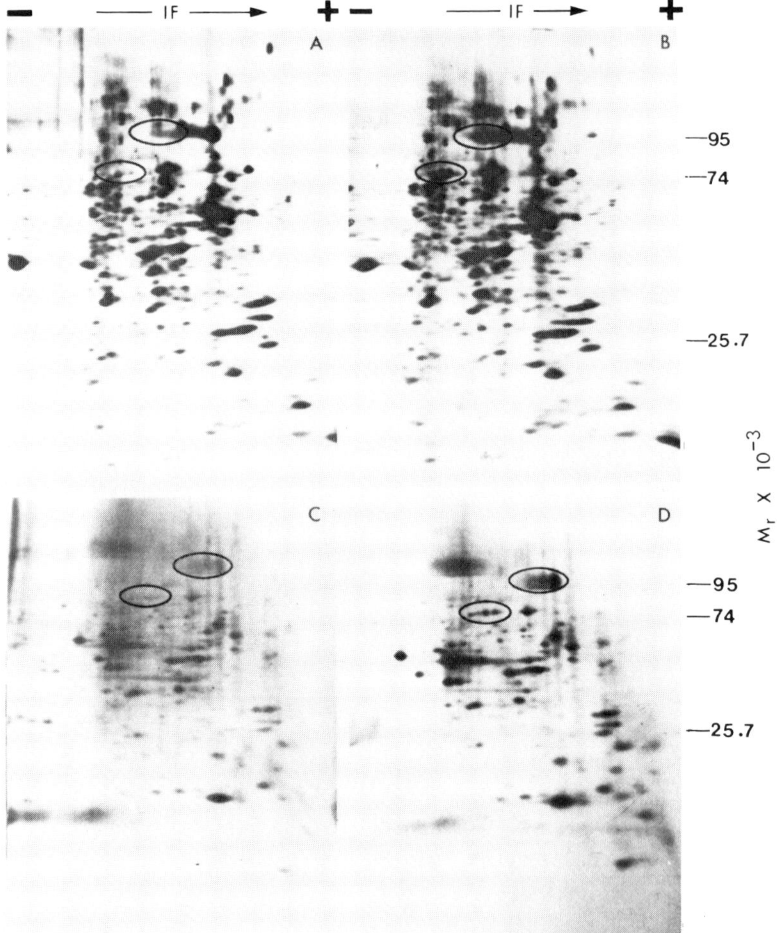

Fig. 5. Induction of heat shock proteins in the rabbit retina after drug-induced hyperthermia. Top: *In vivo*-labeled proteins. Rabbits were injected intravitreally for a 20-min pulse with [^{35}S]methionine 1 hr following intravenous injection of LSD at 100 μg/kg body mass (controls received saline). Labeled retinal proteins in a postmitochondrial supernatant fraction were analyzed as in Fig. 2. (A) Saline control, (B) drug-induced hyperthermia. Bottom: Cell-free translation products. Retinal polysomes were isolated from animals 1 hr following drug injection and translated in an mRNA dependent reticulocyte cell-free system with [^{35}S]methionine. Labeled translation products were analyzed as in Fig. 2. The positions of the 74K and 95K heat shock proteins are encircled. (C) Saline control, (D) drug-induced hyperthermia. (From Clark and Brown, 1982a.)

Our results indicate that the induced 74K heat shock protein but not the 95K protein undergoes axonal transport from neuronal cell bodies in the retina (Clark and Brown, 1985). The 74K protein moves down the optic nerve from retinal ganglion cells, taking 14–30 days to reach synaptic termini in the visual projection areas of the brain. This places the 74K retinal heat shock protein in a class of slowly transported axonal proteins, which includes proteins which are primarily associated with elements of the cytoskeleton (microtubular, microfilament, and neurofilament proteins). This may imply that the 74K heat shock protein in neurons performs a function related to the cytoskeleton, perhaps providing stability during periods of stress.

The retinal system was also useful since it facilitated *in vitro* experiments designed to investigate aspects of the induction of the 74K protein. Elevation of the temperature of incubation of the isolated retina induced synthesis of the 74K heat shock protein, but addition of LSD did not (Clark and Brown, 1982b). Subjecting the isolated retina to heat shock in the presence of an RNA synthesis inhibitor demonstrated that induction of the heat shock protein was independent on the synthesis of new RNA. A portion of the newly synthesized 74K protein moved rapidly into retinal nuclei; however, the 95K protein remained in the cytoplasm.

The cerebral microvascular system is a critical cellular control system in the brain since it is the site of the blood–brain barrier, which regulates the transit of molecules between brain cells and the circulatory system. Protein synthesis in the cerebral microvascular (i.e., capillary) system was inhibited following elevation of body temperature by 2–2.5°C (Inasi and Brown, 1982). Analysis of *in vivo*-labeled proteins demonstrated that during this phase of overall decrease in protein synthesis, hyperthermia induced a marked increase in synthesis of a 74K heat shock protein.

VII. INDUCTION OF mRNA CODING FOR HEAT SHOCK PROTEIN

Changes in the expression of the 74K heat shock gene in maternal and fetal rabbit tissues have been analyzed by Northern blot analysis (Brown *et al.*, 1983). The recombinant DNA probe was derived from the coding region of the major heat shock gene. After a $1\frac{1}{2}$-hr hyperthermic episode in which body temperature increased by 3°C, a heat shock mRNA species approximately 2.7 kilobases (kb) in size was induced in massive amounts in fetal and maternal brain and kidney, with fetal kidney demonstrating the greatest response. Following a $6\frac{1}{2}$-hr recovery period in which body temperature returned to normal, this induced heat shock mRNA virtually disappeared in adult brain and kidney; however, in fetal tissues, elevated levels of heat shock mRNA persisted. Within 24 hr the induced mRNA species was no longer apparent in maternal or fetal tissues. These experiments suggest that *in vivo* induction of the major heat shock gene in intact organs is controlled at the level of transcription.

Our results also revealed that in the absence of hyperthermia, a constitutively expressed heat shock mRNA species of slightly smaller size (i.e., 2.5 kb) was present in all fetal and adult tissue examined. Synthesis of this heat shock-related mRNA species was not affected by hyperthermia. These results suggest that the rabbit genome may contain a family of 74K heat shock genes. Some members of this gene family (''cognates'') are expressed in unstressed organs, while other members are inducible by hyperthermia.

VIII. CONCLUSIONS

The studies discussed in this review indicate that elevation of body temperature to levels similar to that attained during fever reactions induces the synthesis of heat shock protein in intact mammalian organs. In tissue culture systems and unicellular organisms, an increase of ambient temperature induces the synthesis of a set of heat shock proteins (Schlesinger *et al.*, 1982); however, in intact organs a marked increase in synthesis of only a 74K protein is observed after elevation of body temperature by 2 to 3°C. Retinal tissue was an exception in that induction of a 95K protein was also observed. Since mammalian cell lines contain genes coding for a set of heat shock proteins (Kelley and Schlesinger, 1978; Wang *et al.*, 1981; Hightower and White, 1981), it is of interest that primarily only one of the major heat shock proteins is induced in intact organs. It would appear that constraints exist on the induction of the full set of heat shock proteins in intact organs at physiologically relevant temperatures. This may suggest an adaptation of the 74K heat shock protein to a functional role in the cellular reaction of intact organ systems to thermal stress conditions which are experienced during fever.

Organs of the newborn mammal appear to be particularly sensitive to induction of the 74K heat shock protein. As early postnatal development proceeds, higher body temperatures are required to induce synthesis of the 74K protein. The developmental change in the inducibility of the heat shock protein parallels developmental changes in the sensitivity of the rat to hyperthermia-induced seizures; that is, in the newborn animal less increase in body temperature is required to induce seizures.

After hyperthermic treatment, transcription of the 74K heat shock gene is induced in adult and fetal brain and kidney. The rate of appearance and subsequent degradation of this induced mRNA species in intact organs is similar to that seen in mammalian cells grown in tissue culture and reflects the transient nature of the heat shock response. One of the objectives of our studies on mammalian organs was to investigate the expression of ''cognate'' genes in addition to inducible heat shock genes. An advantage of studies on the intact animal over experiments using cells in culture is that one can be more certain that the control cells have not been inadvertently shocked prior to assay. Our observations suggest that transcripts of ''cognate'' heat shock genes are present in all tissues of the adult and fetal rabbit which were examined.

Factors such as amino acid analogs, heavy metal ions, sulfhydryl reagents, and chelating agents have been reported to induce proteins similar in molecular weight to heat shock proteins (Ashburner and Bonner, 1979; Kelley and Schlesinger, 1978; Hightower and White, 1981). We have recently demonstrated that perturbations other than hyperthermia can induce synthesis of a major heat shock protein in organs of the intact mammal. A toxic agent such as sodium arsenite, which does not elevate body temperature of rabbits at the dosage employed, induced synthesis of a 74K protein in the liver, heart, and kidney within 1 hr of its intravenous injection (Brown and Rush, 1984). Organ-specific differences in the ability of sodium arsenite to induce synthesis of the 74K protein were apparent since this protein was not detected in the translation product of brain polysomes. Sodium arsenite rapidly accumulates in the liver and kidney; however, there is a very slow passage of the compound through the blood–brain barrier into the brain (Vahter *et al.*, 1982). A trauma-induced protein of 71K has been reported in incubated organ slices and in tissues of the hyperthermic rat (White, 1980, 1981; Currie and White, 1981; Hightower and White, 1981). It is possible that a common protein may be induced by diverse perturbations in intact mammalian organs as a general cellular response to stress. The observations cited in this review suggest that the heat shock response is a physiologically relevant phenomenon in intact organs of the adult and fetal mammal and that *in vivo* induction of the major heat shock gene is at the level of transcription.

Acknowledgments

I thank research technician Sheila Rush, graduate students Bruce Clark and John Heikkila, and postdoctoral fellow James Cosgrove for their contributions to experiments cited in this review article. These studies were supported by grants from the Medical Research Council of Canada.

REFERENCES

Ashburner, M., and Bonner, J. J. (1979). The induction of gene activity in *Drosophila* by heat shock. *Cell* **17,** 241–254.

Brown, I. R. (1983). Hyperthermia induces the synthesis of a heat shock protein by polysomes isolated from the fetal and neonatal mammalian brain. *J. Neurochem.* **40,** 1490–1493.

Brown, I. R., and Rush, S. J. (1984). Induction of a "stress" protein in intact mammalian organs after the intravenous administration of sodium arsenite. *Biochem. Biophys. Res. Commun.* **120,** 150–155.

Brown, I. R., Cosgrove, J. W., and Clark, B. D. (1982a). Physiologically relevant increases in body temperature induce the synthesis of a heat-shock protein in mammalian brain and other organs. *In* "Heat Shock, from Bacteria to Man" (M. J. Schlesinger, M. Ashburner, and A. Tissières, eds.), pp. 361–367. Cold Spring Harbor Lab., Cold Spring Harbor, New York.

Brown, I. R., Heikkila, J. J., and Cosgrove, J. W. (1982b). Analysis of protein synthesis in the mammalian brain using LSD and hyperthermia as experimental probes. *In* "Molecular Approaches to Neurobiology" (I. R. Brown, ed.), pp. 221–253. Academic Press, New York.

Brown, I. R., Lowe, D. G., and Moran, L. A. (1983). Effect of hyperthermia on the expression of a heat shock gene in intact mammalian organs. *J. Cell Biol.* **97,** 152a.

Clark, B. D., and Brown, I. R. (1982a). Protein synthesis in the mammalian retina following the intravenous administration of LSD. *Brain Res.* **247,** 97–104.

Clark, B. D., and Brown, I. R. (1982b). Effects of LSD and hyperthermia on gene expression in the retina. *In* "Basic and Clinical Aspects of Molecular Neurobiology" (A. M. Giuffrida Stella, G. Gombos, G. Benzi, and H. S. Bachelard, eds.), p. 274. Fond. Int. Menarini, Milan.

Clark, B. D., and Brown, I. R. (1985). Axonal transport of a heat shock protein in the rabbit visual system. *Proc. Natl. Acad. Sci. U.S.A.* (in press).

Cosgrove, J. W., and Brown, I. R. (1983). Heat shock protein in mammalian brain and other organs after a physiologically relevant increase in body temperature induced by D-lysergic acid diethylamide. *Proc. Natl. Acad. Sci. U.S.A.* **80,** 569–573.

Cosgrove, J. W., and Brown, I. R. (1984). Effect of intravenous administration of LSD on initiation of protein synthesis in a cell-free system derived from brain. *J. Neurochem.* **42,** 1420–1426.

Cosgrove, J. W., Clark, B. D., and Brown, I. R. (1981). Effect of intravenous administration of D-lysergic acid diethylamide on subsequent protein synthesis in a cell-free system derived from brain. *J. Neurochem.* **36,** 1037–1045.

Currie, R. W., and White, F. P. (1981). Trauma-induced protein in rat tissues: A physiological role for a heat shock protein. *Science* **214,** 72–73.

Freedman, M. S., Clark, B. D., Cruz, T. F., Gurd, J. W., and Brown, I. R. (1981). Selective effects of LSD and hyperthermia on the synthesis of synaptic proteins and glycoproteins. *Brain Res.* **207,** 129–145.

Heikkila, J. J., and Brown, I. R. (1979a). Disaggregation of brain polysomes after LSD *in vivo*: Involvement of LSD-induced hyperthermia. *Neurochem. Res.* **4,** 763–776.

Heikkila, J. J., and Brown, I. R. (1979b). Hyperthermia and disaggregation of brain polysomes induced by bacterial pyrogen. *Life Sci.* **25,** 347–352.

Heikkila, J. J., and Brown, I. R. (1981). Comparison of the effect of intravenous administration of LSD on free and membrane-bound polysomes in the rabbit brain. *J. Neurochem.* **36,** 1219–1228.

Heikkila, J. J., Holbrook, L. A., and Brown, I. R. (1978). Stress-accentuation of the LSD-induced disaggregation of brain polysomes. *Life Sci.* **22,** 757–766.

Heikkila, J. J., Holbrook, L. A., and Brown, I. R. (1979). Disaggregation of polysomes in fetal organs and maternal brain after administration of D-lysergic acid diethylamide *in vivo*. *J. Neurochem.* **32,** 1793–1799.

Hightower, L. E., and White, R. P. (1981). Cellular responses to stress: Comparison of a family of 71–73 kilodalton proteins rapidly synthesized in rat tissue slices and canavanine-treated cells in culture. *J. Cell Physiol.* **108,** 261–275.

Holbrook, L. A., and Brown, I. R. (1976). Disaggregation of brain polysomes after administration of D-lysergic acid diethylamide (LSD) *in vivo*. *J. Neurochem.* **27,** 77–82.

Holbrook, L. A., and Brown, I. R. (1977a). Disaggregation of brain polysomes after D-lysergic acid diethylamide administration *in vivo*: Mechanisms and effect of age and environment. *J. Neurochem.* **29,** 461–467.

Holbrook, L. A., and Brown, I. R. (1977b). Antipsychotic drugs block LSD-induced disaggregation of brain polysomes. *Life Sci.* **21,** 1037–1044.

Holtzman, D., Obana, K., and Olson, J. (1981). Hyperthermia-induced seizures in the rat pup: A model for febrile convulsions in children. *Science* **213,** 1034–1036.

Inasi, B. S., and Brown, I. R. (1982). Synthesis of a heat shock protein in the microvascular system of the rabbit brain following elevation of body temperature. *Biochem. Biophys. Res. Commun.* **106,** 881–887.

Kalant, H., and Khanna, J. (1975). *In* "Principles of Medical Pharmacology" (P. Seeman and E. Sellers, eds.), pp. 279–284. Univ. of Toronto Press, Toronto.

Kelley, P. M., and Schlesinger, M. J. (1978). The effect of amino acid analogues and heat shock on gene expression in chicken embryo fibroblasts. *Cell* **15,** 1277–1286.

Mahony, J. B., and Brown, I. R. (1979). Fate of brain mRNA following disaggregation of brain polysomes after administration of LSD *in vivo*. *Biochim. Biophys. Acta* **565,** 161–172.

Rossi, G. V. (1971). LSD: A pharmacological profile. *Am. J. Pharm.* **2,** 117–123.

Schlesinger, M. J., Ashburner, M., and Tissières, A., eds. (1982). "Heat shock from bacteria to man." Cold Spring Harbor Lab., Cold Spring Harbor, New York.

Tytell, M., Balck, M. M., Garner, J. A., and Lasek, R. J. (1981). Axonal transport: Each major rate component reflects the movement of distinct macromolecular complexes. *Science* **214,** 179–181.

Vahter, M., Marafante, E., Lindgren, A., and Deneker, L. (1982). Tissue distribution and subcellular binding of arsenic in marmoset monkeys after injection of [^{74}As]arsenite. *Arch. Toxicol.* **51,** 65–77.

Wang, C., Gomer, R. H., and Lazarides, E. (1981). Heat shock proteins are methylated in avian and mammalian cells. *Proc. Natl. Acad. Sci. U.S.A.* **78,** 3531–3535.

White, F. P. (1980). Differences in proteins synthesized *in vivo* and *in vitro* by cells associated with the cerebral microvasculature: A protein synthesized in response to trauma? *Neuroscience* **5,** 1793–1799.

White, F. P. (1981). The induction of "stress" proteins in organ slices from brain, heart and lung as a function of postnatal development. *J. Neurosci.* **1,** 1312–1319.

Willard, M., Cowan, W. M., and Vagelos, P. R. (1974). The polypeptide composition of intra-axonally transported proteins: evidence for four transport velocities. *Proc. Natl. Acad. Sci. U.S.A.* **71,** 2183–2187.

12

Thermotolerance in Mammalian Cells: A Possible Role for Heat Shock Proteins

GLORIA C. LI AND ANDREI LASZLO

I. INTRODUCTION

One of the more interesting aspects of thermal biology in mammalian systems is the response of heated cells to a subsequent heat challenge. Gerner and Schneider (1975) first reported and Henle and Leeper (1976) confirmed that

Changes in Eukaryotic Gene Expression
in Response to Environmental Stress

ISBN 0-12-066290-6

mammalian cells exposed to a nonlethal heat shock acquired transient resistance to subsequent exposures at elevated temperatures. This phenomenon has been termed thermotolerance (Henle and Dethlefsen, 1978). Thermotolerance is induced by a short heat treatment at temperatures above 43°C, provided that this exposure is followed by incubation at 37°C. It can also be induced during continuous heating below 43°C. Several excellent reviews discuss this in considerable detail (Henle and Dethlefsen, 1978; Hahn, 1982; Dewey *et al.*, 1980). The molecular mechanism(s) for the development of thermotolerance has not been well understood, but the experimental evidence suggests that protein synthesis plays a role in its manifestation (Henle and Leeper, 1982; Li, *et al.*, 1982a).

In recent years it has been shown that heat or other environmental stresses (chemical or physical challenges) induce the enhanced synthesis of a family of proteins, the so-called heat shock proteins (hsp's), in a wide variety of cells varying from *Drosophila* to yeast to mammalian systems (Kelly and Schlesinger, 1978; Miller *et al.*, 1979; Ashburner and Bonner, 1979; Loomis and Wheeler, 1980; Johnston *et al.*, 1980; Li and Werb, 1982; Li *et al.*, 1982b; Schlesinger *et al.*, 1982). It has been suggested that hsp's may perform a function that is related to heat-induced thermal resistance in *Drosophila* and yeast (Mitchell *et al.*, 1979; McAlister and Finkelstein, 1980). In mammalian cells, because of the important role that protein synthesis seems to play in the development of thermotolerance, many investigators have performed experiments to determine the relationship of the appearance of thermotolerance and the enhanced synthesis of hsp's (Li *et al.*, 1982b; Li and Werb, 1982; Landry *et al.*, 1982; Subjeck *et al.*, 1982; Tomasovic *et al.*, 1983). These studies have shown that there is temporal correlation between the induction of increased synthesis of hsp's and the acquisition of thermotolerance. The positive correlations suggest that hsp's may play a pivotal role in protecting mammalian cells from thermal damage. Thus, while the exact function of hsp's is unclear at present, the phenomenon of thermotolerance may provide us with a tool to determine the function(s) of hsp's.

II. THERMOTOLERANCE IN MAMMALIAN SYSTEMS

Development of thermotolerance in plateau-phase Chinese hamster HA-1 cells is shown in Fig. 1 (Li and Hahn, 1980). In these experiments, monolayers of cells were first exposed to 43°C for 30 min; some of the cells were challenged immediately by a second treatment consisting of graded periods at 43°C. At the end of the second heat treatment, cell survivals were assayed by the standard colony formation technique (Puck and Marcus, 1956). The other cells were returned to 37°C incubation for 4, 6, 8, or 25 hr before the second 43°C heating. The parameter varied in these experiments, in addition to incubation time at

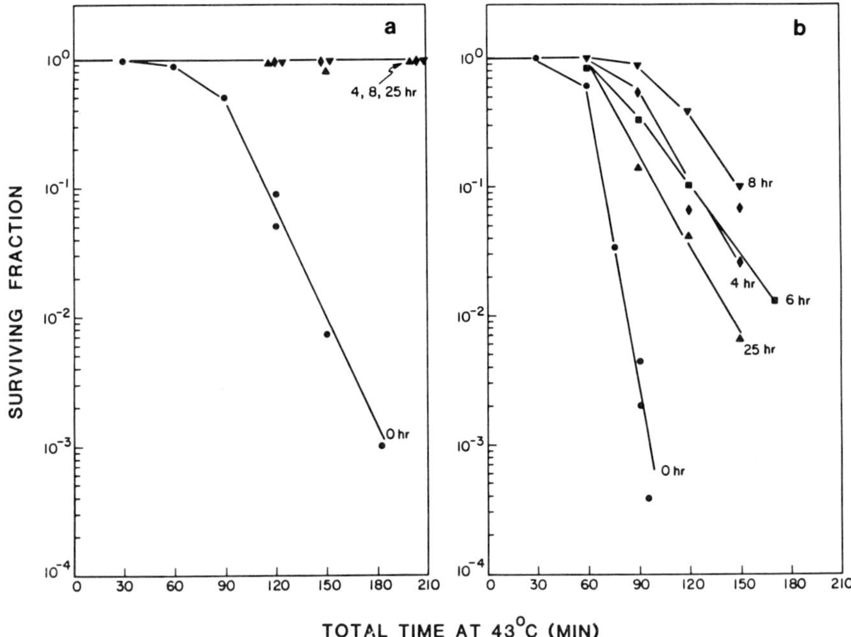

Fig. 1. Development of thermotolerance in plateau-phase Chinese hamster HA-1 cells. Cells were initially exposed to 43°C for 30 min. Some of these cells were challenged immediately by a second heat treatment consisting of graded periods at 43°C. The other cells were incubated at 37°C for 4, 8, 16, or approximately 24 hr, and the cells were then given the second exposure at 43°C. The survival after the first treatment was 85–100%. (a) The initial treatment, the 37°C incubation time, and the second heat treatment were all performed in full medium supplemented with 15% fetal calf serum; (b) the initial treatment, the 37°C incubation time, and the second heat treatment were all performed in Hanks' balanced salt solution.

37°C, was the nutrient environment: during the initial heating, during the development period at 37°C, and during the second heating. These results showed that thermotolerance was induced independently of the nutrient conditions of the first treatment. The time course of decay of thermotolerance was largely determined by the nutrient conditions of the initial as well as the second heat treatment, although the milieu during the development of thermotolerance was also of some importance. Survival at the time of maximum expression of development of thermotolerance depended primarily on the duration and the temperature of the initial treatment (Li and Hahn, 1980).

Thermotolerance can be demonstrated in a somewhat different split dose experiment (Fig. 2). In these experiments, monolayers of cells were first exposed to a predetermined temperature (e.g., 41°C for 1 hr or 45°C for 20 min). Some of the cells were challenged immediately by a second heat treatment at 45°C for 45

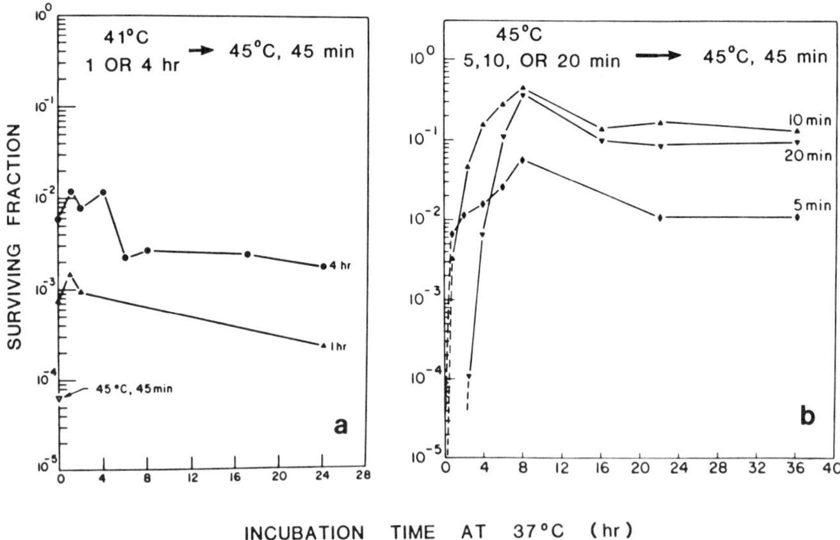

Fig. 2. Kinetics of induction of thermotolerance in plateau-phase Chinese hamster HA-1 cells. Cells were initially exposed to designated temperatures 41°C (a) or 45°C (b) for the durations noted. In each case, the second heat treatment was 45°C for 45 min. The surviving fractions are plotted as a function of the duration of 37°C incubation between the first and second heat treatments. Control cells were exposed to only one treatment at 45°C for 45 min, and the kinetics of development of thermotolerance were measured by the gradual increases in survival above the control values ($4–8 \times 10^{-5}$).

min. The other cells were returned to 37°C incubation for various lengths of time. The cells were then given a similar second treatment at 45°C for 45 min and survival was determined. Cell survival after the two heat treatments was then plotted as a function of the length of 37°C incubation between the two heat treatments. The increase in survival, when compared to the survival of control cells (no preheat treatment), is largely a manifestation of thermotolerance. The increase in cell survival values and the gradual return to the control survival values were used to measure the kinetics of the development and the decay of thermotolerance, respectively; one set of examples is shown in Fig. 2. Note that survival values at 0 time are at (or near) maximum if the initial treatment is 41°C. This is in contrast with the survival kinetics following the initial treatment at 45°C for which survival values at 0 time are minimal; an incubation interval at 37°C is required before thermotolerance manifests itself. The effects of varying the temperature (41–47°C) and the duration of the first heat treatment on the subsequent expression of thermotolerance were studied by Li *et al.* (1982c). These sets of experiments indicated that a temperature of 43°C or higher inhibited the development of thermotolerance during the first heat exposure. In contrast, if the initial exposure was at 41°C, thermotolerance was almost fully expressed by the end of the initial treatment.

On the basis of these and other data, Li and Hahn (1980) proposed an operational model of thermotolerance. The authors suggested that thermotolerance can be divided into three complementary and sometimes competing processes: an initial event (''trigger''), the expression of resistance (''development''), and the gradual disappearance of resistance (''decay''). Each of these components may have its own temperature dependence as well as dependence on other factors, such as pH and nutrients. Conceptually, the three components of thermotolerance may be considered to be independent processes. However, independent measurement of each component is not always possible with the colony formation assay. For example, to measure the kinetics of triggering, development of thermotolerance must be permitted to proceed.

III. CORRELATION BETWEEN SYNTHESIS OF HEAT SHOCK PROTEINS AND DEVELOPMENT OF THERMOTOLERANCE

Our working hypothesis over the past several years has been that the elevated levels of some or one of the hsp's may play a role in conferring heat resistance to cells. Li and Werb's work has shown that, indeed, in Chinese hamster fibroblasts there is a good temporal correlation between the induction of hsp's and the development of thermotolerance (Li and Werb, 1982; Li *et al.*, 1982b). Several lines of evidence imply that hsp's may play a pivotal role in the development of thermotolerance: (i) heat treatment enhances the synthesis of hsp's and induces a transient thermotolerance at 41–46°C, under conditions in which the initial heat treatment either did not suppress total protein synthesis or drastically inhibited it; (ii) when thermotolerance is fully developed, the rates of synthesis of most hsp's return to control values; (iii) the delay of onset of thermotolerance correlates well with the delay in the induction of enhanced synthesis of hsp's (iv) there is good correlation between the persistence of some hsp's and thermotolerance; and (v) agents known to induce thermotolerance also induce synthesis of hsp's, and agents known to induce synthesis of hsp's induce thermotolerance. Therefore, it seems reasonable to hypothesize that hsp's may play a role in providing cells with an additional measure of heat resistance. However, until the functions of hsp's are determined and the basic mechanisms of thermal killing are well understood, a causal relationship cannot be claimed.

A good correlation between the induction of hsp's and the development of thermotolerance has been observed in many different mammalian cell systems that have been examined for both phenomena in our laboratory (Table I). Similar studies by other investigators also showed a temporal correlation between the

TABLE I

Relationship of Development of Thermotolerance and Induced Synthesis of Heat Shock Proteins

Cell line	Organism of origin	Cell type	Thermotolerance development	Increased synthesis of HSP
HA-1	Chinese hamster	Fibroblast	Yes	Yes
BHK	Syrian hamster	Fibroblast	Yes	Yes
SQ-1	Mouse	Squamous cell carcinoma	Yes	Yes
M 8013	Mouse	Adenocarcinoma	Yes	Yes
NNN	Rat	Normal revertant of the B-31 Rous sarcoma virus-transformed cell line	Yes	Yes
B-31	Rat	Rous sarcoma virus-transformed fibroblast cell line	Yes	Yes

induction of hsp's and the acquisition of thermotolerance in other mammalian cell lines (Subjeck *et al.*, 1982; Landry *et al.*, 1982; Tomasovic *et al.*, 1983).

IV. KINETICS OF HEAT SHOCK PROTEIN SYNTHESIS DURING DEVELOPMENT OF THERMOTOLERANCE: EFFECTS OF TEMPERATURE AND DURATION OF INITIAL HEAT TREATMENT

We have examined the effects of temperature and duration of initial heat treatment on the time patterns of hsp synthesis (Li, 1983a). Plateau-phase Chinese hamster HA-1 cells were first exposed to elevated temperature, for example, 41 or 45°C, for various lengths of time. After the initial heating, cells were incubated at 37°C (the normal growth temperature) for 0–24 hr and then pulse labeled with [^{35}S]methionine for 1 hr at 37°C. Electrophoretically separated proteins were quantified by densitometry of autoradiograms. Areas under each protein peak were used as a measure of the amount of labeled protein present (Figs. 3a,b). In parallel experiments, the effects of temperature and duration of initial treatment on the induction of thermotolerance were examined (Figs. 2a,b).

Similar to earlier reports (Li and Werb, 1982), heat treatment (41 or 45°C) induced an increase in production of three major hsp's (70K, 87K, and 110K). If the initial heat treatment was at 41°C (or 42°C), hsp synthesis was greatly enhanced while normal protein synthesis was not suppressed. The rate of hsp synthesis was at its maximum value immediately after 41°C heating; the rate of repression of hsp synthesis was not significantly influenced by the duration of the 41°C exposure. Initial exposure of cells to 45°C (or temperatures above 43°C) resulted in very different kinetics of hsp synthesis and recovery of normal protein synthesis. Immediately after heating, total protein synthesis was inhibited. A subsequent incubation at 37°C was required for induction of enhanced synthesis of hsp's and restoration of normal protein synthesis. For a specific temperature, the time interval at 37°C required for the rate of hsp synthesis to reach its maximum value depended on the duration of the initial treatment (Li, 1983a). Neither the rate of induction nor the rate of repression of hsp synthesis was affected significantly by the duration of 45°C exposure. The kinetics of enhanced synthesis of hsp's and subsequent repression of synthesis varied for different hsp's. For example, hsp70 was always the first protein to reach its maximum value for synthesis and the first to be repressed; hsp87 was always the last to reach its maximum value for synthesis and the last to be repressed. The magnitude of the total accumulation of each hsp depended on the duration of the priming treatment. Interestingly, when the kinetics of thermotolerance development and the kinetics of hsp synthesis were compared (Figs. 2 and 3), we found that the kinetics of the rate of induction of hsp synthesis mirrored the development, but not the decay, of thermotolerance.

V. RELATIONSHIP BETWEEN LEVELS OF HEAT SHOCK PROTEINS AND CELLULAR SURVIVAL DURING DECAY OF THERMOTOLERANCE

During induction and development of thermotolerance, both the survival response and the rate of hsp synthesis change very rapidly (Sections III and IV; G. C. Li and J. Y. Mak, unpublished observations). It is therefore difficult to establish precisely the quantitative relationship between cellular survival and the accumulated levels of hsp's. On the other hand, the decay of thermotolerance is a relatively slow and stable process; thermotolerance maintains itself at constant levels for about 24 hr, then decays slowly, and disappears only after 120 hr (Li, 1984b). Therefore, more precise measurements of cell survival and levels of hsp's can be undertaken at accurately known time points during the decay period. Plateau-phase Chinese hamster HA-1 cells were chosen for these sets of experiments because experiments with such cells were relatively free of complications introduced by cell progression and cell proliferation (Hahn and Little, 1972).

Monolayers of cells were first exposed to 45°C for 20 min and then incubated at 37°C for 6 hr. The 6-hr incubation was chosen because thermotolerance was

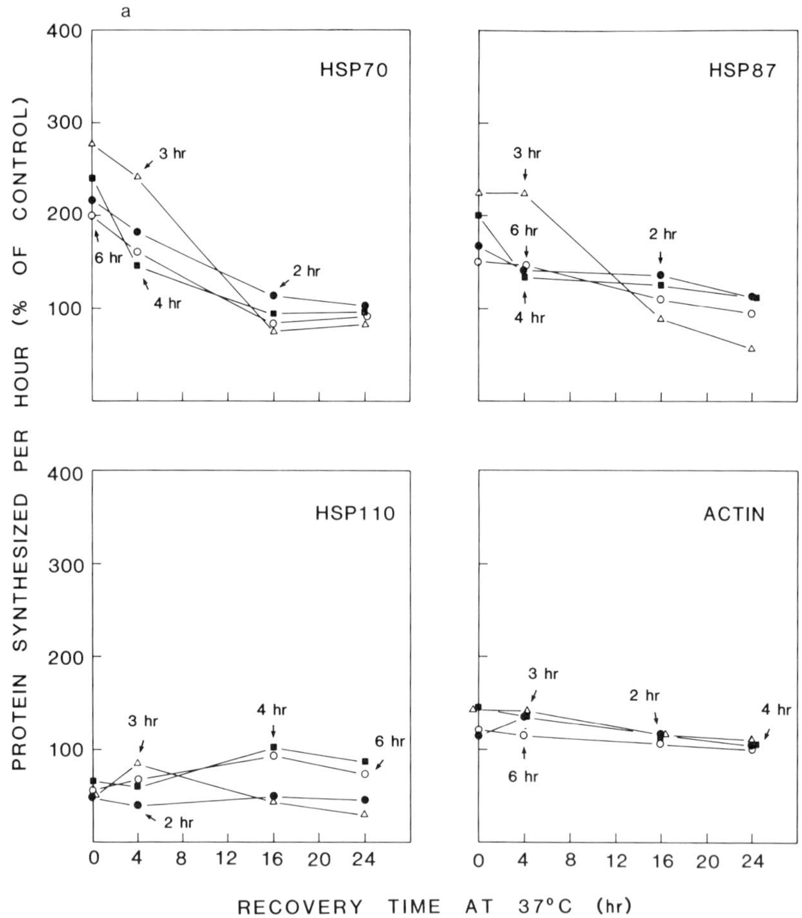

Fig. 3. Kinetics of heat shock protein synthesis in plateau-phase Chinese hamster HA-1 cells during development of thermotolerance: effects of temperature and duration of the initial heat treatment. Plateau-phase Chinese hamster HA-1 cells were first exposed to elevated temperature, for example, 41 or 45°C, for various lengths of time as indicated for each curve. After the initial heating, cells were incubated at 37°C (the normal growth temperature) for 0–24 hr and then pulse labeled with

then close to its maximum value. Cells were then labeled with [^{35}S]methionine for 2 hr at 37°C and incubated afterward at 37°C for 0–120 hr in complete medium containing nonradioactive methionine (medium was changed daily). Electrophoretically separated proteins were quantified by densitometry of autoradiograms (Fig. 4). The accumulation of individual proteins was calculated as

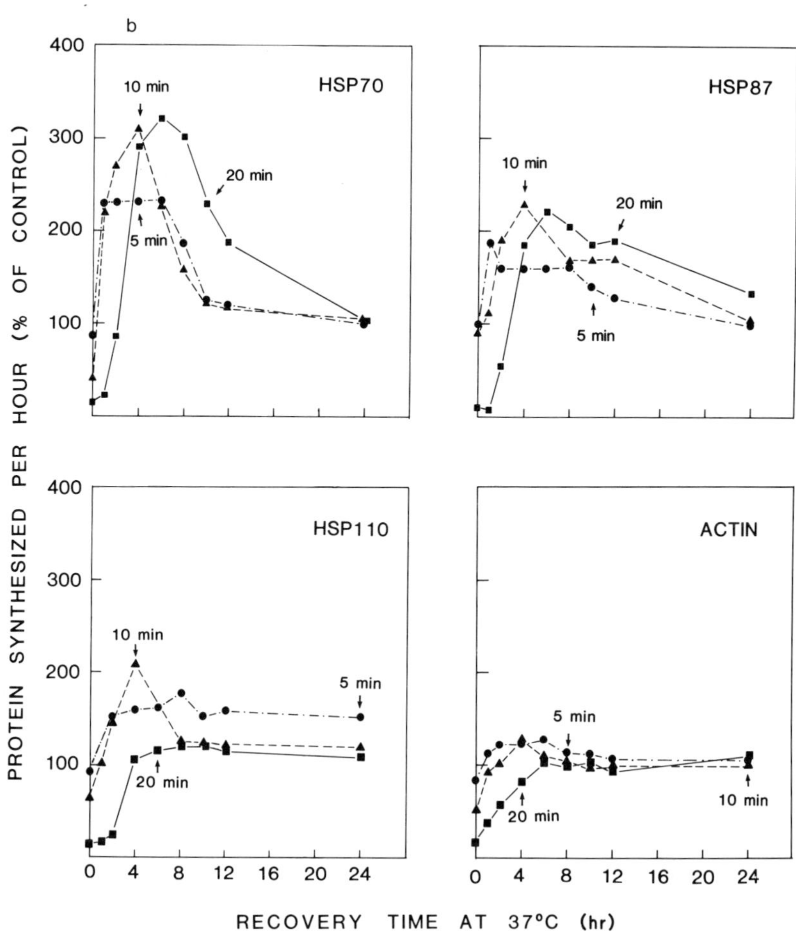

[^{35}S]methionine for 1 hr at 37°C. Electrophoretically separated proteins were quantified by densitometry of autoradiograms. Areas under each protein peak were used as a measure of the amount of labeled protein present. Proteins synthesized per hour (percentage of unheated control) after heat shock at 41°C (Fig. 3a) or 45°C (Fig. 3b) are plotted as a function of recovery time at 37°C (normal growth temperature).

the integrated area under an individual peak divided by the total area under the scan that included all protein bands in the gel lane. In parallel experiments, the induction and decay of thermotolerance were examined. HA-1 cells were first heated at 45°C for 20 min and then incubated at 37°C for 0–120 hr before a second treatment (45°C for 45 min), and survival was assayed. It is clearly

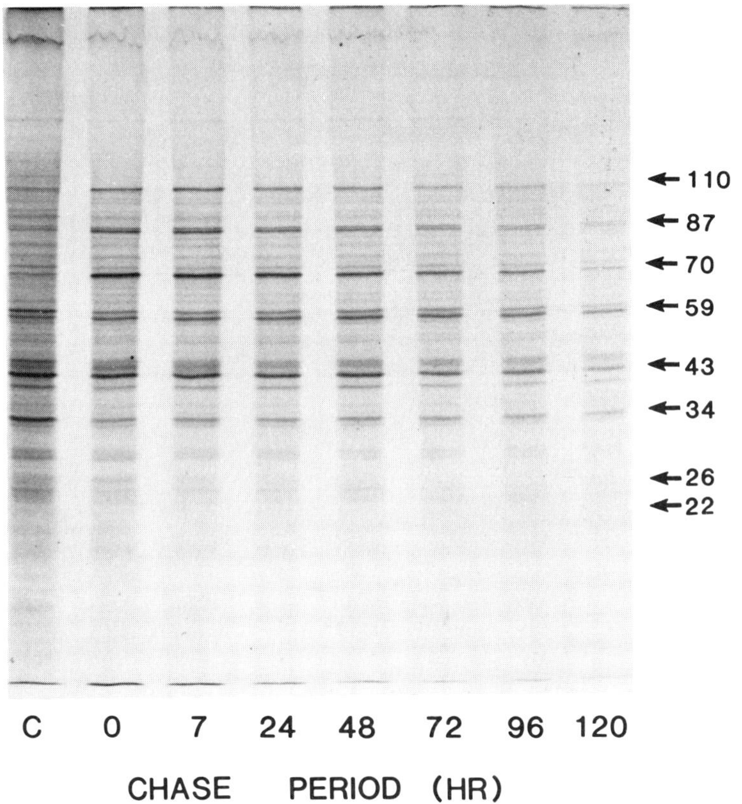

Fig. 4. Persistence of heat shock proteins in plateau-phase Chinese hamster HA-1 cells after 45°C heat shock. Cells were first exposed at 45°C for 20 min and then incubated at 37°C for 6 hr. Cells were labeled with [^{35}S]methionine for 2 hr at 37°C and then incubated at 37°C for additional chase periods of 0–120 hr in complete medium containing nonradioactive methionine. Lysates from equal numbers of cells were analyzed by SDS–gel electrophoresis. Unheated control (C) and chase period (in hours from 1 to 120) at 37°C after labeling. Molecular weights are shown on the right ($\times 10^{-3}$). The molecular weight of actin is indicated as 43,000.

shown in Fig. 5 that the rate of decay of thermotolerance was much slower than the rate of induction. The results showed that (i) at 6 hr after 45°C heat shock, synthesis of at least three proteins with molecular weights of 70,000 (hsp70), 87,000 (hsp87), and 110,000 (hsp110) were greatly enhanced over that seen in unheated controls; (ii) these induced hsp's were degraded gradually during the decay of thermotolerance, the half-times for the degradation of hsp70, hsp87, and hsp110 being between 72 and 96 hr; (iii) actin, a major cellular protein,

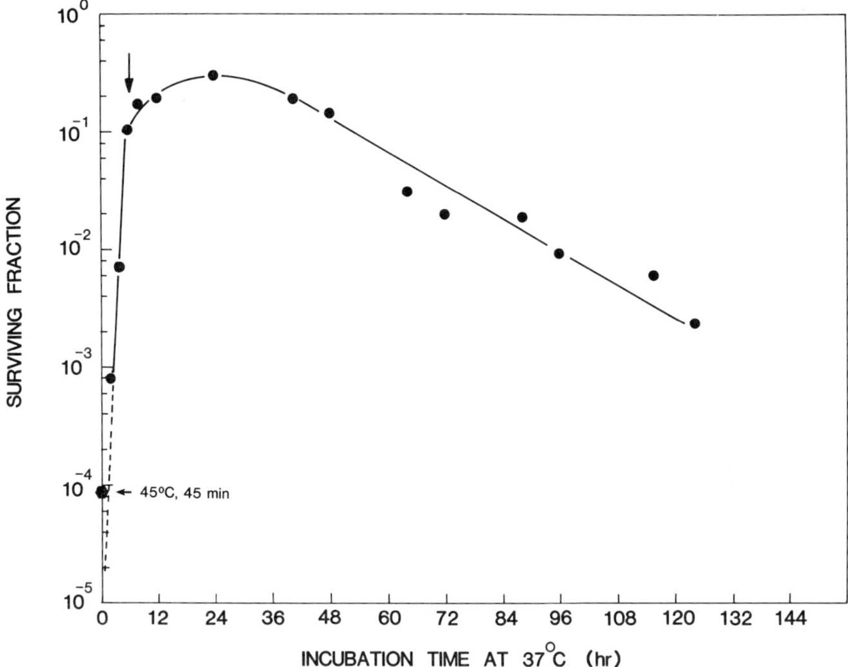

Fig. 5. Kinetics of induction and decay of thermotolerance in plateau-phase HA-1 cells. Cells were initially exposed to 45°C for 20 min followed by incubation at 37°C for 0–124 hr before a second treatment at 45°C for 45 min. The surviving fractions are plotted as a function of incubation time at 37°C between the first and second heat treatments. The control treatment (45°C, 45 min) is also indicated. The arrow indicates the start of pulse labeling as described in Fig. 4.

remained at a constant level in heat-shocked cells, and this level was indistinguishable from that of the unheated control; and (iv) in contrast to the lower eukaryotes, the levels of the lower-molecular-weight proteins (22,000–27,000) did not exceed control values.

The logarithm of survival values resulting from the second heat treatment (45°C, 45 min) delivered during the decay of thermotolerance was plotted as a function of residual radioactivity of [^{35}S]methionine incorporated into hsp's and actin as observed in the pulse–chase experiment. Comparison of results suggested that a decrease in the levels of high-molecular-weight hsp's (e.g., hsp70, hsp87, and hsp110) correlated well with the decrease of cell survival during the decay of thermotolerance (Fig. 6). In contrast, the levels of actin and lower-molecular-weight proteins (22,000–27,000) did not correlate with cell survival during the decay of thermotolerance (Li, 1984b).

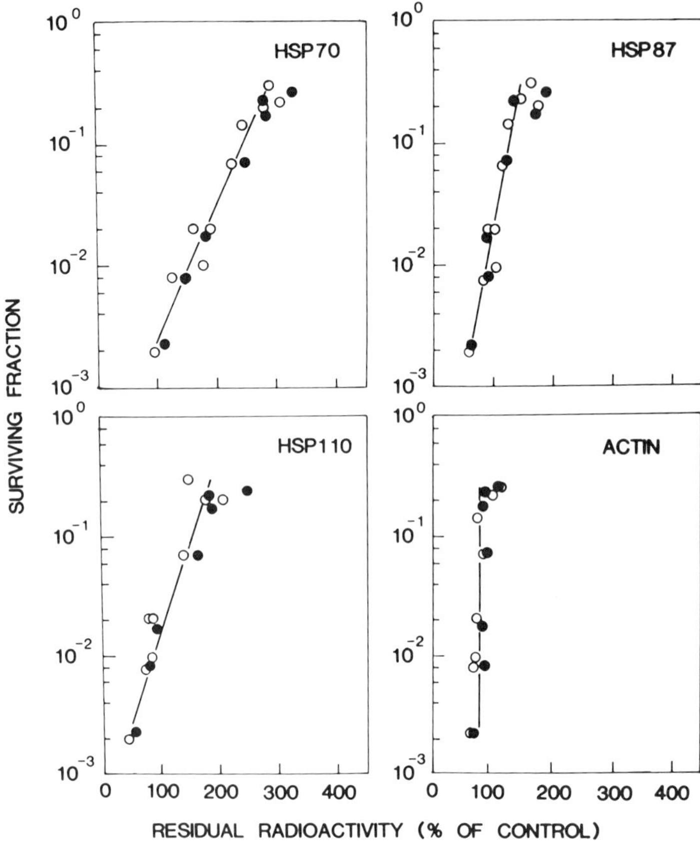

Fig. 6. Relationship between cell survival and residual levels of heat shock proteins during the decay of thermotolerance. At each time point during the decay of thermotolerance, the surviving fraction of thermotolerant cells after a 45°C, 45-min heat challenge (data from Fig. 5) is plotted against the residual radioactivity in hsp70, hsp87, hsp110, and actin. Open and closed circles represent data from two independent experiments.

VI. INDUCTION OF THERMOTOLERANCE AND ENHANCED SYNTHESIS OF HEAT SHOCK PROTEINS BY AGENTS OTHER THAN HEAT

A large number of physical or chemical agents other than heat have been shown to cause preferential synthesis of hsp's in various biological systems (Ashburner and Bonner, 1979; Johnston *et al.*, 1980; Kelley and Schlesinger, 1978; Li, 1983b; Hahn and Li, 1982). It has also been shown that various agents other than heat can induce thermotolerance in mammalian cells (Li and Hahn, 1978; Li, 1983b; Li and Shrieve, 1982). A reasonable question is: Can all these hsp inducers induce thermotolerance? Conversely, can all these thermotolerance inducers induce enhanced synthesis of hsp's? Table II suggests that there is a positive correlation. Here we have listed all the chemical and physical agents that have been examined for both phenomena using the same cell lines in our laboratory. The only exceptions to this positive correlation are the amino acid analogs. The amino acid analog data suggest that proteins synthesized in the presence of amino acid analogs may be nonfunctional. This is discussed in detail in Section VII.

Heat induces thermotolerance, which is a form of heat protection. Thus, it is reasonable to ask whether other inducers of hsp's induce protection against themselves. As already mentioned in a mini-review article by Hahn and Li (1982), this is not necessarily the case (Table III). For example, heat, arsenite, or cadmium induces thermotolerance, but neither heat nor arsenite induces protection against arsenite or cadmium. Thus, heat may represent a unique stress, and the hsp's may represent a unique set of stress proteins.

TABLE II

Heat Shock Protein Inducers' Induction of Thermotolerance

Agent	Induction of tolerance	Induction of heat shock proteins	References
Heat	+	+	Li and Werb (1982); Li (1983b)
Chronic hypoxia	+	+	Li and Werb (1982); Li and Shrieve (1982)
Ethanol	+	+	Li and Hahn (1978); Li (1983b)
Lidocaine	+	+	G. M. Hahn, B. R. West, E. C. Shiu, and G. C. Li (unpublished observations)
Procaine	–	–	G. M. Hahn, B. R. West, E. C. Shiu, and G. C. Li (unpublished observations)
Sodium arsenite	+	+	Johnston *et al.* (1980); Li (1983b)
Cadmium chloride	+	+	Li (1983b); Hahn and Li (1982)
CCP[a]	+	+	G. C. Li and J. Haveman (unpublished observations)
Disulfiram	+	+	G. C. Li and J. Haveman (unpublished observations)
Canavanine (arginine analog)	–	+	Li and Laszlo (1984)
Azetidine (proline analog)	–	+	Li and Laszlo (1984)

[a] CCP: carbonylcyanide-3-chlorophenylhydrazone.

TABLE III

Heat Shock Protein Inducers' Induction of Self-Tolerance

Inducer	Self-tolerance	References
Heat	+	Li *et al.* (1982b); Henle and Leeper (1976); Gerner and Schneider (1975)
Ethanol	+	Li *et al.* (1980)
Lidocaine	+	G. M. Hahn, B. R. West, E. C. Shiu, and G. C. Li (unpublished observations)
Sodium arsenite	−	G. C. Li (unpublished observations)

VII. EFFECT OF AMINO ACID ANALOGS ON THERMAL SENSITIVITY AND DEVELOPMENT OF THERMOTOLERANCE

Exposure of cells in culture to amino acid analogs, such as canavanine or L-azetidine-2-carboxylic acid (azetidine), has been reported to induce the synthesis of hsp's in avian and mammalian cells (Kelley and Schlesinger, 1978; Hightower, 1980; Thomas *et al.*, 1982; Hightower and White, 1982). In our laboratory, we have found that treatment of exponentially growing HA-1 cells with either canavanine or azetidine also induces the enhanced synthesis of hsp's, especially the higher-molecular-weight class of 70,000, 87,000, and 110,000 hsp's (Fig. 7). The hsp's induced by amino acid analogs have migration properties identical with those induced by a mild heat treatment.

Interestingly, we found that amino acid analog treatment did not induce thermotolerance. On the contrary, our results showed that exposure of HA-1 cells to either canavanine or azetidine for various lengths of time at 37°C sensitized cells to a subsequent heat challenge (Fig. 8). This amino acid analog-induced "sensitization" of the cells' thermal response was observed at all elevated temperatures tested (42–45°C) (Li and Laszlo, 1984). The amino acid analog-induced "sensitization" to heat required the incorporation of the analog into cellular proteins (Li and Laszlo, 1984). Similarly, the induction of the enhanced synthesis of hsp's by an amino acid analog also required the incorporation of the analog into cellular proteins (Kelly and Schlesinger, 1978; Hightower, 1980; A. Laszlo and G. C. Li, unpublished observations). One interpretation of these results is that the incorporation of amino acid analogs into the induced hsp's as well as other cellular proteins rendered them nonfunctional, and the analog-substituted hsp's could not protect cells from thermal damage.

We then determined and compared the thermal sensitivities of thermotolerant and control cells after exposure to amino acid analogs. Exponentially growing

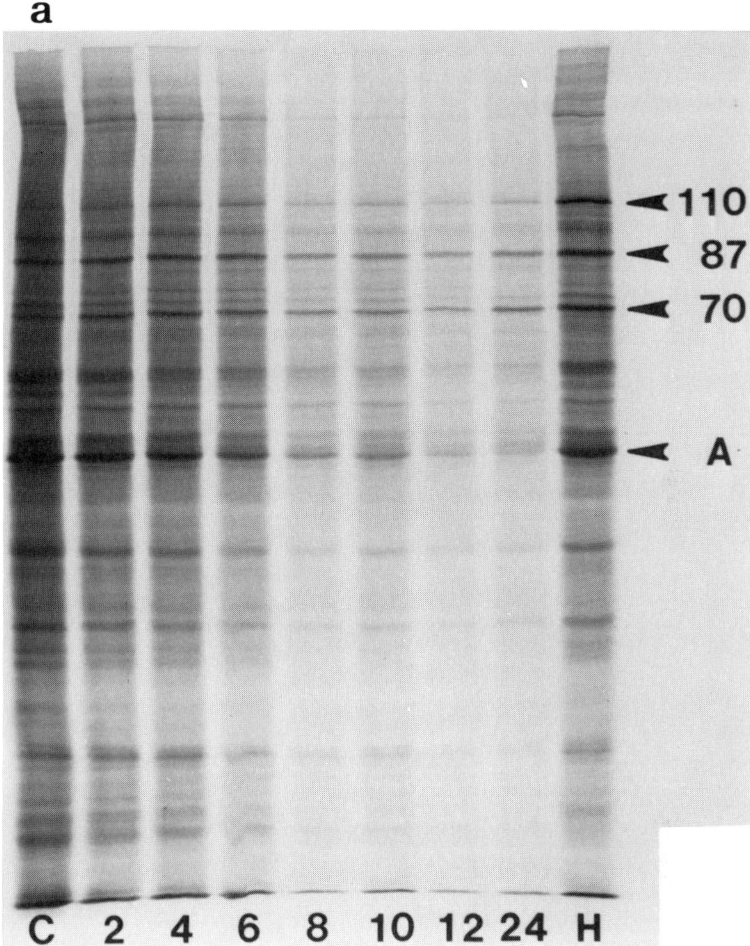

Fig. 7. The effect of amino acid analogs on protein synthesis of Chinese hamster HA-1 cells. Exponentially growing Chinese hamster HA-1 cells were exposed to 0.6 of canavanine in arginine-free medium (a) or 2.5 mmoles of azetidine in complete medium (b). In both cases, the media were supplemented with 5% dialyzed fetal calf serum. After various times of incubation (in hours) at 37°C, cells were pulse labeled with [^{35}S]methionine for 2 hr at 37°C. Lysates from equal numbers of cells were analyzed by SDS–gel electrophoresis. Arrows indicate the location of the three major hsps and actin (A); control (C); heat-shocked cells (H). (Cells labeled 6 hr after a 45°C, 15-min treatment.) (*continued*)

HA-1 cells were exposed to 45°C for 20 min and incubated at 37°C for 12 or 24 hr to allow maximal thermotolerance development. The cells were then exposed to medium containing canavanine or azetidine for 2–12 hr. Such treatments did not sensitize the thermotolerant cells to a subsequent heat challenge (Laszlo and Li, 1983a). This is in contrast with the control, nonthermotolerant cells, which

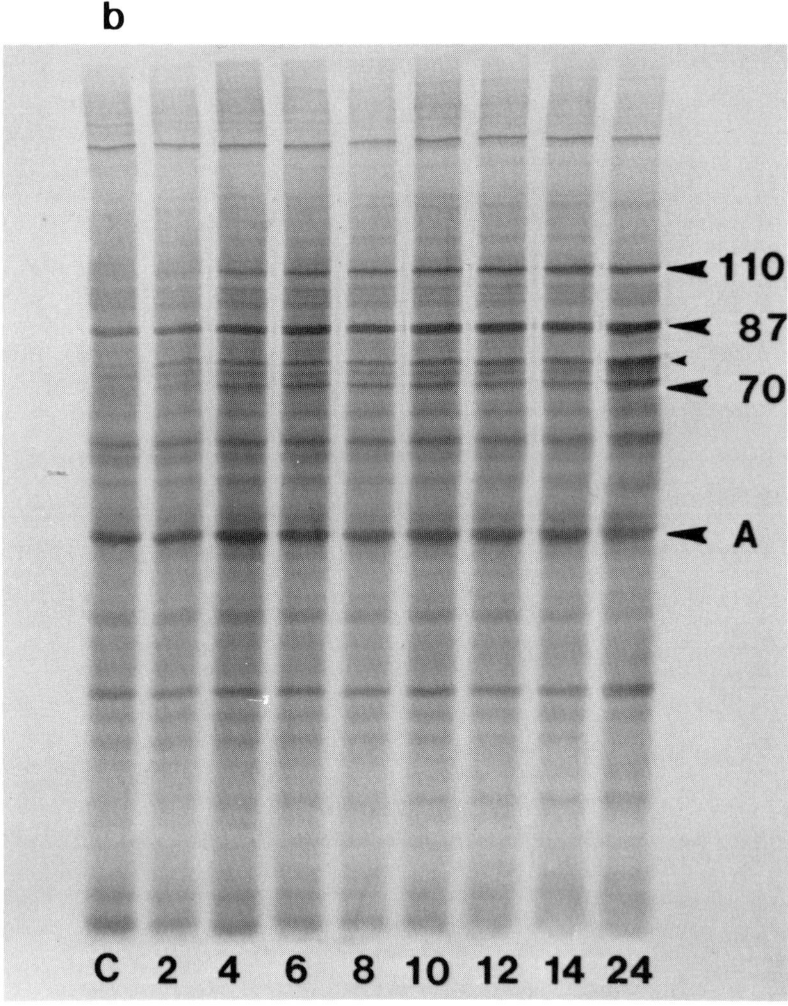

Fig. 7. (*Continued*)

became extremely sensitive to heat (200-fold decrease in survival at 45°C for 45 min) after 6 hr of incubation in the presence of amino acid analogs.

In separate experiments, the effects of amino acid analog treatment on the thermal sensitivity of cells at various stages of thermotolerance development were examined (Fig. 9). HA-1 cells were exposed to a mild heat treatment (45°C, 20 min) and then returned to 37°C incubation for various lengths of time. Some of the cells were then challenged by a 45°C, 45-min heat treatment, and the

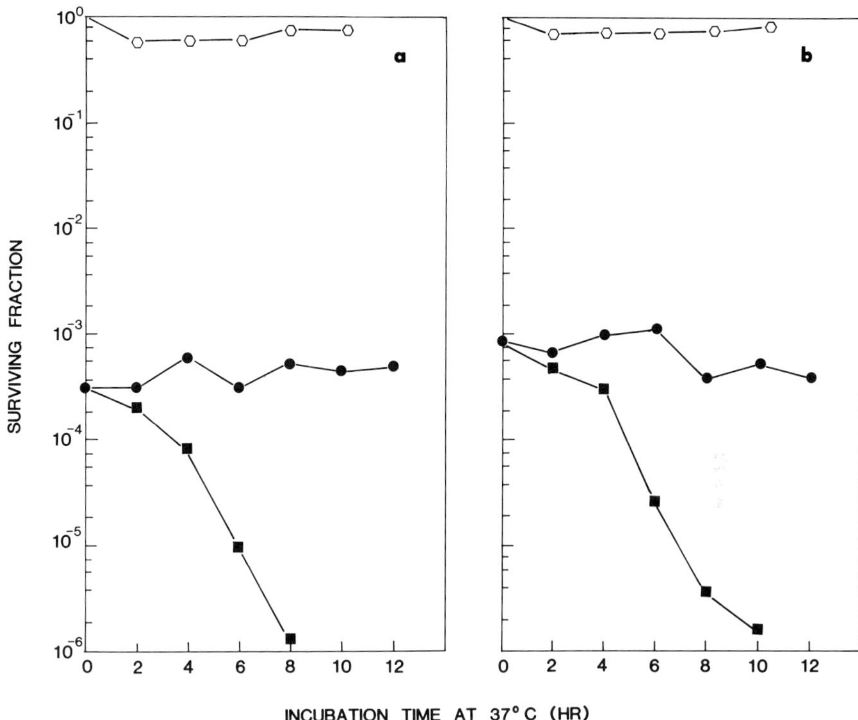

Fig. 8. The effect of amino acid analogs on the thermal response of Chinese hamster HA-1 cells. Exponentially growing HA-1 cells were exposed to 0.6 mmoles of canavanine or 2.5 mmoles of azetidine at 37°C as described in Fig. 7. After various times of exposure to the amino acid analogs, cells were challenged with a heat treatment at 45°C for 45 min. (a) Canavanine-treated cells; (b) azetidine-treated cells. Amino acid analog-treated cells, unheated (○); amino acid analog-treated cells, heated (■); control cells, heated but not exposed to the analogs (●).

increased survival values were used as a measure of the development of thermotolerance. Other cells were transferred to amino acid analog-containing medium for 0–6 hr before a similar heat challenge. Exposure to amino acid analogs immediately after a "triggering" heat or chemical treatment inhibited the development of thermotolerance and led to thermal sensitization (Laszlo and Li, 1983a). However, we found that as the cells became more thermotolerant, for example, 2–8 hr after the first heat treatment, their thermal sensitivity was less affected by the subsequent 6-hr exposure to amino acid analogs. Because heat treatment is known to alter membrane properties, including changes in permeability and the ability to transport small molecules, we examined the uptake of [^{14}C]canavanine into the acid-soluble pool and acid-insoluble macromolecules in normal and thermotolerant cells. We found no significant differences in these

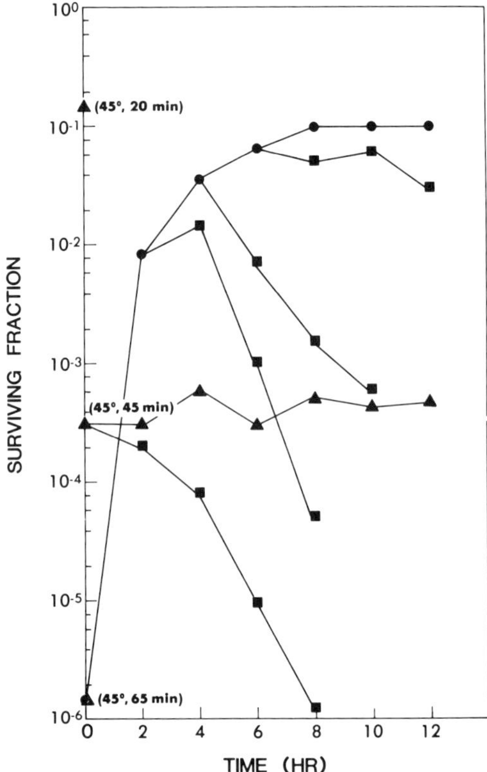

Fig. 9. The effect of an amino acid analog on the thermal response of transiently thermotolerant Chinese hamster HA-1 cells. Exponentially growing HA-1 cells were first heated at 45°C for 20 min and then returned to 37°C incubation for various lengths of time to allow the development of thermotolerance. At various stages of thermotolerance development (e.g., 2, 4, or 6 hr after the first heat treatment), cells were transfered to medium containing 0.6 mmoles of canavanine (for 0–6 hr). Cells were then challenged (no canavanine) with a 45°C, 45-min heat treatment. Control cells exposed to one heat treatment (45°C, 45 min) (▲); cells exposed to 45°C for 20 min and incubated at 37°C for 0–12 hr before the second heat treatment at 45°C for 45 min (●); control cells, or thermotolerant cells, exposed to 0.6 mmoles of canavanine for 0–6 hr before the 45°C, 45-min heat treatment (■). Note that as thermotolerance develops, cells become resistant to the thermal sensitization action of canavanine treatment.

parameters between the normal and thermotolerant HA-1 cells (Laszlo and Li, 1983a).

We interpret our observations as follows. The exposure of both control and thermotolerant cells to amino acid analogs leads to the increased synthesis of hsp's, but such hsp's may be nonfunctional since they contain amino acid analogs. The analog-substituted hsp's cannot protect normal cells from thermal stress as indicated by the enhanced cell killing after preincubation of control cells with amino acid analogs. A significant difference between control and thermotolerant cells is that the latter already contain increased levels of functional hsp's. Since the uptake and the incorporation of amino acid analogs into cellular proteins are unaltered in thermotolerant cells, it is most likely that the already elevated levels of functional hsp's in thermotolerant cells can protect them from thermal stress even in the presence of some additional synthesis of analog-substituted cellular proteins and analog-substituted hsp's. These results strongly support the hypothesis that the hsp's play a direct role in thermotolerance.

VIII. STABLE HEAT-RESISTANT VARIANTS OF CHINESE HAMSTER FIBROBLASTS

Heat-resistant variants can be useful in the investigation of the role that hsp's may play in thermotolerance. There are only a few reports demonstrating the existence of such variants of mammalian cells in culture (Selawry *et al.*, 1957; Harris, 1967, 1980). However, neither the development of thermotolerance and its relationship to hsp synthesis nor the relationship of the heat-resistant phenotype and the levels of hsp's has been examined in those cell lines.

We have isolated a series of heat-resistant (HR) variants from the HA-1 cell line by selecting survivors after repeated heat treatments (Laszlo and Li, 1983b, 1984). Significantly increased cell survival was observed in the HR variants exposed to elevated temperatures (Fig. 10).

The profile of proteins synthesized in HR variants was examined by one-dimensional SDS–PAGE. Densitometric analysis of the autoradiograms indicated that the 70K hsp is expressed at greater rates in the HR variants (Li, 1984; Laszlo and Li, 1983b). We obtained similar results whether the cells were labeled with a short pulse or labeled for up to 3 days with [^{35}S]methionine. Thus, the higher levels of the 70K hsp found constitutively in the HR variants were probably due to altered rates of synthesis, but not altered rates of degradation, of this polypeptides (Laszlo and Li, 1984).

Two-dimensional electrophoretic analysis of the extracts of [^{35}S]methionine-labeled HA-1 cells and the HR strain 3012 indicated that a member of the 70K hsp family which was greatly enhanced in heat-shocked HA-1 cells was expressed at much higher levels in the unheated HR variants (Fig. 11). Since the isoelectric point of the 70K hsp polypeptide that was overexpressed in the HR variant under normal growth conditions is identical to that induced in heat-shocked HA-1 cells, it is likely that these two polypeptides are identical.

Split dose heat experiments showed that thermotolerance also developed in the

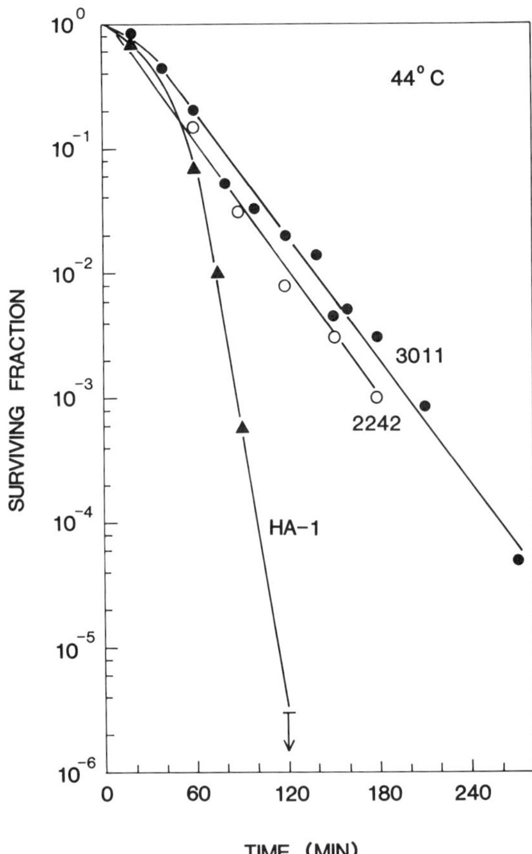

Fig. 10. The thermal response of Chinese hamster HA-1 cells and their stable heat-resistant variants. Exponentially growing HA-1 cells (▲) and the heat-resistant strains 3011 line (●) and 2242 line (○) were exposed to 44°C for various lengths of time. After the heat treatment, cells were trypsinized and survival was assayed.

HR variants. While the kinetics of development of thermotolerance were similar, there was a significant difference in the kinetic patterns of proteins synthesized after the initial heat treatment in HA-1 cells and proteins synthesized after heat treatment in their HR variants (Laszlo and Li, 1983b). In the HR line, the rate of induction of the 70K hsp occurred at a faster rate, the maximum levels achieved were relatively higher and occurred sooner, and the return to control levels occurred sooner. These differences in the kinetics of the rate of synthesis of the 70K hsp were observed in the HR lines subjected to various treatments, including various doses of heat, treatment with arsenite, and exposure to amino acid

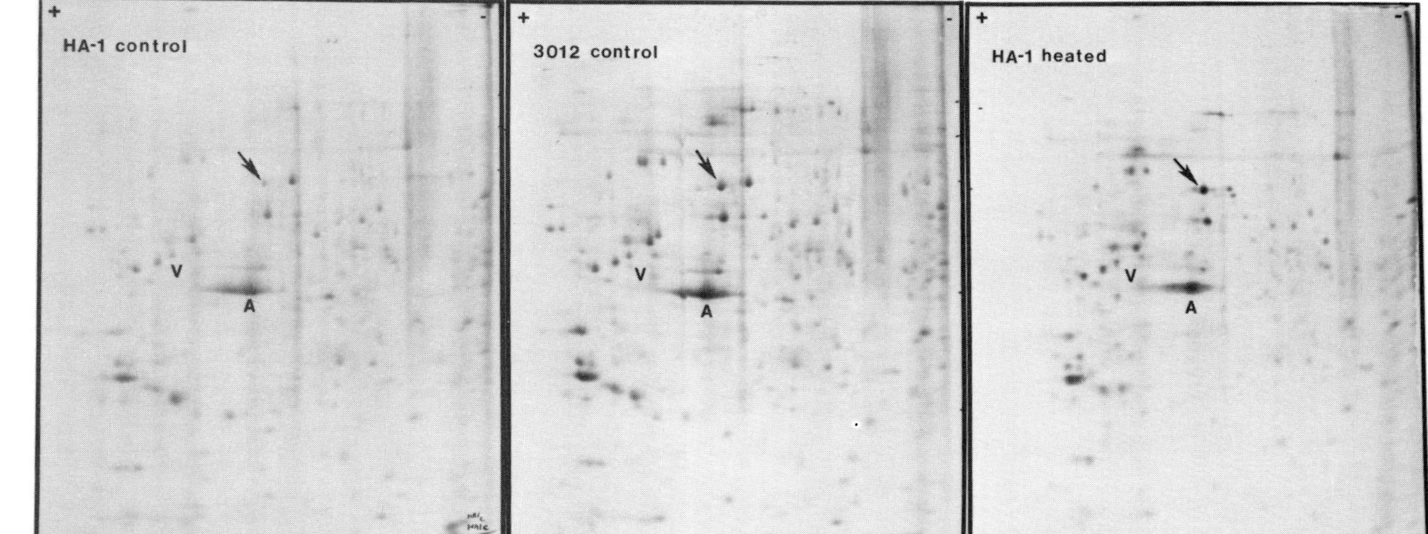

Fig. 11. Increased synthesis of the 70K heat shock protein in a heat-resistant variant derived from Chinese hamster HA-1 cells. Exponentially growing Chinese hamster HA-1 cells and their heat-resistant variant 3012 were labeled with [^{35}S]methionine for 4 hr. Total cellular extracts were analyzed by isoelectric focusing in the first dimension (from right to left) and SDS–PAGE in the second dimension (from top to bottom). For comparison, the heat-shocked HA-1 cells were labeled from 4 to 8 hr after an initial exposure to 45°C for 15 min. Actin (A); vimetin (V). The arrow indicates the 70K hsp induced in heated HA-1 cells, which is overexpressed in the 3012 heat-resistant variant under normal, unheated conditions.

analogs (A. Laszlo and G. C. Li, unpublished observations). The altered induction kinetics of the 70K hsp in the HR lines seem to be associated with the heat-resistant phenotype: the loss of heat resistance in some of our unstable lines was associated with a return to a kinetic pattern of 70K hsp induction similar to that of HA-1 cells (Laszlo and Li, 1984).

The results of *in vitro* translation of total cellular RNA isolated from HA-1 and HR variants indicated that the increased levels of synthesis of the 70K hsp in HR cells is a consequence of increased levels of mRNA coding for this polypeptide (Laszlo and Li, 1984). Our current working hypothesis is that the genetic alteration of the expression of the 70K hsp associated with the HR phenotype may be of a regulatory nature.

In this series of stable variants of the HA-1 line, the heat-resistant phenotype was associated with altered constitutive and inducible expression of the 70K hsp. Increased levels of the 70K hsp correlated well with increased heat survival in the HR variants (Li, 1984). The log-linear relationship between cell survival and levels of hps's in thermotolerant HA-1 cells and their stable heat-resistant variants is shown in Fig. 12. Here the survival of thermotolerant HA-1 cells and the permanently heat resistant strains (all exposed to 45°C for 45 min) was plotted

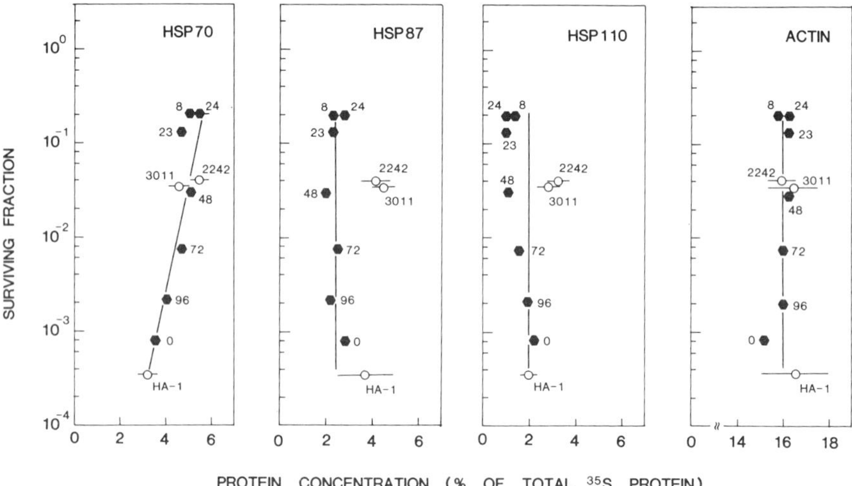

Fig. 12. Relationship between cell survival and levels of hsp's in thermotolerant HA-1 cells and their stable heat-resistant variants. Survival of exponentially growing, thermotolerant HA-1 cells and heat-resistant cells (strains 3011 and 2242) exposed to 45°C for 45 min are plotted against levels of hsp's, both induced and constitutive. The levels of hsp's are expressed as a percentage of total [^{35}S]metionine labeled proteins. To measure the total accumulated levels of hsp's, cells were labeled continuously throughout the experiments. Unheated HA-1, 3011, and 2242 cells (○); transiently thermotolerant HA-1 cells (●). Cells were made thermotolerant by exposure to 45°C for 15 min followed by 37°C incubation for 8, 23, 24, 48, 72, or 96 hr.

against the total levels of hsp's (both induced and constitutive). These results again reinforce the hypothesis that hsp's play a pivotal role in the protection of cells from thermal damage. Detailed molecular studies using these variants may shed light on the mechanisms of regulation of expression of the hsp gene products; for instance, we can examine whether the self-regulatory model proposed for Drosophila cells (DiDomenico *et al.*, 1982) for the control of expression of the 70K hsp is also applicable to mammalian cells.

IX. HEAT-INDUCED PROTECTION OF MICE AGAINST THERMAL DEATH

The possibility that the exposure of organisms to whole-body hyperthermia may provide protection against subsequent thermal exposures is intriguing and may play an important role in the clinical scheduling of fractionated hyperthermia. Li *et al.* (1983) used C3H mice to investigate whether whole-body heating can be used as a conditioning treatment to induce protection of mice against thermal death from subsequent heat treatment (Fig. 13). The data clearly showed that a conditioning whole-body heat dose (41°C for 40 min), by itself nonlethal, could give substantial protection to animals against a later heat treatment. The heat-induced protection was transient in nature: it reached a maximum by 6–24 hr following the 41°C conditioning dose and decayed by approximately 60% by 72 hr.

The kinetics of induction of this transient heat resistance differ from the kinetics of thermotolerance induction in mammalian cells in culture exposed to 41.0°C. In the latter, thermotolerance develops during the course of exposure to 41.0°C heat, with subsequent decay of thermotolerance over time (Fig. 2a). In this study, no protective benefit of the conditioning exposure was evident immediately on its completion. The protective benefit was not in evidence until 3 hr following the conditioning exposure and only reached maximum at 6–24 hr following exposure. Interestingly, these kinetics are similar to the kinetics of thermotolerance induction for mammalian cells exposed to temperatures of 42.5°C and higher (Fig. 2b).

The data presented do not shed any light on the cause of death following whole-body hyperthermia. Whether the prior hyperthermia dose causes physiological rather than cellular changes remains unanswered. However, it is intriguing that our data shared some similar kinetics with the accumulation of a 71K protein induced in rat tissue after whole-body hyperthermia (Currie and White, 1983). The authors examined the tissues of rats subjected to a brief 42°C hyperthermia shock by two-dimensional gel electrophoresis for the synthesis and accumulation of a 71K stress-induced protein. Their results showed that tissue of 6-week-old rats, killed immediately after heat shock, contained little or no 71K

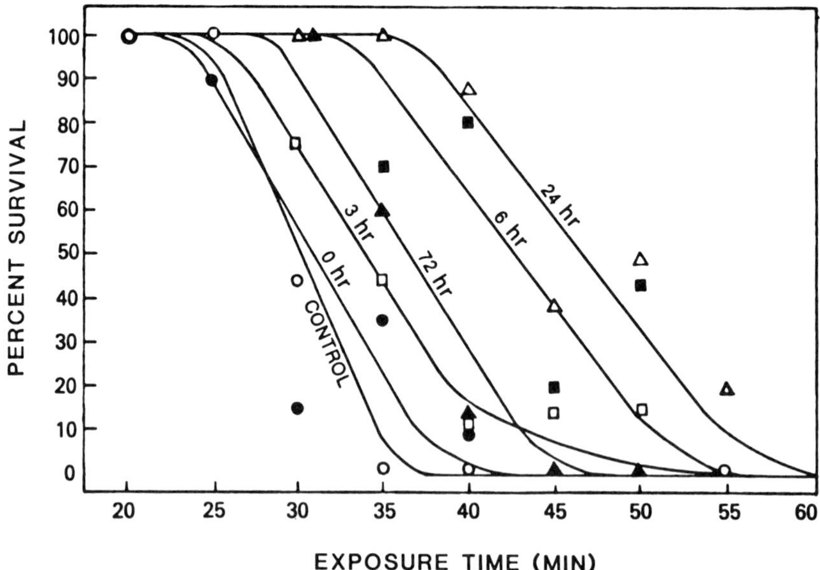

Fig. 13. Heat-induced protection of C3H mice against thermal death at 42.5°C. The experimental groups of C3H mice received a conditioning exposure (41°C for 40 min) for 0–72 hr before the second exposure at 42.5°C for various lengths of time. No deaths occurred after the conditioning exposure alone. The maximum protection against the second heat challenge at 42.5°C occurred when 6 or 24 hr elapsed between the two treatments. The percentage of animals (with or without prior conditioning thermal exposure) surviving increasing challenging thermal doses at 42.5°C is shown. Control group (○); 0 hr (●); 3 hr (□); 6 hr (■); 24 hr (△); and 72 hr (▲) after the conditioning exposure.

stress-induced protein. However, in all tissues tested (e.g, brain, heart, liver, bladder, etc.) synthesis and accumulation of this 71K protein were easily detected as early as 2.5 hr after heat shock. The synthesis of 71K protein was reduced by 1 and 2 days postshock, while the concentration in all tissues remained high up to 2 days postshock. Whether the accumulation of this 71K protein was related to heat-induced protection against thermal death remains to be elucidated.

X. INDUCTION OF THERMAL TOLERANCE AND ENHANCED SYNTHESIS OF HEAT SHOCK PROTEINS IN MURINE TUMORS

A radiation-induced fibrosarcoma (RIF) was implanted in the flank of C3H mice. The tumors were exposed to elevated temperatures (e.g., 43°C for 30 min). Some tumors were excised immediately and then challenged by a second heat shock *in vitro* (44°C, 30 min). The other mice with tumors were returned to

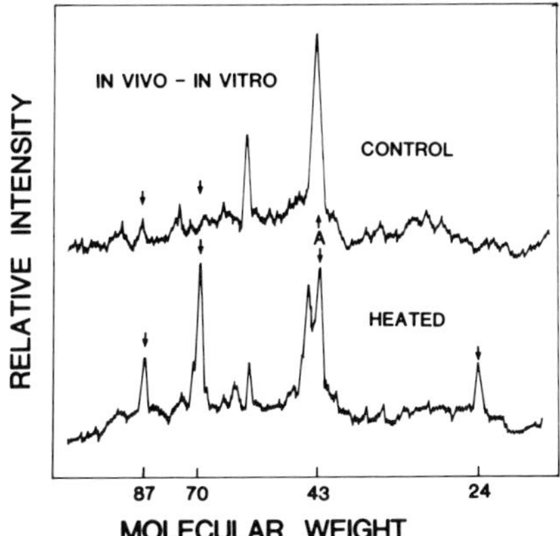

Fig. 14. Induction of enhanced synthesis of heat shock proteins in murine tumors. Flank tumors (RIF) in C3H mice were exposed to 43°C for 30 min. Some tumors were excised immediately, and the cell suspensions were labeled for 2 hr at 37°C with [^{35}S]methionine. The other mice were returned to their cages and allowed to recover *in situ* before being sacrificed. The tumors were then excised, and the cell suspensions were labeled. The cell extracts were prepared for SDS–PAGE. Densitometric scans of the autoradiograms are shown. Heat shock proteins were identified by comparing proteins synthesized by heat-shocked tumors with proteins synthesized by non-heat-shocked control tumors. Molecular weights ($\times 10^{-3}$) and actin (A) are shown.

their cages for 0–24 hr before being sacrificed and then exposed to a second heat shock. The tumor response to the combined heat treatments was measured using an *in vivo*/*in vitro* cloning assay. We found that this mild heat treatment induced thermotolerance, and our results agreed well with Hahn's previous observations (Hahn, 1982). In parallel experiments, the tumors were first heated at 43°C for 30 min and allowed to recover *in situ*. The tumors were then excised and the cell suspensions were labeled for 2 hr at 37°C with [^{35}S]methionine. The cell extracts were prepared for SDS–gel electrophoresis. The enhanced synthesis of hsp's was identified by comparing proteins synthesized in heat-shocked tumors with proteins synthesized in non-heat-shocked controls. Our results showed that mild heat shock at 43°C also enhanced the rate of synthesis of some of the high-molecular-weight hsp's, especially the 70K hsp (Fig. 14) (Li and Mak, 1984). These data suggest that the quantification of 70K hsp can be used as an assay to predict the thermal response of RIF tumors and, presumably, other tumors during fractionated hyperthermia treatments.

XI. CLINICAL RELEVANCE

The principles derived from these studies may be of practical value in planning clinical hyperthermia treatments of cancer.

Information on the kinetics of induction and disappearance of thermotolerance in mammalian cells, normal tissues, and tumors is essential when designing treatment protocols of fractionated hyperthermia in cancer therapy. By inducing tolerance in selected normal tissues at the proper time but not in tumors, the therapeutic efficacy of localized hyperthermia might be maximized while damage to treatment-limiting normal tissue might be minimized.

Since the levels of heat shock protein, especially hsp70, and the kinetics of thermotolerance development and cells' thermal sensitivity are closely correlated in many different cultured mammalian cell lines, thermal sensitivity of normal tissues and tumors might be predictable based on measurements of the levels of hsp70 in those tissues. Ultimately, one could even foresee the application of such an assay in clinical hyperthermia, for example, in deciding on the appropriate fractionation schedules.

Acknowledgments

We thank J. Y. Mak for his expert technical assistance. This work was supported by Grants CA 31397 and CA 09215 from the U.S. Public Health Service.

REFERENCES

Ashburner, M., and Bonner, J. J. (1979). The induction of gene activity in *Drosophila* by heat shock. *Cell* **17,** 241–254.

Currie, R. W., and White, F. P. (1983). Characterization of the synthesis and accumulation of a 71-kilodalton protein induced in rat tissue after hyperthermia. *Can. J. Biochem. Cell Biol.* **61,** 438–446.

Dewey, W. C., Freeman, M. L., Raaphorst, G. P., Clark, E. P., Wong, R. S. L., Highfield, D. P., Spiro, I. J., Tomasovic, S. P., Denman, D. L., and Coss, R. A. (1980). Cell biology of hyperthermia and radiation. *In* "Radiation Biology in Cancer Research" (R. E. Meyn and H. R. Withers, eds.), pp. 589–621.

DiDomenico, B. J., Bugaisky, G. E., and Lindquist, S. (1982). The heat shock response is self-regulated at both the transcriptional and post-transcriptional level. *Cell* **31,** 593–603.

Gerner, E. W., and Schneider, M. J. (1975). Induced thermal resistance in HeLa cells. *Nature (London)* **256,** 500–502.

Hahn, G. M., ed. (1982). "Hyperthermia and Cancer." Plenum, New York.

Hahn, G. M., and Li, G. C. (1982). Thermotolerance and heat shock proteins in mammalian cells. *Radiat. Res.* **92,** 452–457.

Hahn, G. M., and Little, J. B. (1972). Plateau phase cultures of mammalian cells: An *in vitro* model for human cancer. *Curr. Top. Radiat. Res. Q.* **8,** 39–83.

Harris, M. (1967). Temperature-resistant variants in clonal populations of pig kidney cell. *Exp. Cell Res.* **46,** 301–314.

Harris, M. (1980). Stable heat resistant variants in populations of Chinese hamster cells. *J. Nat. Cancer Inst. (U.S.)* **64,** 1495–1501.

Henle, K. J., and Dethlefsen, L. A. (1978). Heat fractionation and thermotolerance: A review. *Cancer Res.* **38,** 570–574.

Henle, K. J., and Leeper, D. B. (1976). Interaction of hyperthermia and radiation in CHO cells: Recovery kinetics. *Radiat. Res.* **66,** 505–518.

Henle, K. J., and Lepper, D. B. (1982). Modification of heat response and thermotolerance by cycloheximide, hydroxyurea and lucanthone in CHO cells. *Radiat. Res.* **90,** 339–347.

Hightower, L. E. (1980). Cultured animal cells exposed to amino acid analogues or puromycin rapidly synthesize several polypeptides. *J. Cell. Physiol.* **102,** 407–427.

Hightower, L. E., and White, F. P. (1982). Cellular responses to stress: comparison of a family of 71–73 kilodalton proteins rapidly synthesized in rat tissue slices and canavanine treated cells in culture. *J. Cell. Physiol.* **108,** 261–275.

Johnston, D., Oppermann, H., Jackson, J., and Levinson, W. (1980). Induction of four proteins in chicken embryo cells by sodium arsenite. *J. Biol. Chem.* **225,** 6975–6980.

Kelley, P. M., and Schlesinger, M. J. (1978). The effect of amino acid analogues and heat shock on gene expression in chicken embryo fibroblasts. *Cell* **15,** 1277–1286.

Landry, J., Bernier, D., Cretien, P., Nicole, L. M., Tanguay, R. M., and Marceau, N. (1982). Synthesis and degradation of heat shock proteins during development and decay of thermotolerance. *Cancer Res.* **42,** 2457–2461.

Laszlo, A., and Li, G. C. (1983a). Thermotolerant HA-1 cells are resistant to the thermosensitization action of amino acid analogues. *In* "Proceedings of the 7th International Congress of Radiation Research, Tumor Biology and Therapy" (J. J. Broerse, G. W. Barendsen, H. B. Kal, and A. J. van der Kogel, eds.), pp. D6–27. Martinus Nijhoff, Amsterdam.

Laszlo, A., and Li, G. C. (1983b). Heat resistant variants of Chinese hamster cells are altered in heat shock protein expression. *J. Cell Biol.* **97,** 151a.

Laszlo, A., and Li, G. C. (1984). Heat resistant variants of Chinese hamster cells altered in heat shock protein expression. *Proc. Natl. Acad. Sci. U.S.A.* (in press).

Li, G. C. (1983a). Time patterns of synthesis of heat shock proteins during development of thermotolerance. *In* "Proceedings of the 7th International Congress of Radiation Research, Tumor Biology and Therapy" (J. J. Broerse, G. W. Barendsen, H. B. Kal, and A. J. van der Kogel, eds.), pp. D6–32. Martinus Nijhoff, Amsterdam.

Li, G. C. (1983b). Induction of thermotolerance and enhanced heat shock protein synthesis in Chinese hamster fibroblasts by sodium arsenite and by ethanol. *J. Cell. Physiol.* **115,** 116–122.

Li, G. C. (1984). Elevated levels of 70,000 dalton heat shock protein in transiently thermotolerant Chinese hamster fibroblasts and in their stable heat resistant variants. *Int. J. Radiat. Oncol. Biol. Phys.* (in press).

Li, G. C., and Hahn, G. M. (1978). Ethanol-induced tolerance to heat and adriamycin. *Nature (London)* **1274,** 699–701.

Li, G. C., and Hahn, G. M. (1980). A proposed operational model of thermotolerance based on the effects of nutrients and the initial treatment temperature. *Cancer Res.* **40,** 4501–4508.

Li, G. C., and Laszlo, A. (1984). Amino acid analogues, while inducing heat shock proteins, sensitize cells to heat. *J. Cell. Physiol.* (in press).

Li, G. C., and Mak, J. Y. (1984). Induction of thermal tolerance and enhanced synthesis of heat shock proteins in murine tumors. *Cancer Res.* (in press).

Li, G. C., and Shrieve, D. C. (1982). Thermal tolerance and specific protein synthesis in Chinese hamster fibroblasts exposed to prolonged hypoxia. *Exp. Cell Res.* **142,** 464–468.

Li, G. C., and Werb, Z. (1982). Correlation between synthesis of heat shock proteins and development of thermotolerance in Chinese hamster fibroblasts. *Proc. Natl. Acad. Sci. U.S.A.* **79,** 3219–3222.

Li, G. C., Shiu, E. C., and Hahn, G. M. (1980). Similarities in cellular inactivation by hyperthermia or by ethanol. *Radiat. Res.* **82,** 257–268.

Li, G. C., Fisher, G. C., and Hahn, G. M. (1982a). Modification of the thermoresponse by D_2O. II. Thermotolerance and the specific inhibition of development. *Radiat. Res.* **92,** 541–551.

Li, G. C., Petersen, N. S., and Mitchell, H. K. (1982b). Induced thermal tolerance and heat shock protein synthesis in Chinese hamster ovary cells. *Int. J. Radiat. Oncol. Biol. Phys.* **8,** 63–67.

Li, G. C., Fisher, G. C., and Hahn, G. M. (1982c). Induction of thermotolerance and evidence for a well defined, thermotropic cooperative process. *Radiat. Res.* **89,** 361–368.

Li, G. C., Meyer, J., Mak, J. Y., and Hahn, G. M. (1983). Heat induced protection of mice against thermal death. *Cancer Res.* **43,** 5758–5760.

Loomis, W. F., and Wheeler, S. (1980). Heat shock response in *Dictyostelium. Dev. Biol.* **79,** 399–408.

McAlister, L., and Finkelstein, D. B. (1980). Heat shock proteins and thermal resistance in yeast. *Biochem. Biophys. Res. Commun.* **93,** 819–824.

Miller, M. J., Xuong, N. H., and Geidescheck, E. P. (1979). A response of protein synthesis to temperature shift in the yeast *Saccharomyces cerevisiae. Proc. Natl. Acad. Sci. U.S.A.* **76,** 5222–5225.

Mitchell, H. K., Moller, G., Petersen, N. S., and Lipps-Sarmiento, L. (1979). Specific protection from phenocopy induction by heat shock. *Dev. Genet. (Amsterdam)* **1,** 181–192.

Puck, T. T., and Marcus, P. I. (1956). Action of X-rays on mammalian cells. *J. Exp. Med.* **103,** 653–666.

Schlesinger, M. J., Ashburner, M., and Tissières, A., eds. (1982). "Heat Shock, from Bacteria to Man." Cold Spring Harbor Lab. Cold Spring Harbor, New York.

Selawry, O., Goldstein, M., and McCormick, T. (1957). Hyperthermia in tissue-culture cells of malignant origin. *Cancer Res.* **17,** 785–791.

Subjeck, J. R., Sciandra, J. J., and Johnson, R. J. (1982). Heat shock proteins and thermotolerance: A comparison of induction kinetics. *Br. J. Radiol.* **55,** 579–584.

Thomas, G. P., Welch, W. J., Mathews, M. B., and Feramisco, J. R. (1982). Molecular and cellular effects of heat shock and related treatments on mammalian tissue culture cells. *Cold Spring Harbor Symp. Quant. Biol.* **46,** 985–996.

Tomasovic, S. P., Steck, P. A., and Heitzman, D. (1983). Heat-stress proteins and thermal resistance in rat mammary tumor cells. *Radiat. Res.* **95,** 399–413.

Yang, S. J., Hahn, G. M., and Bagshaw, M. A. (1966). Chromosome aberrations induced by thymidine. *Exp. Cell Res.* **42,** 130–135.

II

Plants and Fungi

13

Heat Shock Genes of *Dictyostelium*

ELLIOT ROSEN, ANNEGRETHE SIVERTSEN, RICHARD A. FIRTEL, STEVEN WHEELER, AND WILLIAM F. LOOMIS

I. INTRODUCTION

Until the recent evolution of homeotherms, cells of most organisms have had to survive the temperature extremes of their environments. A small set of genes arose enabling cells to survive at the upper extreme of their temperature range, and these genes have been conserved to a surprising extent through subsequent evolution. While the temperatures which induce these genes vary depending on the species, many aspects of the gene products have remained constant. Perhaps most telling is the immunological cross-reactivity of one of the major heat shock-induced proteins, hsp70, isolated from yeast, *Dictyostelium, Drosophila,* and man (Kelley and Schlesinger, 1982). Among the heat shock-induced proteins is a small group from 25,000 to 35,000 daltons which are found preferentially in the nucleus shortly after their synthesis (Velasquez *et al.*, 1980; Arrigo *et al.*, 1980; Loomis and Wheeler, 1982). These proteins appear to be essential for protection

Changes in Eukaryotic Gene Expression
in Response to Environmental Stress

ISBN 0-12-066290-6

from lethal temperatures since heat shock of cells in which protein synthesis has been inhibited does not protect the cells. Moreover, a mutant strain of *Dictyostelium* (HL122) was isolated which is impaired in heat shock protection, and cells of this strain fail to synthesize the low-molecular-weight (LMW) nuclear proteins following heat shock induction (Loomis and Wheeler, 1982).

The conservation of the heat shock response in a wide variety of divergent organisms suggests that the response provides some important cellular function. In addition to its role in thermal protection, it has been observed that several heat shock genes are induced in some systems by a variety of environmental insults such as water deprivation, anoxia, heavy metal ions, and metabolic inhibitors (see Ashburner and Bonner, 1979). Thus, it may be that the heat shock response is part of a general cellular reaction protecting cells from the potentially lethal effects of environmental stress (Gerner and Schneider, 1975; Li and Hahn, 1978, McAlister and Finkelstein, 1980; Mitchell *et al.*, 1979; Petersen and Mitchell, 1981).

A considerable amount is now known about the conditions which induce heat shock proteins, the nature of the induction, the characteristics of the proteins and the genes which code for them, and the physiological consequences of the accumulation of these proteins. We will discuss the information available in *Dictyostelium* and relate it to similar analyses in other organisms.

II. PHYSIOLOGICAL ROLE OF HEAT SHOCK PROTEINS

Dictyostelium discoideum is a species of phagocytic amoeba which lives in the soil and can grow at temperatures up to 27°C (Loomis, 1969). If a growing culture is shifted to 30°C, the cells neither grow nor divide but die slowly with a half-life of approximately 4 hr (Figs. 1 and 2). At this temperature a few heat shock genes are induced, resulting in the rapid accumulation of a few heat shock proteins (Loomis and Wheeler, 1980). Furthermore, the expression of most genes which were active at the time of the temperature shift is now repressed. If a growing culture is shifted to 34°C, the cells rapidly die ($t_{1/2} \sim 20$ min). At this temperature the cells die before there is an appreciable accumulation of heat shock protein. However, if cells are first incubated at 30°C and allowed to accumulate heat shock proteins, the cells will survive at 34°C (Fig. 1). Thus it appears that induction of the heat shock response at 30°C protects the cells from the lethal effects of subsequent treatment at the higher temperature. The acquisition of thermal protection at 30°C requires *de novo* protein synthesis. Cells treated with cycloheximide at 30°C fail to survive a subsequent shift to 34°C

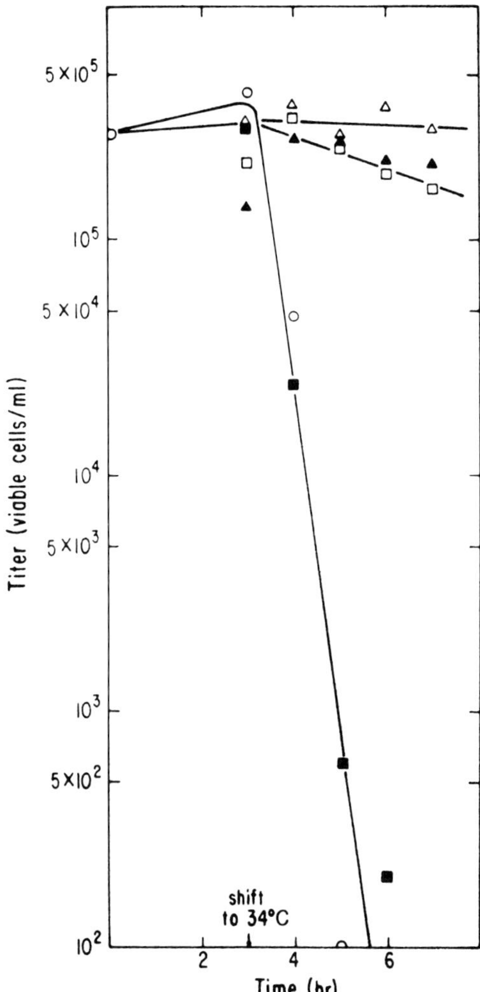

Fig. 1. Heat shock protection. Cells of strain Ax-3 growing exponentially at 22°C in HL-5 medium were shifted to 30°C. After 3 hr, some were kept at 30°C (△, ▲) and others were shifted to 34°C (□, ■). Cells which had been kept at 22°C were shifted directly to 34°C (○). Cycloheximide (500 μg/ml) was added to some of the cultures at the start of the experiment (▲, ■). At various times samples were withdrawn, diluted, and plated in association with *Klebsiella aerogenes* on SM agar plates which were incubated at 22°C. Plaques which arose in 4 days were counted to calculate the viable titer. (From Loomis and Wheeler, 1980.)

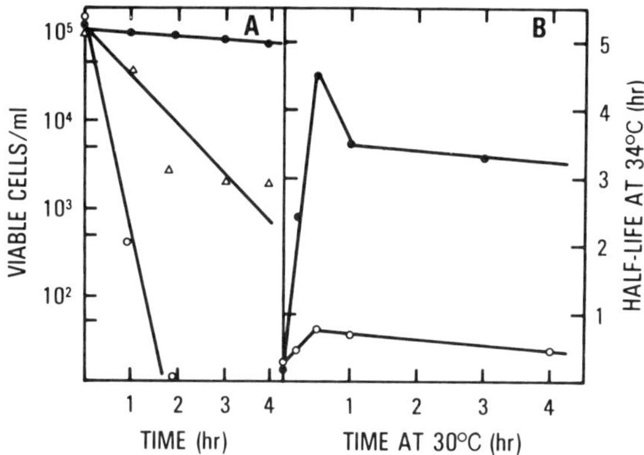

Fig. 2. Thermal protection in a heat shock mutant. (A) Cells of strain HL122 growing at 22°C were heat shocked at 30°C for 1 hr before being transferred to 34°C (△) or were directly transferred to 34°C (○) or 30°C (●). (B) The half-life of cells of strain Ax-3 (●) or strain HL122 (○) at 34°C was determined after various periods of heat shock at 30°C. (From Loomis and Wheeler, 1982.)

(Fig. 1). Moreover, the mutant strain of *Dictyostelium* HL122 fails to acquire thermal protection at 30°C following heat shock (Fig. 2; Loomis and Wheeler, 1982). Interestingly, this strain fails to accumulate heat shock proteins following a shift to 30°C.

To delineate the biochemical processes involved in thermotolerance, it is first necessary to know how high temperature kills the cells. Unfortunately, the available data either fail to support or contradict many of the hypotheses used to explain thermal killing. Although it has been suggested that high temperature might cause thermal denaturation of enzymes, no convincing data have pinpointed a specific thermolabile function. It should be remembered that *Dictyostelium* cells die rapidly at 34°C unless there has been an accumulation of heat shock proteins. This is a permissive temperature for growth of many cells and does not inactivate a series of different *Dictyostelium* enzymes even in highly purified preparations (Loomis, 1975). Another possibility is that high temperature affects the integrity of the cell membrane. Although *Dictyostelium* cells continue to exclude trypan blue for several hours after the irreversible events leading to cell death, we cannot eliminate the possibility that high temperature either causes the membrane to leak essential ions and small molecules or disrupts some other essential membrane function.

The nuclear localization of LMW proteins leads us to consider the susceptibility of DNA replication and chromosomal segregation to thermal damage. However, killing at 34°C in non-heat-shocked cells is rapid, occurring in less than a tenth of a generation. Thus, it is unlikely that the initial lethal event is related to replication. Furthermore, the thermosensitivity of heat-shocked cells in which protein synthesis is blocked suggests that thermal death does not result from the

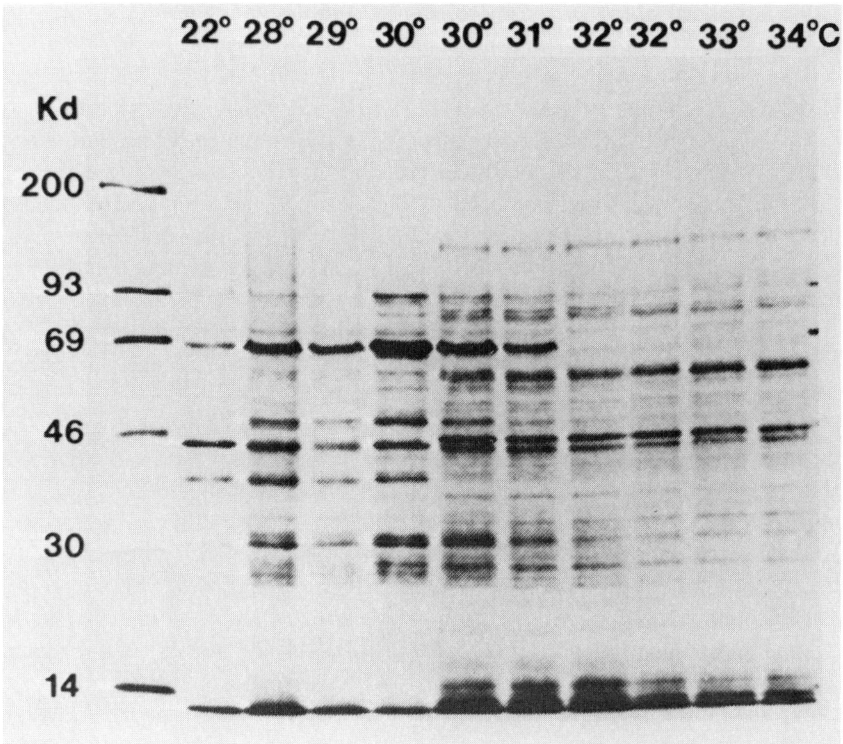

Fig. 3. Protein synthesis during heat shock at various temperatures. Autoradiograms of a gradient SDS–polyacrylamide gel of proteins (5×10^3 cpm) from cultures of developing *D. discoideum* cells shifted to the temperature (°C) indicated for 2 hr and labeled for 1 hr. (From Loomis and Wheeler, 1980.)

accumulation of errors in macromolecular synthesis. Although the elucidation of the mechanism of thermal killing is important to the determination of the function of the heat shock proteins, at present the mechanism remains obscure.

III. INDUCTION OF HEAT SHOCK GENES

As observed in *Drosophila,* the pattern of protein synthesis in *Dictyostelium* following heat shock varies with the temperature (Fig. 3). Heat shock proteins are observed at 28°C, but their expression is maximal at 30°C. At the latter temperature hsp70 is the major protein synthesized by the cell, while at higher temperatures its expression is repressed. At 34°C the cells die before they accumulate significant amounts of heat shock protein.

We have tried to induce these proteins by a variety of other stress treatments including 1 m*M* arsenite, 1 mg/ml L-canavanine, 5% ethanol, 10 m*M* butyrate, 40 μg/ml cycloheximide, saturated antimycin A, and 2 m*M* 2,4-dinitrophenol, all of which slow or stop growth of *Dictyostelium*. Many of these treatments induce puffing at the sites of heat shock genes in dipteran cells (Ashburner and Bonner, 1979) or induce heat shock proteins in vertebrate cells (Kelley and Schlesinger, 1978). However, none of these compounds specifically induced heat shock proteins (W. F. Loomis and S. A. Wheeler, unpublished observations). Thus, the only stress we have found so far to affect *Dictyostelium* is high temperature. Just as the absolute temperature necessary to induce these genes varies from species to species, the range of environmental stimuli coupled to the genes appears to be relatively free to adapt to the specific challenges facing one organism or another.

It is useful to make a distinction between the environmental signals triggering the heat shock response and the molecular mechanisms activating particular heat shock genes. Transformation studies, in which cloned heat shock genes from one organism are introduced into the cells of another, indicate that the mechanisms activating the endogenous heat shock genes in the host cell also act on the heterologus heat shock gene (Corces *et al.*, 1981; Pelham, 1982; Mirault *et al.*, 1982; Voellmy and Rungger, 1982). It has been found that the *Drosophila* hsp70 gene introduced into mouse cells (Corces *et al.*, 1981), monkey cells (Pelham, 1982; Mirault *et al.*, 1982), or *Xenopus* oocytes (Voellmy and Rungger, 1982; Bienz and Pelham, 1982) is induced after heat shock of the host cells. Although the particular environmental signal triggering the heat shock response varies among organisms, the results indicate that the molecular mechanisms activating the heat shock genes have been conserved.

IV. CONTROL OF TRANSCRIPTION

Following a heat shock there are rapid and dramatic changes in the pattern of accumulation of RNA. These changes include a block in the synthesis of mature 17 S and 26 S ribosomal RNA, the cessation of synthesis of most mRNA found in the cells at 22°C, and the rapid accumulation of transcripts for a few heat shock genes (Rosen *et al.*, submitted for publication). Some of these changes can be observed by comparing the incorporation of [^{32}P]phosphate into RNA of cells incubated at 22 and 30°C (Fig. 4). At 30°C a small amount of the label is incorporated into mature 17 S and 26 S rRNA (Fig. 4). Interestingly, at 30°C the accumulation of the label into a larger poly(A)$^-$ species occurs, suggesting that the failure to accumulate 17 S and 26 S rRNA may result from the failure to process ribosomal RNA precursors into stable mature forms, although a decrease in the transcription rate of ribosomal genes is not excluded.

The difference in the pattern of mRNA synthesis at 22 and 30°C can be seen in a comparison of *in vivo*-labeled poly(A)$^+$ RNA from cells incubated at 22 and 30°C (Fig. 4). The few poly(A)$^+$ bands present at 22°C disappear at 30°C while new prominent RNA species appear. Most prominent of the heat shock-inducible bands are those at 2.5 kb and 0.9 kb. The induction of heat shock poly(A)$^+$ RNA is more strikingly demonstrated in Fig. 5; the heat shock mRNA's accumulate to the point of producing prominent bands when stained with ethidium bromide.

In order to examine the expression of particular transcripts after heat shock, it is necessary to use specific cloned genes as hybridization probes. To this end we have isolated and characterized *Dictyostelium* genomic fragments containing heat shock genes. These clones were identified by probing recombinant λ phage containing *Dictyostelium* DNA with *in vivo*-labeled poly(A)$^+$ RNA from heat-shocked cells in the presence of cold competitor RNA from cells incubated at 22°C. Three clones encoding heat shock-induced transcripts have been identified (Rosen *et al.*, submitted for publication). One clone, hs70, hybridizes to a 2.5-kb heat shock-inducible mRNA (Fig. 6). DNA sequence analysis as well as *in vitro* translation of hybrid-selected mRNA indicates that the clone encodes the hsp70 gene of *Dictyostelium* (Fig. 7). In addition, we have isolated two other clones, HS1.4 and D1, which hybridize to heat shock-induced messages of 0.9 and 2.5 kb, respectively (Fig. 6). At present the heat shock proteins encoded by these two genes are unknown. Preliminary sequence analysis of clone D1 suggests that it does not encode the hsp70 protein although it hybridizes to an mRNA of the appropriate size.

In addition, we have isolated a number of clones containing the same transcribed reiterated sequence which hybridizes to two heat shock-inducible transcripts (Rosen *et al.*, submitted for publication). Subsequent analysis of these clones indicates that they belong to the M4-repeat family initially characterized by Kimmel and Firtel (1979, 1982). The M4-repeat contains the reiterated sequence $(\text{A-A-C})_n$ that is found about 100 times in the genome and in approximately 50 transcripts (Kimmel and Firtel, 1982; A. R. Kimmel and R. A. Firtel, manuscript in preparation). Although the M4 clones which we isolated probably do not contain heat shock structural genes, they hybridize to two heat shock-inducible messages of 0.9 and 1.2 kb as well as a population of mRNA's present at 22°C (Fig. 6). The use of *in vivo*-labeled heat shock RNA, which includes these two transcripts containing the reiterated sequence, resulted in the isolation of a large number of genomic clones containing the M4-repeat.

Analysis of Northern blots of RNA from control and heat-shocked cells reveals that there are two patterns of accumulation of heat shock mRNA (Fig. 6) (Rosen *et al.*, submitted for publication). Clones D1, hs70, and HS1.4 hybridize to messages which are induced within 20 min after heat shock and accumulate for 3–4 hr after the temperature shift. After 3–4 hr these messages begin to disap-

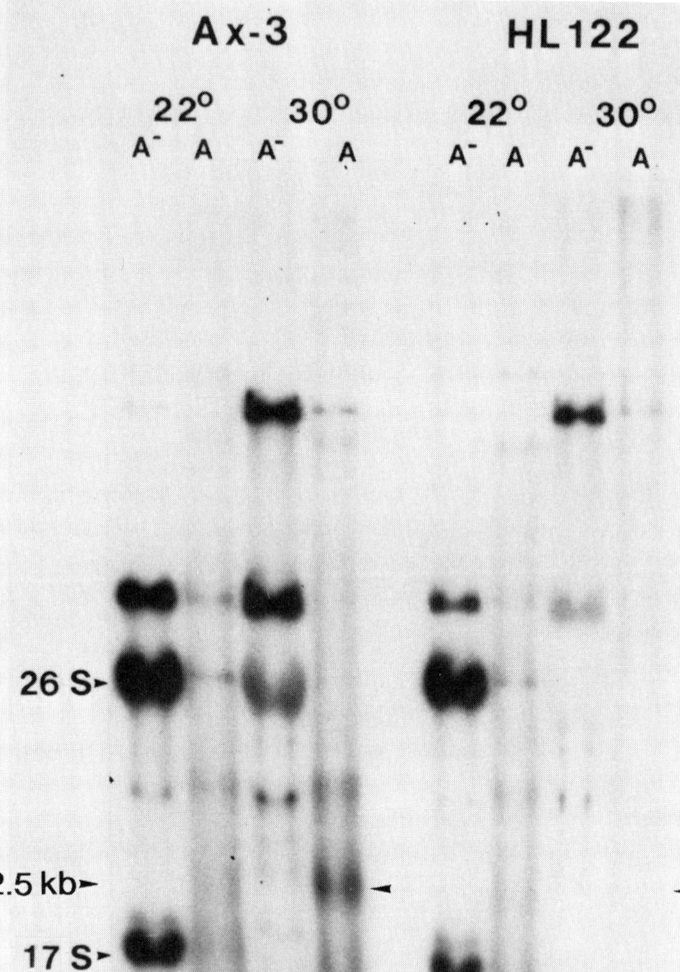

Ax-3
HL122
22°
30°
22°
30°
A⁻
A
A⁻
A
A⁻
A
A⁻
A
26 S
2.5 kb
17 S

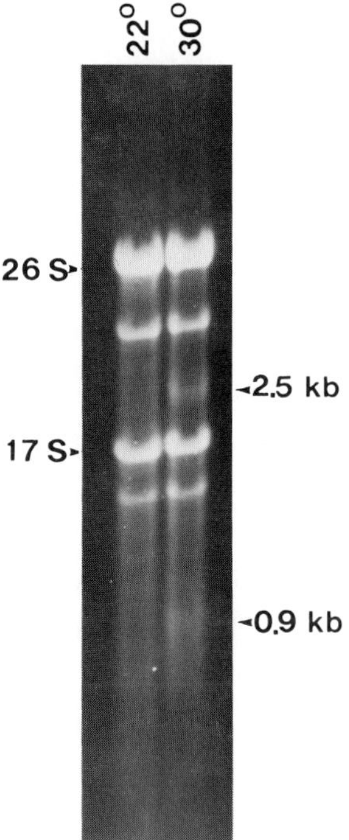

Fig. 5. Ethidium bromide staining of heat shock RNA. Poly(A)+ RNA extracted from *Dictyostelium* Ax-3 cultures grown at 22°C or heat shocked at 30°C for 2 hr were size fractionated on denaturing formaldehyde–agarose gels (Firtel and Lodish, 1973; Lebrach *et al.*, 1977). Gels were stained with 0.5 μg/ml of ethidium bromide and photographed under UV illumination. Arrows indicate the major poly(A)+ RNA's which accumulate after heat shock.

pear. In contrast, the heat shock-inducible messages hybridizing to the M4-repeat only accumulate for the first hour and then are lost. It is interesting to note that the more transient messages contain a transcribed repeat sequence.

In contrast to the heat shock-induced RNA's, mRNA from non-heat-shock genes is rapidly lost after heat shock. As an example, Northern blots of total

---

Fig. 4. *In vivo* labeling of RNA during heat shock. A log phase culture of strains Ax-3 or HL122 growing at 22°C washed and resuspended in 1 : 10 volume of MES–PDF are shown (Firtel and Lodish, 1973). After a 1-hr incubation at 22°C half of each culture was transferred to 30°C. After 15 min [^{32}PO]phosphate was added to the 22 and 30°C cultures. After 2 hr the cells were harvested, RNA was extracted, and poly(A)+ RNA and poly(A)− RNA were separated on poly(U)–Sepharose columns (Firtel *et al.*, 1973). Equal numbers of counts were loaded on denaturing formaldehyde–agarose gels (Lebrach *et al.*, 1977). Left, RNA from Ax-3 cultures; right, RNA from the mutant HL122; A− and A, 22°C and 30°C denote poly(A)− or poly(A)+ RNA isolated from cells incubated at 22 or 30°C. The mature 17 S and 26 S ribosomal RNA's are denoted. The arrows indicate the position of the 2.5-kb mRNA for hsp70 that accumulates in Ax-3 after heat shock.

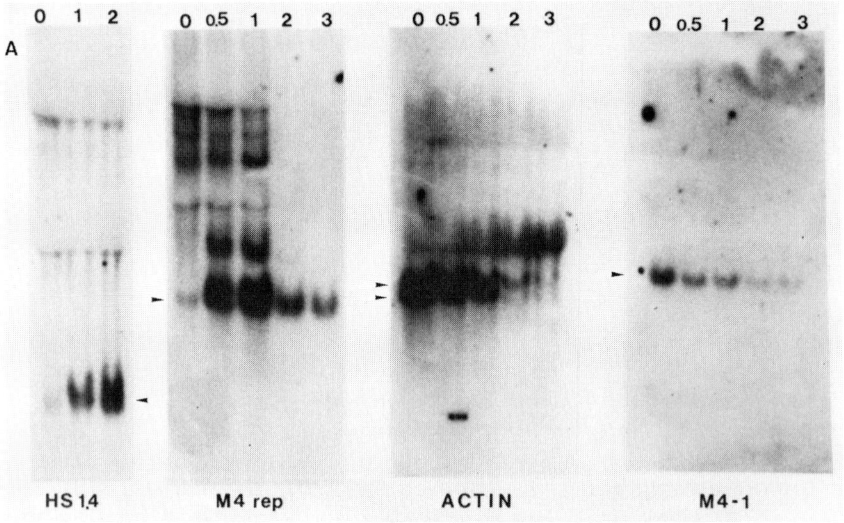
A
0
1
2
0
0.5
1
2
3
0
0.5
1
2
3
0
0.5
1
2
3
HS 1.4
M4 rep
ACTIN
M4-1

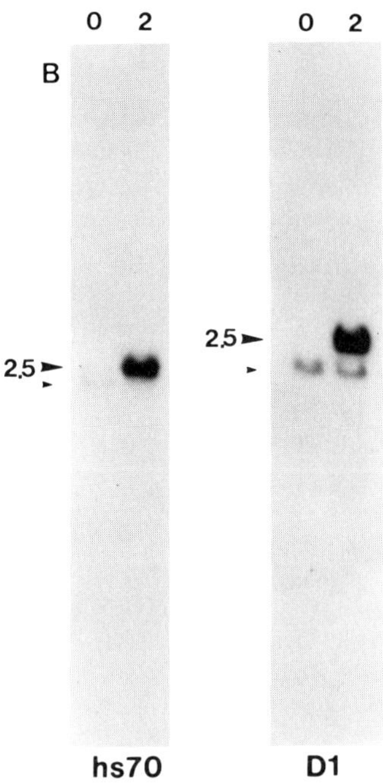
B
0
2
0
2
2.5
2.5
hs70
D1

RNA from control and heat-shocked cells probed with a cloned actin gene or M4-1 (a single-copy gene expressed during both vegetative growth and development) shows that these mRNA's are lost quickly after the temperature shift (Fig. 6). Similar results are obtained with probes to other non-heat-shock genes. These results are in contrast to what is seen in *Drosophila* where non-heat-shock messages present in the cell at the time of the heat shock are stably maintained even though they are not translated (Mirault *et al.*, 1978; Storti *et al.*, 1980).

The loss of non-heat-shock RNA is quite rapid. Estimates of the half-life of non-heat-shock transcripts based on Northern blots suggest these messages decay with a half-life of approximately 30 min (Fig. 6). Although there is some disagreement from the work of different groups, estimates of messenger half-life during *Dictyostelium* vegetative growth or development are considerably longer. Margolskee and Lodish (1980a,b) observed a total mRNA half-life of 3.5–4 hr in vegetative and developing cells. On the other hand, Palatnik *et al.* (1980) and Casey *et al.* (1983) reported that there are two populations of vegetative mRNA, one which decays with a half-life of 2 hr and the other with a half-life of 10 hr. The significance of the very rapid decay of non-heat-shock messages after a temperature shift is unclear. Perhaps *Dictyostelium* amoebas can only regulate the pattern of protein synthesis by controlling the level of specific classes of mRNA. To assure the rapid accumulation of heat shock protein, non-heat-shock messengers are rapidly degraded.

We have examined RNA synthesis after heat shock in the mutant HL122. The mutant HL122 was isolated by its failure to acquire thermal protection after heat shock at 30°C. Analysis of proteins made after heat shock indicated that none of the heat shock proteins were induced following a temperature shift. This failure to induce heat shock protein synthesis results from the failure to accumulate heat shock mRNA (Figs. 3 and 8). Although the mutant failed to accumulate heat shock mRNA, other changes in the pattern of RNA accumulation appeared

---

Fig. 6. Pattern of accumulation of specific heat shock mRNA's. A log phase culture of Ax-3 ($\sim 2 \times 10^6$ cell/ml) was resuspended in MES–PDF. The culture was incubated at 22°C for 1 hr and then transferred to 30°C. At various times the cells were removed and total RNA prepared as described in Firtel and Lodish (1973). RNA was size fractionated in formaldehyde–agarose gels (Lebrach *et al.*, 1977) and transferred to a gene screen according to the manufacturer's specifications. The lanes 0, 0.5, 1, 2, and 3 represent RNA isolated from cells incubated at 30°C for 0, 0.5, 1, 2, and 3 hr, respectively. The RNA blots were hybridized with the following nick-translated *Dictyostelium* genes: (A) filter-labeled HS1.4, probed with the respective heat shock genes; filter-labeled actin, probed with the actin cDNA pcDd actin B1, which hybridizes to the two size classes of actin mRNA (McKeown *et al.*, 1978; Kindle and Firtel, 1978); filter-labeled M4-1, probed with the constitutive single-copy gene M4-1 (Kimmel and Firtel, 1979); and filter-labeled M4-repeat (M4-rep), probed with clone pC2.4, which belongs to the M4-repeat family. (B) filter-labeled D1 and hs70, probed with the respective heat shock genes. Arrows indicate RNA's of interest.

```
                aa104
Drosophila       AT   G       G T C   G AGA   T G T   C   G   G     AG
Dictyostelium   TTC AAA GGT GAA ACT AAA GTT TTC TCA CCA GAA GAA ATC TCT
Dictyostelium   phe lys gly glu thr lys val phe ser pro glu glu ile ser
Drosophila      tyr             ser     arg     ala

                118
Drosophila        G       G   G ACC   G       G   G   G   G   G   G
Dictyostelium   TCA ATG GTA CTC TTA AAA ATG AAA GAA ACC GCT GAA GCT TAT
Dictyostelium   ser met val leu leu lys met lys glu thr ala glu ala tyr
Drosophila                      thr

                132
Drosophila        G   C G G  G    C  CG G       A   C   C   A
Dictyostelium   CTT GGT AAA ACC ATT AAT AAT GCT GTA ATT ACC GTT CCA GCT
Dictyostelium   leu gly lys thr ile asn asn ala val ile thr val pro ala
Drosophila              glu ser     thr asp

                146
Drosophila        C       C   C TC    G   C   G   T           C   C  G
Dictyostelium   TAT TTC AAT GAT AGT CAA CGT CAA GCA ACC AAA GAT GCT GCT
Dictyostelium   tyr phe asn asp ser gln arg gln ala thr lys asp ala ala
Drosophila                                                          gly

                160
Drosophila      CAC   C G C GGC C G       G  TC   C   C   C       G
Dictyostelium   ACA ATT TCA AAA TTA AAT GTT CAA CGT ATT ATT AAT GAA
Dictyostelium   thr ile ser lys leu asn val gln arg ile ile asn glu
Drosophila      his     ala gly             leu


                aa278
Drosophila        C A C   C   G   C   C G   T G       G   C CAA   C
Dictyostelium   GCT TCA ATT GAA ATT GAT TCA CTC TTT GAA GGT ATT GAT TTC
Dictyostelium   ala ser ile glu ile asp ser leu phe glu gly ile asp phe
Drosophila          thr                 ala                 gln

                292
Drosophila        C   C AA  G A  GC C T   C A G       G   G   G   C   G
Dictyostelium   TAT ACT TCA ATT ACT AGA GCA CGT TTT GAA GAA CTC TGT GCT
Dictyostelium   tyr thr ser ile thr arg ala arg phe glu glu leu cys ala
Drosophila              lys val ser

                306
Drosophila      A C C C       C AAC ACC C G C G   T   G   G   G  CC C C
Dictyostelium   GAT GTT TTC CGT GGT TGT TTA GAT CCA GTT GAA AAA GTA TTA
Dictyostelium   asp val phe arg gly cys leu asp pro val glu lys val leu
Drosophila      asn leu         asn thr     gln                 ala

                320
Drosophila        C     GCC   G A       C     GGT CAG   C   C   C   C   G
Dictyostelium   AAA GAT AGT AAA TTG GAT AAG AAA TCA ATT CAT GAA ATT GTT
Dictyostelium   lys asp ser lys leu asp lys lys ser ile his glu ile val
Drosophila      asn     ala     met         gly gln         asp

                334
Drosophila      C C   C   C   A   C       C       C
Dictyostelium   TTA GTT GGT GGT TCA ACT CGT ATT CCA
Dictyostelium   leu val gly gly ser thr arg ile pro
Drosophila
```

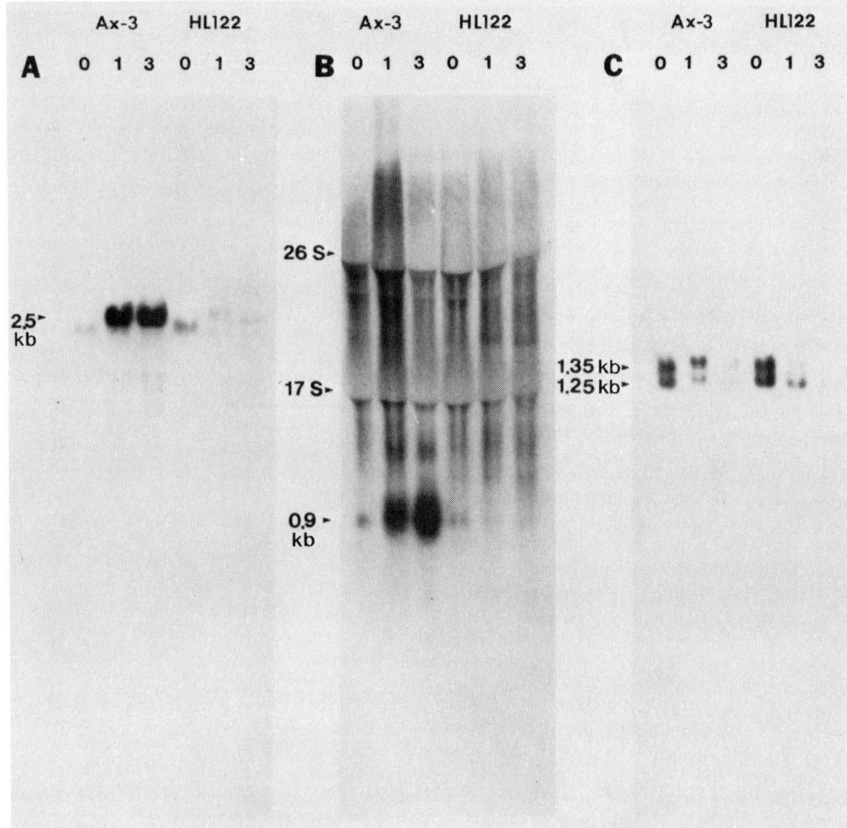

Fig. 8. Heat shock mRNA's in strain HL122. Northern blots of RNA from Ax-3 and HL122 cultures were prepared as in Fig. 6. Lanes marked 0, 1, and 3 contain RNA from cells incubated for 0, 1, or 3 hr at 30°C. The Northern blots were probed with either (A) hsp70 gene insert, (B) plasmid pC24 which belongs to the M4-repeat family, or (C) an actin cDNA, pcDd actin B1 (Kindle and Firtel, 1978; McKeown *et al.*, 1978). There are two actin mRNA size classes, 1.25 and 1.35 kb.

Fig. 7. DNA sequence of portions of cloned *Dictyostelium* hsp70 gene. The sequences of portions of the *Dictyostelium* hsp70 gene are compared to corresponding regions of one of the *Drosophila* hsp70 genes (Ignolia *et al.*, 1980). The derived amino acid sequences are shown. The numbers refer to the positions of the amino acids in the *Drosophila* sequence. Only the nucleotide and amino acid differences between the two sequences are shown for *Drosophila*.

normal. First, non-heat-shock mRNA present in the cell was degraded as seen in the loss of actin mRNA (Fig. 8). In addition, the accumulation of 17 S and 26 S ribosomal RNA was blocked as observed in wild-type cells (Fig. 4). Thus, HL122 is not a "thermometer" mutant that fails to register the temperature shift. The fact that the mutant normally regulates rRNA synthesis and the degradation of non-heat-shock mRNA suggests that the defect is involved in the expression of heat shock genes.

The defect in HL122 resulting in the failure to induce heat shock mRNA synthesis remains obscure. Conceivably, the mutation affects some mechanism specifically regulating heat shock gene expression. Alternatively, the mutation might cause a general temperature-sensitive defect in the machinery of messenger synthesis or maturation. Consistent with this interpretation are the presence of RNA species that hybridize to the M4-repeat and hs70 probes found in the mutant which are larger than those found in the wild type (Figs. 6B and 8). These RNA's might represent incorrect or immature unstable transcripts which accumulate in the mutant at high temperature.

V. THE HEAT SHOCK PROTEIN 70 GENE OF *DICTYOSTELIUM*

One of the components of heat shock response that is conserved in a wide variety of species is the major heat shock protein hsp70. DNA sequence analysis (Fig. 7) of segments of the clone hs70 indicates that it encodes the part of the hsp70 gene corresponding to amino acid 104 of the *Drosophila* gene and extending toward the carboxyl terminus (Rosen *et al.*, submitted for publication).

Sequence comparisons between the corresponding *Dictyostelium* and *Drosophila* sequences indicate that there is a 70% homology in the amino acid sequence of the two genes in the regions examined. This is similar to the degree of homology found between *Drosophila* and yeast hsp70 genes (Ignolia *et al.*, 1980, 1982). The nucleotide sequences are less conserved; there is only 55% homology between *Dictyostelium* and *Drosophila* DNA sequences. *Dictyostelium* genomic DNA has a very low G + C content (22%) with much of the nonprotein coding sequence having a G + C content of only 10–20%. The coding region has a G + C content of 38%, close to the lower limit possible to encode standard proteins (Kimmel and Firtel, 1982, 1983). As expected, the codon utilization is highly skewed to codons having a high A + U content (Kimmel and Firtel, 1983). Interestingly, almost half of the differences in the sequence of the genes are substitutions of A–T base pairs in *Dictyostelium* for G–C base pairs in *Drosophila* at the third position of codons.

Analysis of Northern blots of heat shock RNA indicates that a probe from within the hsp70 gene hybridizes to two mRNA's of ~2.5 kb in length (Fig. 6B).

The larger message is present at 22°C but is induced approximately 50-fold after heat shock. The slightly smaller message is also present at 22°C but is lost after the temperature shift. Conceivably, there are cognate hsp70 genes in *Dictyostelium* as there are in yeast and *Drosophila* (Ignolia *et al.*, 1982; Ignolia and Craig, 1982). Alternatively, the hsp70 gene may produce two different transcripts during incubation at 22 and 30°C. During growth at 22°C both transcripts may be synthesized at a low constitutive level, whereas after heat shock the large message is induced while the smaller one is lost.

VI. A HEAT SHOCK-INDUCED MESSAGE IS ENCODED BY A TRANSPOSABLE ELEMENT

The clone HS1.4 contains a 4.2-kb *Eco*RI fragment that hybridizes to a 0.9-kb heat shock-inducible poly(A)$^+$ RNA (Fig. 6). Analysis of genomic fragments hybridizing to HS1.4 indicates that the 4.2-kb insert forms the central portion of a 4.9-kb transposon, which has been named Tdd-1 (Fig. 9) (Rosen *et al.*, 1983). Analysis of genomic Southern blots indicates there are ~50 copies of the conserved element in the genome as well as 100 diverse fragments containing Tdd-1 homologous sequences. The Tdd-1 transposon contains 313 base-pair inverted repeats near the end of the element, but, unlike other eukaryotic transposons, one end of Tdd-1 extends 35 base pairs (bp) beyond the repeat (Rosen *et al.*, 1983). The pattern of hybridization of sequences flanking Tdd-1 are different in different wild-type and laboratory *Dictyostelium* strains. This is in contrast to the regions flanking the 17–20 members of the actin multigene family, suggesting that Tdd-1 is a mobile genetic element.

We have localized the heat shock RNA coding region to the right *Hae*III·-*Eco*RI fragment (Fig. 9) by probing RNA blots of heat shock RNA with labeled subfragments of HS1.4. Furthermore, using probes made with subclones in which the 2.1-kb *Hind*III·*Eco*RI fragment is cloned in both orientations in M13, we have determined the polarity of the heat shock message as shown in Fig. 9

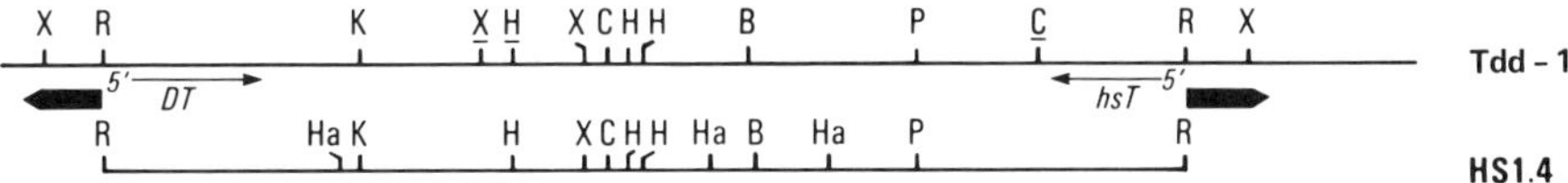

Fig. 9. Restriction map of Tdd-1. The restriction map of the 4.2-kb insert of HS1.4 is shown on the bottom line. The top line presents the composite map of the transposon Tdd-1 (see Rosen *et al.*, 1983). The heavy bullet-shaped lines beneath the map of Tdd-1 indicate the inverted repeats. The arrows DT and hsT indicate the polarity of the developmental and heat shock-induced transcripts hybridizing to the element. The restriction site symbols are: *Bgl* (B); *Cla*I (C); *Hind*III (H); *Hae*III (Ha); *Kpn*I (K); *Pvu*II (P); *Eco*RI (R); and *Xba*I (X). Partially conserved restriction sites of Tdd-1 are underlined. The *Hae*III sites are shown for the clone HS1.4 but not for the composite Tdd-1 element.

(Rosen *et al.*, 1983). The element also hybridizes to several developmentally regulated transcripts which have the opposite polarity as the heat shock-inducible transcript.

VII. HEAT SHOCK PROTEINS

Since the physiological mechanisms of the heat shock response are only poorly understood, it has not been possible to define heat shock proteins by their functional roles. It then becomes a bit arbitrary as to which proteins should be considered specifically in context with heat shock.

In *Drosophila,* the heat shock response was first recognized by the puffing of seven major chromosomal bands (Ritossa, 1962). Heat shock induced proteins which could be shown to be coded for at these bands were clearly heat shock proteins. These include hsp 82, 70, 68, 27, 26, 23, and 22 (see Chapter 1, this volume). However, four or five distinct genes at two separate puffs turned out to code for hsp70 while all the low-molecular-weight heat shock proteins (hsp 27, 26, 23, and 22) have genes at a single puff. The products of several heat shock puffs have yet to be recognized.

Because of the conservation of immunological determinants on hsp70 among *Dictyostelium, Drosophila,* and chicks (Kelley and Schlesinger, 1982), there is little question that this protein is similar among species. Comparison of the predicted amino acid sequence from the nucleic acid sequence of portions of cloned hsp70 genes in yeast, chick, *Drosophila,* and *Dictyostelium* shows that at least 70% of the residues are conserved among different pairs of species (Fig. 10). In the region of amino acids 297 to 322 of *Drosophila,* four residues can be seen to be unique to *Dictyostelium* while this region is conserved in three or more of the other species. Conversely, residues unique to *Drosophila,* yeast, or chick are still conserved in *Dictyostelium.* When the degree of conservation is calculated over this region, *Dictyostelium* falls about equidistant from yeast and chick but significantly further from *Drosophila* (Fig. 11).

	297			300											310											320		
Dictyostelium	arg	ala	arg	phe	glu	glu	leu	cys	ala	asp	val	phe	arg	gly	cys	leu	asp	pro	val	glu	lys	val	leu	lys	asp	ser	lys	leu
Drosophila	.	.	.	.	.	.	.	.	.	asn	leu	.	.	asn	thr	.	gln	.	.	.	.	ala	.	asn	.	ala	.	met
chick	.	.	.	.	.	.	.	asn	.	.	leu	.	.	.	thr	.	ser	gln	ser	.	.	ala	.	arg	.	ala	.	.
yeast	.	.	.	.	.	.	.	.	.	.	leu	.	.	ser	thr	.	.	.	.	.	.	.	.	arg (lys)	.	ala	.	.
E. coli	302 .	.	lys	leu	.	ser	.	val	glu	.	leu	val	asn	arg	ser	ile	glu	.	leu	lys	val	ala	.	gln	.	ala	gly	.

Fig. 10. Comparison of hsp70 sequences. The sequence between amino acids 297 and 320 of *Drosophila* hsp70 encoded by homologous DNA sequences is compared to the *Dictyostelium* sequence. Where no amino acid is indicated, the residue is the same as in *Dictyostelium.* The chick sequence is from Morimoto and Hunt (1984); the *dnaK* sequence of *E. coli* is from Bardwell and Craig (1984); and the yeast sequence is from Ignolia *et al.* (1982).

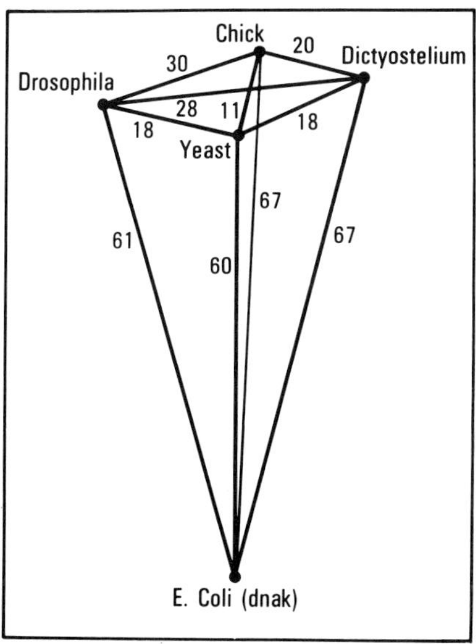

Fig. 11. Evolutionary divergence of hsp70. The amino acid sequences shown in Fig. 10 (amino acids 297 to 320) were analyzed for the proportion of divergence calculated as amino acid changes per 100 residues. The distances between the sequences in diverse organisms are roughly to scale. The degree of divergence in this region is similar to that calculated in other regions.

A heat shock protein of *Escherichia coli* is coded for by the *dnaK* gene (Bardwell and Craig, 1984). When the sequence of this gene is compared to the eukaryotic hsp70 gene, an amazing homology can be seen in the sequence near amino acid 300 (Fig. 10). There is little doubt that these genes shared a common ancestor before evolutionary divergence. The *E. coli* sequence is about equally distant from each of the eukaryotic sequences (Fig. 11).

The position of hsp70 on two-dimensional gels is very similar in material from yeast, *Dictyostelium,* and *Drosophila,* indicating that the size and isoelectric point of these proteins have been conserved (Fig. 12). In each case hsp70 separates into several distinct spots. For *Drosophila,* this may be the consequence of a slight divergence of the multiple hsp70 genes as well as the fact that hsp68 is a closely related protein. For *Dictyostelium,* there is evidence from analysis of partial proteolytic cleavage products of the isolated spots that they comprise two related but distinct proteins, each with a major phosphorylated member (Loomis *et al.*, 1982). Moreover, there is an indication on many two-dimensional gels for another heat shock-induced protein of M_r of about 68,000 running just below hsp70 (Fig. 13). Heat shock protein 70 of *Dictyostelium* is phosphorylated on threonime residues, which makes it somewhat unusual among the phosphoproteins, most of which carry serine phosphate. The phosphate is

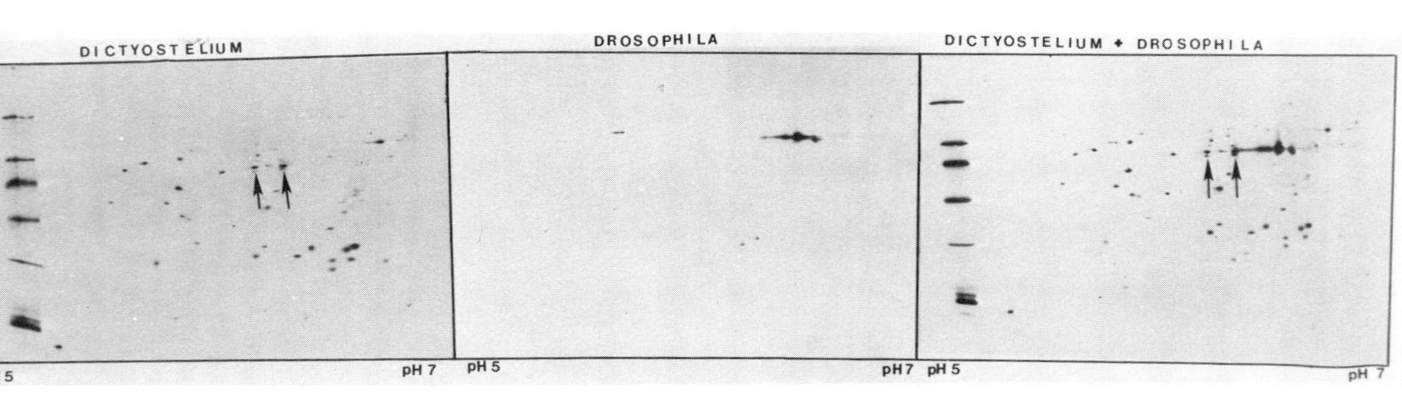

Fig. 12. Electrophoretic mobility of hsp70 from *Dictyostelium* and *Drosophila*. Cells of either *D. discoideum* or *D. melanogaster* were heat shocked at 30 and 37°C, respectively, before being labled with [^{3}H]leucine. Proteins were separated on two-dimensional gels (acidic end shown to the left). Samples from *Dictyostelium* and *Drosophila* were analyzed separately and after being mixed. Arrows indicate the hsp 70 proteins of *Dictyostelium*. The *Drosophila* samples were kindly provided by Dr. Susan Lindquist.

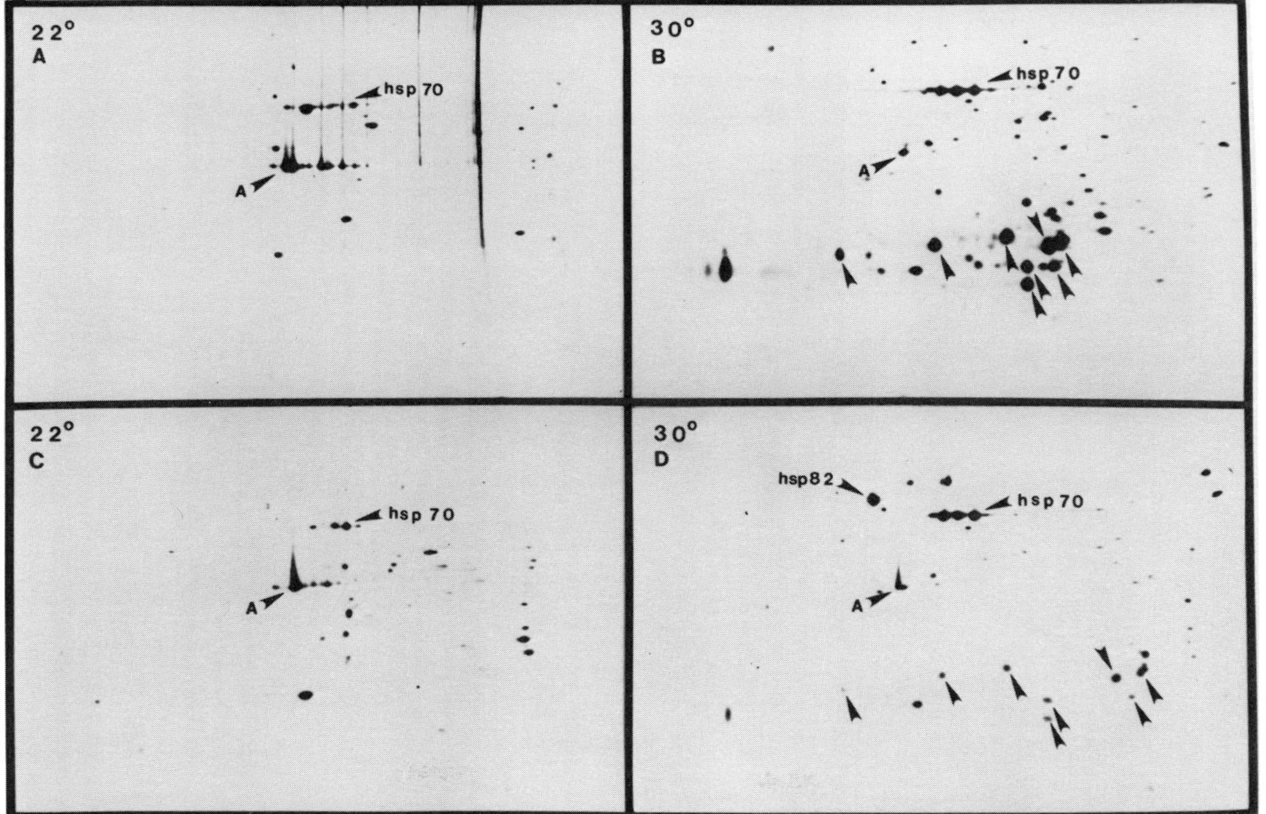

Fig. 13. Subcellular localization of heat shock proteins. Cells of strain Ax-3 growing at 22°C were transferred to 30°C. After 1 hr at 30°C [^{35}S]methionine was added, and the newly made proteins were labeled for 2 hr (B and D). Control cells were labeled for 2 hr at 22°C (A and C). The cells were broken and the nuclei (A and B) and supernatant fractions (C and D) were separated. Proteins containing 10^5 cpm from each fraction were analyzed by two-dimensional gel electrophoresis. The acid end (pH 3.5) is on the left. Arrow A indicates actin. Other arrows indicate low-molecular-weight proteins induced by heat shock. (From Loomis and Wheeler, 1982.)

rapidly turned over ($t_{1/2} \sim 20$ min). This dynamic state suggests that phosphorylation of hsp70 may modulate its role.

Heat shock induces the synthesis of eight proteins ranging in molecular weight from 26,000 to 32,000 in *Dictyostelium* (Fig. 13). These low-molecular-weight heat shock proteins rapidly enter the nucleus and become associated with the chromatin (Loomis and Wheeler, 1982). On shift-down from 30 to 22°C, the LMW heat shock proteins disappear from the nucleus within 3 hr, coincident with the loss of protection conferred by heat shock. The LMW heat shock proteins of *Drosophila* are also preferentially found in the nucleus and exit on shift-down (Velasquez *et al.*, 1980; Arrigo *et al.*, 1980). While they are roughly the same size as the LMW heat shock proteins of *Dictyostelium,* they have distinct isoelectric points. These LMW proteins, therefore, do not seem to be as highly conserved between species as the larger heat shock proteins.

While hsp70 has been conserved to a surprising extent in all cells analyzed so far, no direct evidence has been found concerning its physiological role. Its abundance in many cell types including *Dictyostelium* suggests that it may play a structural role. Likewise, no direct functions have been indicated for hsp82 or the LMW nuclear heat shock proteins in any organism. Perhaps by insertional mutagenesis using cloned sequences in yeast or other systems the physiological roles can be determined.

Acknowledgments

We would like to thank Drs. E. Craig, R. Morimoto, and S. Lindquist for providing us with fascinating unpublished data. During part of the period of this work, E.R. was supported by a postdoctoral research training award from NIH. A.S. is supported by the NATO Science Fellowship program and the Danish Science Research Council. During part of this work, R.A.F. was the recipient of a Faculty Research Award from the ACS. This work was supported by grants from the National Institute for Health to R.A.F. and from the ACS to W. F. L.

REFERENCES

Arrigo, A. P., Fakan, S., and Tissières, A. (1980). Localization of the heat shock induced proteins in *Drosophila melanogaster* tissue culture cells. *Dev. Biol.* **78,** 84–103.

Ashburner, M., and Bonner, J. J. (1979). The induction of gene activity in *Drosophila* by heat shock. *Cell* **17,** 241–254.

Bardwell, J. A. C., and Craig, E. (1984). Major heat shock gene of *Drosophila* and the *Escherichia coli* heat inducible *dnak* gene are homologous. *Proc. Natl. Acad. Sci. U.S.A.* **81,** 848–852.

Bienz, M., and Pelham, H. (1982). Expression of a *Drosophila* heat shock protein in *Xenopus* oocytes: Conserved and divergent regulatory signals. *EMBO J.* **1,** 1583–1588.

Casey, L., Palatnik, C. M., and Jacobson, A. (1983). Messenger RNA half-life in *Dictyostelium discoideum. Dev. Biol.* **95,** 239–243.

Corces, V., Pellicer, A., Axel, R., and Meselson, M. (1981). Integration, transcription and control of a *Drosophila* heat shock gene in mouse cells. *Proc. Natl. Acad. Sci. U.S.A.* **78,** 7038–7042.

Firtel, R. A., and Lodish, H. F. (1973). A small nuclear precursor RNA in the cellular slime mold *Dictyostelium discoideum*. *J. Mol. Biol.* **79,** 295–314.

Firtel, R. A., Baxter, L., and Lodish, H. F. (1973). Actinomycin D and the regulation of enzyme biosynthesis during development of *Dictyostelium discoideum*. *J. Mol. Biol.* **79,** 315–327.

Gerner, E. W., and Schneider, M. J. (1975). Induced thermal resistance in HeLa cells. *Nature* **256,** 500–502.

Ignolia, T. D., and Craig, E. A. (1982). *Drosophila* gene related to the major heat shock induced gene is transcribed at normal temperatures and not induced by heat shock. *Proc. Natl. Acad. Sci. U.S.A.* **79,** 525–529.

Ingolia, T. D., Craig, E. A., and McCarthy, B. J. (1980). Sequence of three copies of the gene for the major *Drosophila* heat shock induced protein and their flanking regions. *Cell* **21,** 669–679.

Ignolia, T. D., Slater, M., and Craig, E. A. (1982). *Saccharomyces cerevisiae* contains a complex multigene family related to the major heat shock inducible gene of *Drosophila*. *Mol. Cell. Biol.* **2,** 1388–1398.

Kelley, P. M., and Schlesinger, M. J. (1978). The effect of amino-acid analogues and heat shock on gene expression in chicken embryo fibroblasts. *Cell* **15,** 1277–1286.

Kelley, P. M., and Schlesinger, M. J. (1982). Antibodies to two major chicken heat shock proteins cross react with similar proteins in widely divergent species. *Mol. Cell. Biol.* **2,** 267–274.

Kimmel, A. R., and Firtel, R. A. (1979). A family of short, interspersed repeat sequences at the 5′ end of a set of *Dictyostelium* single copy mRNAs. *Cell* **16,** 787–796.

Kimmel, A. R., and Firtel, R. A. (1982). The organization and expression of the *Dictyostelium* genome. *In* "The Development of Dictyostelium discoideum" (W. F. Loomis, ed., pp. 233–324. Academic Press, New York.

Kimmel, A. R., and Firtel, R. A. (1983). Sequence organization in *Dictyostelium:* Unique structure at the 5′-ends of protein coding genes. *Nucleic Acids Res.* **11,** 541–552.

Kindle, K. L., and Firtel, R. A. (1978). Identification and analysis of *Dictyostelium* actin genes, a family of moderately repeated genes. *Cell* **15,** 763–778.

Lebrach, H., Diamond, D., Wozney, J. M., and Boedtken, H. (1977). RNA molecular weight determinations by gel electrophoresis under denaturing conditions: A critical examination. *Biochemistry* **16,** 4743–4751.

Loomis, W. F. (1969). Temperature sensitive mutants of *Dictyostelium discoideum*. *J. Bacteriol.* **99,** 65–99.

Loomis, W. F. (1975). "*Dictyostelium discoideum:* A Developmental System." Academic Press, New York.

Loomis, W. F., and Wheeler, S. A. (1980). Heat shock response of *Dictyostelium*. *Dev. Biol.* **79,** 399–408.

Loomis, W. F., and Wheeler, S. A. (1982). Chromatin associated heat shock proteins of *Dictyostelium*. *Dev. Biol.* **90,** 412–418.

Loomis, W. F., Wheeler, S. A., and Schmidt, J. (1982). Phosphorylation of the major heat shock protein of *Dictyostelium discoideum*. *Mol. Cell. Biol.* **2,** 484–489.

Li, G. C., and Hahn, G. M. (1978). Ethanol-induced tolerance to heat and to adriamycin. *Nature (London)* **274,** 699–701.

McAlister, L., and Finkelstein, D. B. (1980). Heat shock proteins and thermal resistance in yeast. *Biochem. Biophys. Res. Commun.* **93,** 819–824.

McKeown, M., Tayler, W., Kindle, K., Firtel, R. A., Bender, W., and Davidson, N. (1978). Multiple, heterogenous actin genes in *Dictyostelium*. *Cell* **15,** 789–800.

Margolskee, J. P., and Lodish, H. F. (1980a). Half-lives of messenger RNA species during growth and differentiation of *Dictyostelium discoideum*. *Dev. Biol.* **74,** 37–49.

Margolskee, J. P., and Lodish, H. F. (1980b). The regulation of the synthesis of actin and two other proteins induced early in *Dictyostelium discoideum* development. *Dev. Biol.* **74,** 50–64.

Mirault, M.-E., Goldschmidt-Clermont, M., Moran, L., Arrigo, A. P., and Tissières, A. (1978). The effect of heat shock on gene expression in *Drosophila melanogaster. Cold Spring Harbor Symp. Quant. Biol.* **42,** 819–827.

Mirault, M.-E., Delwart, E., and Southgate, R. (1982). A DNA sequence upstream of *Drosophila* hsp70 genes is essential for their heat shock induction in monkey cells. *In Heat Shock, from Bacteria to Man*'' (M. J. Schlesinger, M. Ashburner, and A. Tissières, eds.), pp. 35–42. Cold Spring Harbor Lab., Cold Spring Harbor, New York.

Mitchell, H. K., Moller, G., Petersen, N. S., and Lipps-Sarmiento, L. (1979). Specific protection from phenocopy induction by heat shock. *Dev. Genet. (Amsterdam)* **1,** 181–192.

Morimoto, R., and Hunt, C. (1984). The gene for the major 70,000 dalton heat shock protein is highly conserved in eukaryotes: Cloning of the chicken hsp70 gene. *Proc. Natl. Acad. Sci. U.S.A.* (in press).

Palatnik, C., Storti, R., Capone, A., and Jacobson, A. (1980). Messenger RNA stability in *Dictyostelium discoideum.* Does poly adenylic-acid have a regulatory role? *J. Mol. Biol.* **141,** 99–118.

Pelham, H. R. B. (1982). A regulatory upstream promoter element in the *Drosophila* hsp70 heat shock gene. *Cell* **30,** 517–538.

Petersen, N. S., and Mitchell, H. K. (1981). Recovery of protein synthesis after heat shock: Prior heat treatment affects the ability of cells to translate mRNA. *Proc. Natl. Acad. Sci. U.S.A.* **78,** 1708–1711.

Ritossa, F. (1962). A new puffing pattern induced by temperature shock and DNP in *Drosophila. Experientia* **18,** 571–573.

Rosen, E., Sivertsen, A., and Firtel, R. A. (1983). An unusual transposon encoding heat shock inducible and developmentally regulated transcripts in *Dictyostelium. Cell* **35,** 243–251.

Rosen, E., Sivertsen, A., Nellen, W., Romans, P., and Firtel, R. A.. Heat shock induced changes in gene expression in *Dictyostelium.* Submitted for publication.

Storti, R. V., Scott, M. P., Rich, A., and Pardue, M. L. (1980). Translational control of protein synthesis in response to heat shock in *D. melanogaster* cells. *Cell* **22,** 825–834.

Valazquez, J. M., DiDomenico, B. J., and Lindquist, S. (1980). Intracellular localization of heat shock proteins in *Drosophila. Cell* **20,** 679–689.

Voellmy, R., and Rungger, D. (1982). Transcription of a *Drosophila* heat shock gene is heat induced in *Xenopus* oocytes. *Proc. Natl. Acad. Sci. U.S.A.* **79,** 1776–1780.

14

Plant Productivity, Photosynthesis, and Environmental Stress

DONALD R. ORT AND JOHN S. BOYER

I. INTRODUCTION

Plants frequently face periods of environmental stress sufficient to diminish their reproductive capacity. In natural populations, these periods represent a major force leading to genetic change, and those individuals with superior char-

Abbreviations used: B, Secondary quinone acceptor of photosystem II(Q_B); CF_0 and CF_1, the intrinsic and extrinsic membrane portions of the coupling factor enzyme complex; $[CO_2]_i$, $[CO_2]_a$, intercellular CO_2 concentration, ambient CO_2 concentration; cyt, cytochrome; DAD, DAD_{ox}, diaminodurene, diiminodurene; DCMU, 3-(3,4-dichlorophenyl)-1,1-dimethylurea; DHQ, duroquinol; HEPES, *N*-2-hydroxylpiperazine-*N'*-2-ethanesulfonic acid; MOPS, 3-(*N*-morpholino)propanesulfonic acid; MV, methyl viologen; P_{680}, P_{700}, primary electron donors of photosystem II and photosystem I; PAR, photosynthetically active radiation; PC, plastocyanin; p_{CO_2}, partial pressure of CO_2; PD, PD_{ox}, *p*-phenylenediamine, *p*-phenylenediimine; PQ, plastoquinone/plastoquinol; PS, photosystem; ψ_l, leaf water potential; ϕ, quantum yield; Q, first quinone acceptor of photosystem II(Q_A); R-FeS, Rieske iron–sulfur center; RH, relative humidity; and X, to denote the membrane bound photosystem I electron acceptors.

Changes in Eukaryotic Gene Expression
in Response to Environmental Stress

ISBN 0-12-066290-6

acters leave more progeny in succeeding generations. In agriculture, successful genetic traits are not so closely linked to the composition of succeeding generations because of the intervention of man. Genetic manipulation, nevertheless, plays a large role in improving agricultural productivity, and part of the improvement can be attributed to the plant's coping better with environmental stress (Boyer, 1982).

A clear indication of the magnitude of stress-induced losses in agricultural productivity can be determined by comparing the genetically determined potential for the productivity of a crop under nonlimiting conditions, that is the so-called record yields, with average productivity realized by the farmers. It is worthwhile to specify that the terms "productivity" and "yield" will be used interchangeably in our chapter and that we restrict their meaning to the production of economically desirable plant parts. The record yields and the average yields of eight major United States crops are assembled in Table I. The difference between average yields and record yields for some crops is astonishing, as with wheat and sorghum, in which record yields are greater by a factor of more than seven. Crops having reproductive structures as the sought-after plant part (i.e., corn, wheat, soybean, sorghum, oats, and barley) show greater discrepancies between average and record yields than those with marketable vegetative structures (i.e., potatoes and sugar beets). Even in crops with marketable vegetative structures, however, record yields are still three times larger than average yields.

The explanation for the remarkable difference between the genetic potential

TABLE I

Record Yields, Average Yields, and Yield Losses Due to Unfavorable Physicochemical Environments in Major Crops of the United States as of 1975[a]

Crop	Record yield	Average yield	Average losses due to unfavorable environments
Corn	19,300	4,600	12,700
Wheat	14,500	1,880	11,900
Soybean	7,390	1,610	5,120
Sorghum	20,100	2,830	16,200
Oat	10,600	1,720	7,960
Barley	11,400	2,050	8,590
Potato	94,100	28,200	50,900
Sugar beet	121,000	42,600	61,300

[a] Data are from Boyer (1982). Units: kilograms per hectare.

for crop production and the production actually realized by the farmer lies in the prevalence of biological pests, which cause significant losses, and in the limitations imposed by physicochemical factors (Table I), which cause even larger losses. Those environmental parameters which may conspire to diminish yields are numerous, but the most significant ones can be categorized in two broad areas: inappropriate soils and unfavorable climates. It has been estimated (Boyer, 1982) that only about 12% of the United States land surface can provide a soil environment for plant growth which does not normally place constraints on productivity, and even this 12% provides constraints when plant nutrients are not supplied to replenish those removed as a part of the harvest. The deleterious effects of unfavorable climates are equally widespread. Inadequate water, excessive water, and cold are the most ruinous conditions for crop productivity in this country (Boyer, 1982). These considerations emphasize that research aimed at improvements in productivity is often more appropriately directed at bringing average yields closer to the existing genetic potential than at further increasing the genetic potential. Research directed at this goal should focus on how plants can most successfully deal with the physicochemical factors of the environment that limit plant growth.

Because plant growth is the result of many integrated and regulated physiological processes, it is seldom possible to assign limitations of plant growth to a single process. Certainly in the field, plant growth is controlled simultaneously and to varying extents by a number of limiting processes. However, among the diversity of overlapping factors there are likely to be some central physiological processes that are highly sensitive to the environment and are, therefore, dominant in determining plant response to stress. In many instances the dominant process is photosynthesis. Plant growth as biomass production is in fact simply a measure of net photosynthesis integrated over time, so factors limiting plant growth are unavoidably the same factors that limit net photosynthesis. It is not surprising, as Zelitch (1982) has pointed out, that there is considerable evidence linking the seasonal photosynthetic performance of plant canopies to the production of economic yield.

The efficiency of light energy conversion into photosynthate is an important aspect of this link, and it is worthwhile considering the efficency of photosynthesis that is attained in the field compared to that which is theoretically attainable given the mechanisms of photosynthesis that have evolved. Good and Bell (1980) have treated this subject in a way that leads to an important conclusion regarding improvements in productivity. Four moles of electrons, originating from the photosynthetic oxidation of water, are required to reduce 1 mole of CO_2 to the level of carbohydrate. By virtue of the sequential arrangement of the two photosystems in higher plant photosynthesis, 8 moles of quanta are necessary to drive the reduction process. The average energy content of the quanta can be calculated, and the calculation is simplified by selecting an average wavelength for photosynthetically active light of 550 nm, in which case the 8 moles of quanta are equivalent to 422 kcal. One mole of CO_2 reduced to the level of carbohydrate stores about 114 kcal. The ratio of these two numbers is 0.27 and represents the maximum efficiency with which the energy of short wavelength light can be stored in the products of photosynthesis. This efficiency must be further corrected downward for the spectrum of sunlight at wavelengths beyond 700 nm which are not sufficiently energetic to participate in photosynthesis in higher plants. These factors reduce the theroretical maximum storage of sunlight energy in the products of photosynthesis to about 0.11. Then it must be recognized that rapid rates of reaction, which remove the system appreciably from the equilibrium condition, are always achieved at the expense of efficiency. Good and Bell (1980) point out that irreversibility losses associated with disequilibrium are inescapable for any reaction including those of photosynthesis. Finally, we must not forget the energy costs of building and maintaining the photosynthetic machinery. The theoretical maximum sustained by photosynthesis is thus unable to conserve anything close to 0.11 of the incident light, how much less is difficult to determine. For our purposes, we would like to know how close to the theoretical maximum photosynthesis in the field can operate. The measurements obtained by R. B. Musgrave and E. R. Lemon (from Good and Bell, 1980) for corn fields in midday during the period of maximum growth indicate that 5% of the total energy of incident sunlight is conserved in products of photosynthesis even at high rates for which irreversibility losses should be great. It is clear that this value must be approaching the theoretical maximum, and, therefore, the maximum efficiency actually attainable in the field is unlikely to be improved by anything short of the evolution of a new mechanism for photosynthesis. As with the comparison of record yields to average yields, plants under field conditions seldom achieve 5% conversion of light energy into carbohydrate and never on a sustained, season-long basis. One or more physicochemical, environmental factors usually conspire to lower the energy conservation efficiency, and in some cases the decrease is so drastic that diminished photosynthetic capacity becomes the direct determinant of yield (e.g., McPherson and Boyer, 1977).

Work from our laboratories has been aimed at uncovering the biochemical and physiological basis underlying the inhibition of photosynthesis by two separate environmental stresses, each of which is an important cause of unattained agricultural productivity: water deficiency and exposure of thermophilic plants to chilling temperatures. The literature in each of these areas is abundant, but it is more remarkable for the number of contradictory findings than it is for clues to underlying mechanisms. Our effort here will not be a comprehensive review of the literature but instead the presentation of a research strategy for dealing with the special experimental problems encountered during studies of the mechanisms of inhibition of photosynthesis by unfavorable environmental conditions.

II. RESEARCH STRATEGY

The mechanism of inhibition of photosynthesis brought on by environmental stress cannot be adequately studied at a single level of organization, and attempts to do so often lead to grossly inaccurate conclusions. The photosynthetic performance of a plant relies on certain processes, such as leaf–atmosphere gas exchange and plant or plant canopy regulation of light interception, which perforce must be assessed in the whole plant. At the other extreme of organization, photosynthetic performance also relies on the membrane-associated reactions of electron transfer and ATP formation and the functioning of the Calvin cycle enzymes in chloroplast stroma. The detailed analysis of these requires a variety of *in vitro* subcellular preparations. The special difficulty associated with studying stress-induced perturbations is the difficulty of distinguishing primary inhibitory events from the inevitable secondary consequences which are a result rather than a cause of inhibition. In other cases it is difficult to distinguish between inhibitions which are actually part of the *in vivo* process and inhibitions induced through the manipulation of tissue made labile by the stress. Thus, one can easily study a stress-induced biochemical phenomenon which actually presents no real-life problem for the intact plant. Oftentimes a systematic approach that emphasizes a search for corroborating and complimentary results from experiments with materials ranging in organization from intact plants to isolated thylakoid membrane protein complexes can successfully deal with the very troublesome problems associated with stress-induced inhibition of photosynthesis. This involves demonstrating that effects detectable at the subcellular level also occur at the whole-plant level and, equally important, that they are rate limiting at the whole-plant level. The final step requires that the putative limitation be shown to determine the success of the plant in coping with the stress under field conditions. Such an approach is perhaps the only way to build an integrated framework of knowledge from which additional work can be logically developed. Such an approach is also the best safeguard against being misled by accurate but irrelevant data.

A. Measurement of Whole-Plant Gas Exchange

Gas exchanges between the intercellular air spaces of leaves and the atmosphere occur almost exclusively through stomata. Since the exchange of water vapor and CO_2 between the leaf and the atmosphere unavoidably occurs along a common pathway, the relative rates of diffusion of the two gases in the steady state can be computed from concentration differences and diffusion coefficients according to Fick's first law of diffusion. The H_2O concentration difference underlying the loss of water vapor from the leaf into the atmosphere, from which

CO_2 must be obtained, is often very much greater than the CO_2 concentration difference. In consequence, plants lose a great deal of water in exchange for a modest acquisition of CO_2. The data in Table II reflect this imbalance in four species of plants measured in the laboratory. The imbalance is usually far greater under field conditions. As expected, the C_4 plant (corn), which operates at a lower internal CO_2 concentration, displayed a more favorable ratio, reflecting the greater driving force for CO_2 diffusion into the leaf. Mesophytic plants close stomata to limit water loss and, oftentimes when challenged by stress, do so at the expense of photosynthesis. This occurs because the stomatal resistance to gas fluxes increases and net photosynthesis declines in response to a diminished intercellular CO_2 concentration.

1. The Dependence of the Rate of Photosynthesis on the Intercellular CO_2 Concentration

To begin to understand the basis for stress-induced inhibition of photosynthesis, it is necessary to identify the various components of inhibition and to measure the effects of each separately. The first determination to be made is whether the inhibition is due to an increase in stomatal resistance to gas exchange or is of nonstomatal origin. Since direct control of stomatal aperture over the rate of CO_2 reduction is exerted through the control that stomates have on intercellular CO_2 concentration ($[CO_2]_i$), the contribution of stomatal resistance to stress-induced inhibition of photosynthesis can be evaluated and actually measured by examining the dependence of the rate of photosynthesis on the intercellular CO_2 concentration before, during, and after stress. Intercellular CO_2 concentrations can readily be computed from measurements in a leaf chamber of transpiration (T), water vapor concentration difference between the leaf and the ambient atmosphere, CO_2 concentration of the ambient atmosphere ($[CO_2]_a$), and CO_2 exchange rate (P).

$$[CO_2]_i = [CO_2]_a - (P/T)\,(D_{H_2O}/D_{CO_2})\,([H_2O]_i - [H_2O]_a)$$

In this expression, the subscripts i and a represent the concentration of CO_2 and H_2O vapor in the intercellular air spaces of the leaf and in the ambient atmosphere. The diffusion coefficients for water and CO_2 in still air are D_{H_2O} and D_{CO_2}. Moss and Rawlins (1963) have shown that these computations, based on simple physical considerations, are valid provided that the diffusive pathways for water vapor and CO_2 are identical. The computations have recently been further validated by direct measurement of the internal CO_2 concentrations of leaves (Sharkey *et al.*, 1982).

An understanding of the dependence of the rate of light-saturated photosynthesis on the intercellular CO_2 concentration is facilitated by the generalized situation depicted in Fig. 1 for a plant with C_3 metabolism. Point A reflects a typical situation prior to stress in which the plant is photosynthesizing at atmo-

TABLE II

Water Use Efficiency Calculated from Gas Exchange Data

Plant	P (μmoles CO_2 m^{-2} sec^{-1})	T (mmoles H_2O m^{-2} sec^{-1})	moles H_2O : moles CO_2
Tomato (*Lycopersicon esculentum* Mill.)[a]	19	4.6	242
Soybean (*Glycine max* L. Merr.)[b]	21	2.8	133
Corn (*Zea mays* L.)[c]	35	2.2	63
Sunflower (*Helianthus annus* L.)[d]	38	6.2	163

[a] Measurement of an attached leaf at 25°C, 69% RH, with an incident light intensity of 1800 μeinsteins m^{-2} sec^{-1} of photosynthetically active radiation.

[b] Measurement made on a whole potted plant at 25°C, 77% RH, with an incident light intensity of 1800 μeinsteins m^{-2} sec^{-1} of photosynthetically active radiation.

[c] Measurement made on a whole potted plant at 25°C, 77% RH, with an incident light intensity of 1800 μeinsteins m^{-2} sec^{-1} of photosynthetically active radiation.

[d] Measurement of an attached leaf at 25°C, 70% RH, with an incident light intensity of 1800 μeinsteins m^{-2} sec^{-1} of photosynthetically active radiation.

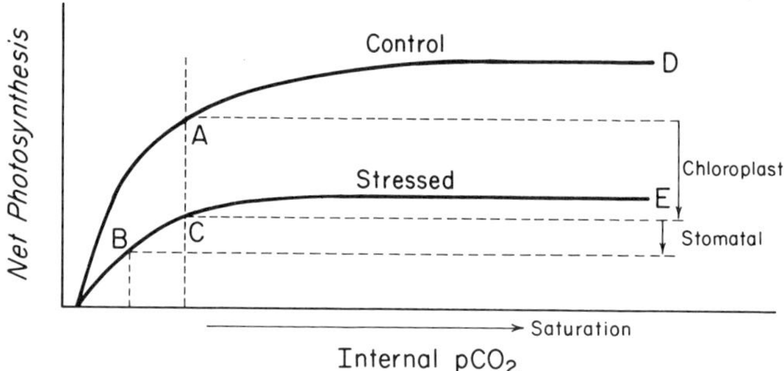

Fig. 1. Generalized diagram depicting the dependence of the rate of light-saturated photosynthesis on the intercellular CO_2 concentration (p_{CO_2}). The interpretation of this diagram is described in the text (Section II,A,1)

spheric CO_2 levels (i.e., 300 μbars CO_2). Even under these optimal conditions the steady-state intercellular concentration is less than the atmospheric level. Following stressed-induced inhibition of CO_2 reduction, an increase in stomatal resistance and a decrease in the intercellular CO_2 concentration often result (point B in Fig. 1). The stomatal contribution to the overall stress-induced impairment of photosynthesis can be overcome, and thereby revealed, when any decreases in the intercellular CO_2 concentration caused by the stress are reversed by increasing the ambient CO_2 level. In other words, any remaining inhibition of photosynthesis, after readjustment of the intercellular CO_2 concentration to the level existing prior to stress (point C in Fig. 1), must be of non-stomatal origin. If the stressed-induced inhibition had entirely nonstomatal causes, then no decrease in the intercellular CO_2 level would be expected and, indeed, some increase would be anticipated. By further elevating the ambient CO_2 level, a saturating CO_2 concentration at the site of CO_2 reduction can be attained, and the rate of photosynthesis is then no longer affected by diffusional resistances. Thus, any differences in rate which persist at high CO_2 concentrations also reflect photosynthesis inhibitions of nonstomatal origin (compare D and E, Fig. 1).

a. Chilling-Induced Inhibition of Photosynthesis in Tomato. In a fairly diverse group of chill-sensitive plants, photosynthesis is inhibited for an extended period after brief exposure in the dark to low temperature ($0°C < T < 12°C$). Dramatic reductions in the rate of whole-plant photosynthesis have been documented in numerous plants (e.g., Drake and Raschke, 1974; Izhar and Wallace, 1967; Martin *et al.*, 1981), including several of the most significant commercial crop plants of temperate North America. Although it is clear that chilling of thermophilic plants in the dark results in inhibition of photosynthesis the subsequent day which is often severe and persistent, there has been no general agreement in the literature regarding the site of this inhibition. As a consequence, a consensus regarding the underlying physiological and biochemical basis of the inhibition has not been possible. It is clearly important to identify the contributions which partial stomatal closure and impaired chloroplast activities make to the overall inhibition of photosynthesis by chilling.

In order to distinguish quantitatively between the stomatal and nonstomatal contributions to the chilling impairment of photosynthesis in tomato (*Lycopersicon esculentum* Mill. cv. Floramerica), we investigated the dependence of photosynthesis by attached leaves on $[CO_2]_i$. Prechilling in the dark at 1°C for 16 hr had the following consequences in the subsequent light period: when the ambient $[CO_2]$ was 300 μbars, photosynthesis was depressed about 53% and the internal $[CO_2]$ was only 180 μbars (Fig. 2, Point B), although in unchilled controls the internal $[CO_2]$ was 250 μbars (Fig. 2, Point A). Thus, at least some of the inhibition of prechilling can be attributed to increased stomatal resistance to CO_2 uptake. However, when the internal $[CO_2]$ was readjusted back to the control level of 250 μbars by raising the ambient $[CO_2]$ (Fig. 2, Point C on the interpolated curve, the closest experimental point, is at $[CO_2]_i$ = 230 μbars), most of the inhibition remained (39% out of 53%). It follows that some steps in the biochemistry were severely limited by the prechilling experience. A similar

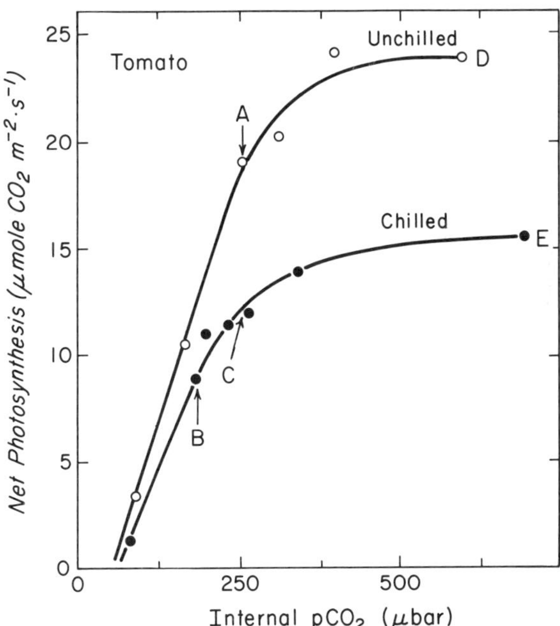

Fig. 2. The dependence of photosynthesis of an attached prechilled and an attached unchilled tomato leaf on the intercellular CO_2 concentration. The ambient temperature was 25°C, the relative humidity was 60%, oxygen was 21%, and the incident irradiation was 1800 μeinsteins m^{-2} sec^{-1}. The highest intercellular CO_2 concentration reported in this figure required ambient CO_2 levels of 700 μbars in the unchilled plant and 1500 μbars in the prechilled plant. See Section II,A,1,a for details.

extent of inhibition of the biochemistry was revealed by simply raising the level of ambient CO_2 until the resulting increases in the internal $[CO_2]$ caused no further increase in the rate of photosynthesis. Under these conditions, where the CO_2-requiring biochemical reactions of the prechilled and unchilled plants alike were provided with all the substrate they could use, the remaining inhibition, independent of stomatal conductance, was about 35% (Fig. 2, compare Points D and E). Clearly in this experiment about two-thirds of the inhibition of photosynthesis due to prechilling resulted from injury to the biochemical processes of the chloroplast and only about one-third resulted from stomatal closure.

b. Inhibition of Photosynthesis by Low Leaf Water Potential. Photosynthetic activity declines as water becomes limited and the leaf water potential decreases. For sunflower plants (*Helianthus annus* L. cv. IS894) grown under well-watered conditions, the response of photosynthesis to a water deprivation that decreased leaf water potential to −16.3 bars is shown in Fig. 3. At atmospheric CO_2 levels, light-saturated photosynthesis was inhibited 75% (cf. Fig. 3, points A and B). Although there was an increase in leaf diffusive resistance from about 2 to about 8 sec cm^{-1} associated with the drop in leaf water potential, this stomatal closure did not result in a decline in the $[CO_2]_i$. If anything, there was a slight increase (Fig. 3, compare experimental point B to point C on the interpolated curve). Thus, in this case of stress-induced inhibition of photosynthesis, the cause lies entirely with nonstomatal processes.

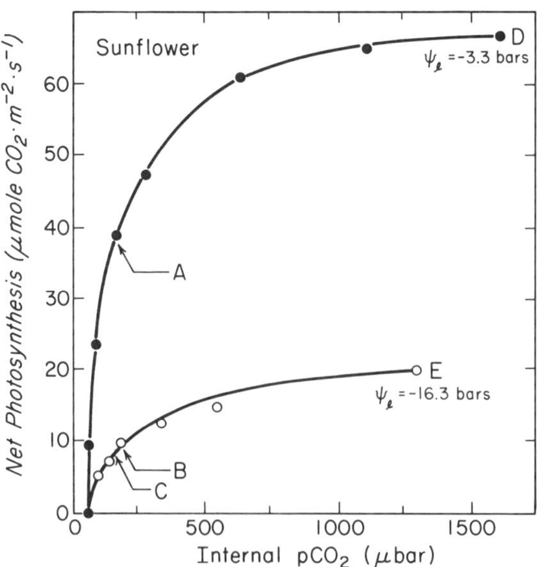

Fig. 3. The dependence of photosynthesis of an attached sunflower leaf on the intercullar CO_2 concentration when measured at high and low leaf water potentials. Conditions of the measurement were as described in Fig. 2. See Section II,A,1,b for details.

c. The Origin of Nonstomatal Inhibition of Photosynthesis: Increased "Mesophyll Resistance" or Impaired Chloroplast Activity. Nonstomatal inhibition of photosynthesis has been attributed to increased resistance to CO_2 diffusion in the liquid phase from the mesophyll walls to the site of CO_2 reduction in the chloroplast (Gaastra, 1959). This notion was based on a calculation of "mesophyll resistance" obtained as a residual effect, after accounting for the resistance to diffusion of CO_2 by the leaf boundary layer and the stomata. However, in order for such an analysis to be valid, photosynthesis must be limited solely by diffusion of CO_2, and this cannot be known without an extensive set of measurements. Our predilection is that "mesophyll resistance" is never great enough to constitute a limitation on the rate of CO_2 reduction since no physical basis for such a severe restriction of CO_2 diffusion at this level is apparent. Indeed, the introduction of the term "mesophyll resistance" was unfortunate since it implies an analogy to diffusion or at least to a linear term, when in fact the greater part of "mesophyll resistance" clearly involves a saturable process and, therefore, is an exceedingly nonlinear function. The analysis discussed here not only recognizes the saturable nature of the process responsible for "mesophyll resistance" but actually makes use of the fact of saturation and, in so doing, clearly distinguishes biochemical limitations from all diffusional limitations. Thus, from the data presented in Figs. 2 and 3 it is clear that the nonstomatal inhibition of photosynthesis brought on by chilling or drought is not caused by mesophyll diffusional resistance. In sunflower, the photosynthesis in the inhibited plants responded nonlinearily even to low $[CO_2]_i$ and virtually not at all to high concentrations. In tomato, there was nearly a linear response at $[CO_2]_i$ less than 280 μbars but little if any response at high $[CO_2]_i$. Such nonlinear or concentration-independent behavior to $[CO_2]_i$ is unlike that expected for any sort of diffusional limitation. By elimination, this implicates the chloroplast as the origin of nonstomatal inhibition of photosynthesis resulting from chilling and drought stress.

2. *In Vivo Measurements of the Quantum Yield of CO_2 Reduction*

Measurement of the quantum yield of CO_2 reduction in attached leaves can be rigorously diagnostic of stress-induced inhibition of photosystem II activity, including water oxidation capability. Therefore, it is an important *in vivo* test of chloroplast activity. In order to understand the relationship of photosystem II activity to the quantum yield of CO_2 reduction, it is necessary to understand the mechanism of electron transfer within the photosystem II reaction center. Photosystem II electron transport, from the oxidation of water to the reduction of the secondary quinone acceptor B, is carried out by an integral membrane polypeptide complex (Satoh and Butler, 1978; Diner and Wollman, 1980). Each of these operates independently with no electron shuttling between complexes but acts to reduce a common plastoquinone acceptor pool (Joliot and Kok, 1975). The association between PSII reaction centers and water-oxidizing enzyme complexes is such that the failure of one water-oxidizing complex leads directly to the loss of photochemical activity of one PSII reaction center. The effect that the loss of photochemical activity of a reaction center will have on the quantum yield will depend on the extent to which energy transfer between PSII centers can occur. Joliot *et al.* (1973) demonstrated that an inactive PSII center will not allow escape of exciton energy if its reaction center is in a "quenching form." The inhibition of water oxidation would cause the reaction center of the PSII complex to be in a quenching state (Joliot *et al.*, 1973). Consequently, loss of photochemical activity due to loss of water oxidation capability should have a direct effect on the quantum yield of PSII turnover, and it is an observed fact that all inhibitions of photosystem II lower the quantum yield of photosynthesis. Furthermore, impaired water oxidation capacity should have a substantially more severe effect on the quantum yield than on the maximum rate of CO_2 reduction, because the rate-limiting step(s) in light- and CO_2-saturated photosynthesis is clearly not associated with PSII activity. Consequently, damage to a small fraction of the PSII centers would immediately be reflected in quantum yield measurements (where light is limiting), whereas any effect on light- and CO_2-saturated photosynthesis would be absent or greatly reduced. This prediction was borne out in experiments with attached tomato leaves in which PSII was partially inhibited by DCMU (Fig. 4). DCMU and a large number of other PSII inhibitors, many of which are agriculturally important herbicides, act by displacing B from its binding site on the PSII reaction center (Crofts and Wraight, 1983). Inasmuch as DCMU acts prior to the rate limiting step(s) of CO_2 reduction, the effect of DCMU was more than twice as large on the quantum yield values as on the light- and CO_2-saturated rates of photosynthesis. Inhibition of water oxidation directly by heat treatment of leaves produced similar results.

Partial inactivation of the intersystem electron transfer carriers, that is, inactivation of the components of the cytochrome b_6/f complex and plastocyanin, is not expected to have an appreciable effect on the quantum yield of CO_2 reduction since there appears to be free sharing of electrons in this region of the chain (e.g., Haehnel *et al.*, 1980). However, inactivation of PSI reaction centers would cause a decline in quantum yield of CO_2 reduction, but, unlike inactivation of PSII centers, this reduction in quantum yield would be in proportion to the reduction in the maximum light-saturated rate. In addition, a change in ratio of the reduction of CO_2 versus the reduction of O_2, by either oxygenase activity of the carboxylase (Ehleringer and Björkman, 1976) or direct reduction of oxygen by PSI, would also affect the quantum yield of CO_2 reduction in a fashion similar to that expected when PSI reaction centers are inactivated.

a. Chilling Inhibition of Photosynthesis and Its Relationship to the Quantum Yield of CO_2 Reduction. There is a good deal of experimental evidence indicating that chloroplasts isolated from prechilled leaves of sensitive plants

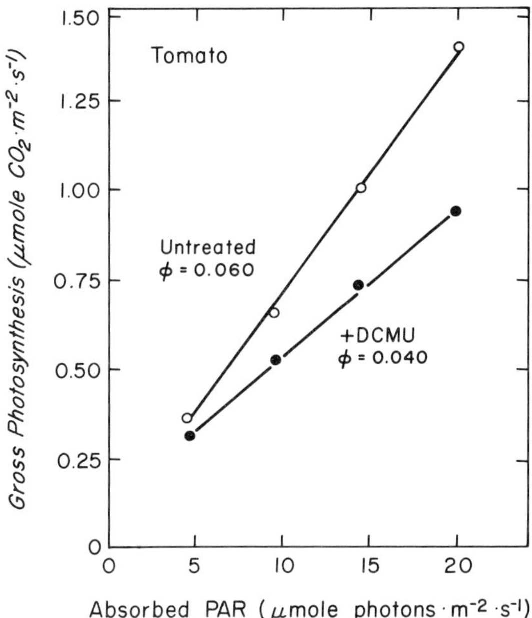

Fig. 4. The effect of DCMU on the quantum yield for photosynthetic CO_2 fixation in an attached tomato leaf. After completion of the control measurements, the attached leaf was immersed in 15 μ*M* DCMU for 5 min at 25°C. It was then stored overnight at 25°C in the dark for measurements the following day. Experimental conditions were identical to those given for Fig. 2. The quantum yield was reduced from 0.060 to 0.040 by the DCMU treatment. The rate of light- and CO_2-saturated photosynthesis was 33 μmoles CO_2 m^{-2} sec^{-1} prior to treatment and 24 μmoles CO_2 m^{-2} sec^{-1} after the treatment with DCMU.

have impaired electron transport. Kaniuga and associates (Kaniuga and Michalski, 1978; Kaniuga *et al.*, 1978) have identified water oxidation as the chill-labile process in tomato chloroplasts, supporting similar conclusions drawn earlier by others for a variety of plants (e.g., Margulies, 1972; Simillie, 1979). Consequently, it is a widely held belief that the reduced capacity for water oxidation observed in isolated chloroplasts is a major cause of the reduced photosynthesis measured in attached, prechilled leaves. However, we have three lines of experimental evidence that have led us to the conclusion that a significant reduction in the capacity of water oxidation does *not* occur in tomato due to prechilling and is not a primary element of the inhibition of photosynthetic CO_2 fixation observed in attached leaves. Furthermore, the data demonstrate that no photosystem II reaction prior to the reduction of plastoquinone can account for the inhibition observed in attached leaves.

The absolute quantum yield of CO_2 reduction was calculated from the slope of the dependence of the CO_2 reduction rate on the amount of light absorbed (Fig. 5). All measurements were made at saturating intercellular CO_2 levels and irra-

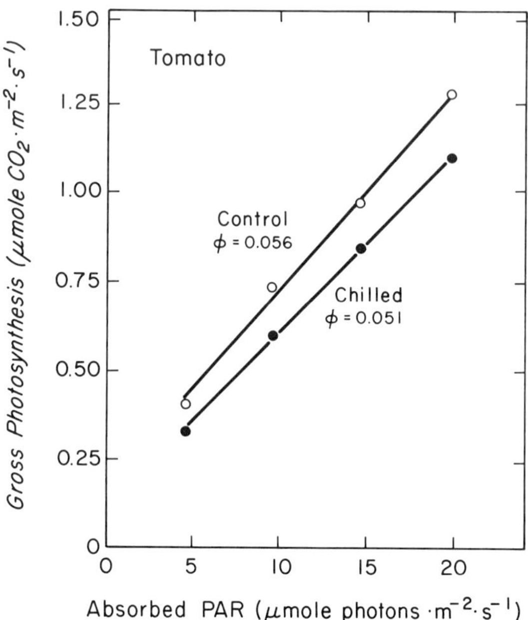

Fig. 5. The effect of dark chilling on the quantum yield for CO_2 reduction in an attached tomato leaf. The ambient CO_2 concentration $(P_{CO_2})_a$ was 1500 μbars, while all other conditions for the measurement were identical to those given for Fig. 2. The light- and CO_2-saturated rates of photosynthesis were 38 μmoles CO_2 m^{-2} sec^{-1} prior to chilling and decreased to 27 μmoles CO_2 m^{-2} sec^{-1} after the chilling treatment.

diances near or below the light compensation point, thus circumventing entirely any involvement of stomata in determining photosynthesis rates. The data show a decrease in quantum yield of only 10% (from 0.056 prior to chilling to 0.051 subsequent to chilling), while the maximum photosynthetic rate (both CO_2- and light-saturated) was inhibited nearly 25% in the same leaf (Fig. 5). Thus, our data show clearly that photosynthesis, when impaired by chilling, does not conform to what is expected of impaired water oxidation or PSII activity, because the effect of chilling on quantum yield was small compared to the effects on maximum rates.

b. The Effect of Low Leaf Water Potential on the Quantum Yield of CO_2 Reduction. In contrast to the effects of chilling, the inhibition of photosynthesis by low leaf water potential in sunflower is associated with a dramatic decline in the quantum yield of CO_2 reduction. As always, quantum yields for CO_2 fixation were obtained at low photon flux densities for which photosynthesis had a linear dependence on absorbed radiation to assure that photochemical conversion was the sole limiting factor. Associated with a decline in leaf water potential to −12.1 bars was a 35% decrease in the quantum yield (Fig. 6) but a somewhat larger decrease in the light-saturated maximum photosynthetic rate. This relationship between the decline in quantum yield and the decline in maximum photosynthetic rate, in which the inhibition of maximum photosynthetic rate is larger than the inhibition of the quantum yield, is different from any situation discussed thus far. Probably more than a single stress-induced malfunction is involved, and, while photosystem II inhibition is probably a major component, it cannot account for the total effect for the reasons described above. The labile and highly integrated nature of the process of photophosphorylation makes it a prime candidate for disruption by unfavorable environmental conditions, and the disruption of photophosphorylation is, therefore, a likely contributor to the overall inhibition of photosynthesis brought on by low leaf water potentials. In this case, the quantum yield and light-saturated photosynthesis measurements have alerted us to the possibility of more than a single malfunction in chloroplast activity at this degree of leaf desiccation. This information has had an impact on the design and the interpretation of subsequent experiments.

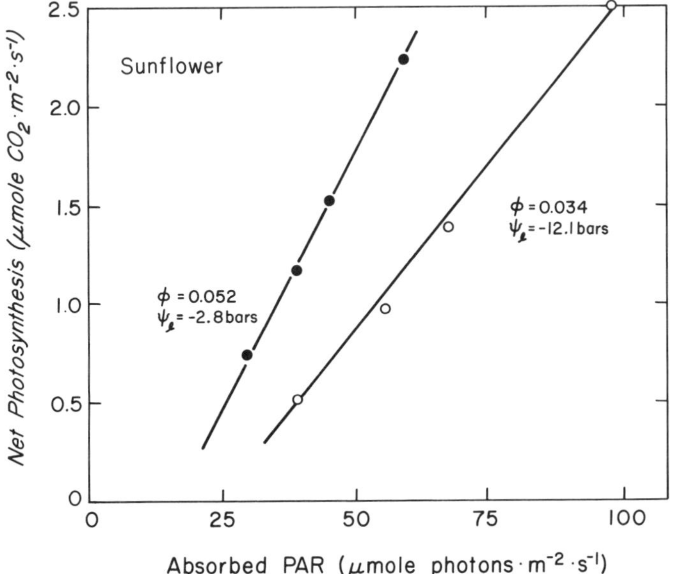

Fig. 6. The effect of low leaf water potential on the quantum yield for CO_2 reduction in an attached sunflower leaf. Leaf water potentials were determined in the experimental leaf immediately on completion of the gas exchange measurements. The conditions of the experiment were as given for Fig. 5. The light- and CO_2-saturated rates of photosynthesis were 40 μmoles CO_2 m^{-2} sec^{-1} at a leaf water potential of −2.8 bars and 16 μmoles CO_2 m^{-2} sec^{-1} at −12.1 bars.

B. *In Vitro* Measurements of Photosynthesis

A detailed analysis of stress-induced inhibition of chloroplast activity must be made at the organelle level. Noted earlier was a special difficulty which is associated with studying stress-induced perturbations of photosynthesis, distinguishing between inhibitions which are actually part of the *in vivo* process and inhibitions induced through the manipulation of tissue made labile by the exposure to the stress. In other words, we must be cautious lest the inhibitions observed *in vitro* are simply a manifestation of unfavorable conditions for organelle isolation induced by the stress, conditions which have no effect whatsoever *in vivo*. Part of the information gained from whole-plant gas exchange measurements is a set of identifying characteristics of stress-impaired chloroplast activity, and these are of vital importance in making this judgment.

1. Measurement of the Electron Transport Capacity of Thylakoid Membranes Isolated from Chill-Stressed Tomato

We have noted previously that three lines of experimental evidence lead us to conclude that inhibition of water oxidation and PSII activity are not the underlying causes of chill-impaired photosynthesis. The first line of evidence has been discussed (the unaffected quantum yield of CO_2 reduction in attached leaves). This finding was corroborated by measuring the quantum yield of electron transfer in isolated thylakoid membranes (Martin and Ort, 1982).

The second line of evidence involves measurement of maximum rates of photosystem II turnover in chloroplasts isolated from unchilled and prechilled leaves (Table III). Two points are demonstrated. First, the rate of photosystem II turnover is only marginally less in chloroplasts from chilled leaves than in chloroplasts from unchilled leaves (Table III). Second, in all cases the rate of photosystem II turnover far exceeds the electron flux required to support the rate of light- and CO_2-saturated CO_2 reduction measured in attached leaves. This second point is an important criterion for judging the suitability of a chloroplast preparation for assessing rate-limiting reactions. It would, for instance, be difficult to have much confidence in data pertaining to the site of a stress-induced electron transport rate-limitation if the maximum rates in chloroplasts isolated from the unstressed control plants were only 25% of that needed to support *in vivo* rates of CO_2 fixation. Clearly, in that case the isolation and assay procedures are, of themselves, causing massive inhibitions.

The third line of support, measurement of the concentration of active photosystem II centers, shows there is vanishingly little difference between chloroplasts isolated from chilled leaves versus chloroplasts isolated from unchilled leaves (Table III). In this experiment (Martin and Ort, 1982), proton release from the oxidation of water was measured. A decrease in the amount of protons

TABLE III

Comparison of the Effect of Prechilling on CO_2 Reduction and on Photosystem II Activity

	Ambient CO_2 (μbars)	Rate of leaf CO_2 reduction[a]	Rate of electron flux required to support CO_2 reduction[b]	Rate of PSII electron transport[c]	Concentration of active PSII centers[d]
Untreated	300	20 ± 3, $n = 6$	115	245 ± 15, $n = 12$	620 ± 90, $n = 46$
	1500	28 ± 4, $n = 10$	170		
Chilled	300	10 ± 2, $n = 8$	60	215 ± 25, $n = 12$	680 ± 40, $n = 6$
	1500	20 ± 2, $n = 7$	115		

[a] Units: μmoles CO_2 m^{-2} sec^{-1}.

[b] Units: mmoles electrons mole $chlorophyll^{-1}$ sec^{-1}.

[c] Photosystem II-dependent electron transport was measured spectrophotometrically as the loss of absorbance at 420 nm due to reduction of ferricyanide mediated by the lipophilic oxidant *p*-phenylenediimine. Dibromothymoquinone (1 μ*M*) was present to prevent PSI turnover. Statistical errors are reported as the standard deviations of the mean, and *n* is the number of experiments included in the calculation of the mean. Units: mmoles electrons mole $chlorophyll^{-1}$ sec^{-1}.

[d] Units: chlorophyll molecules per active reaction center.

released (calculated on the basis of total chlorophyll content) would be a necessary consequence of reduced water oxidation capacity. Thus, the fact that we observed very smiliar values for proton release in the two sorts of chloroplasts indicates again that prechilling has little effect on the water oxidation capacity.

Since photosystem II activity and water oxidation capacity are not significantly impaired in tomato by prechilling in the dark and, therefore, are not elements of the inhibition observed in attached leaves, we extended our study to search for malfunctions which might be present in the intermediate electron transfer reactions. During the past 10 or so years it has become possible to dissect functionally (and in a lesser number of cases, physically) the photosynthetic electron transport chain into discrete segments (e.g., Yamashita and Butler, 1968; Saha *et al.*, 1971; Ort, 1975; Izawa and Pan, 1978). The sites of interaction of numerous oxidizing and reducing compounds with the electron transport chain is governed by their particular midpoint potentials and lipophilicities. Thus, using various electron donors and electron acceptors in conjunction with various specific inhibitors of electron transport, it is now possible to find locations of lesions in the pathway of electron flow. In addition, it is sometimes possible to detect multiple sites of inhibition. The photosynthetic electron transport chain is presented in schematic form in Fig. 7, showing the points at which exogenous electron donors (marked D), electron acceptors (marked A), and specific inhibitors (marked I) interact. Included below the schematic is a catalog of representative compounds for each different type of donor, acceptor, and inhibitor.

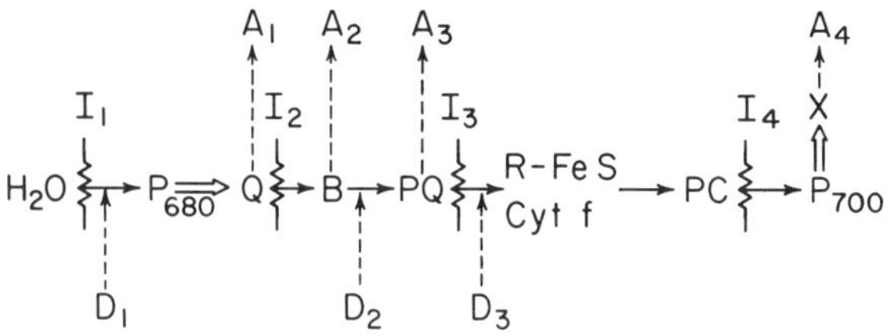

Fig. 7. Schematic drawing of the chloroplast electron transport chain indicating the sites of interaction by exogenous electron donors, electron acceptors, and specific inhibitors. **Sites of interaction of electron donors**: D_1, catechol (Ort and Izawa, 1973); D_2, duroquinol (Izawa and Pan, 1978); and D_3, diaminodurene (Ort, 1975). **Sites of interaction of electron acceptors:** A_1, silicomolybdate (Giaquinta *et al.*, 1974); A_2, oxidized *p*-phenylenediamine (Saha *et al.*, 1971) and 2,3-dimethyl-5,6-methylenedioxy-*p*-benzoquinone (Trebst *et al.*, 1976); A_3, dibromothymoquinone (Izawa *et al.*, 1973); and A_4, methyl viologen, ferricyanide (Saha *et al.*, 1971). **Sites of interaction of specific inhibitors:** I_1, hydroxylamine treatment (Ort and Izawa, 1973); I_2, diuron (Pfister and Arntzen, 1979); I_3, dibromothymoquinone (Trebst, *et al.*, 1970); and I_4, potassium cyanide treatment (Ouitrakul and Izawa, 1973).

Those partial reactions supporting the highest electron transport activities were selected to maximize our chances of detecting even slight inhibitions of endogenous electron transfer reactions. Data from a large number of experiments are compiled in Table IV. The data in this table make two important points. (1) Chloroplasts isolated from prechilled leaves are as competent in electron transport in every region of the chain as are the chloroplasts isolated from control leaves. (2) In every case the rate of electron flux measured in chloroplasts from prechilled leaves is in excess of what is necessary to support the light- and CO_2-saturated rate of CO_2 reduction measured in the attached prechilled leaf. It is worth noting that reactions which include the rate limiting step of electron flow (Avron and Chance, 1967) [i.e., the oxidation of plastohydroquinone by cytochrome *f* or of an iron–sulfur center known to be in this region of the chain (Malkin and Posner, 1978; Whitmarsh and Cramer, 1979)] support rates which are not quite adequate to account for light- and CO_2-saturated photosynthesis of the unchilled attached leaf. This may reflect a limitation of the isolation procedure (see earlier). We conclude that the membrane-bound electron transport reactions (including water oxidation, photosystem II activity, and electron transport between photosystem II and photosystem I) are *not* significantly impaired in tomato by prechilling in the dark and, therefore, seem not to be elements of the inhibition observed in attached leaves.

2. *Measurement of the Electron Transfer Capacity of Thylakoid Membranes Isolated from Leaves of Low Water Potential*

Inasmuch as the whole-plant quantum yield measurements of CO_2 reduction by sunflower leaves at low water potential indicate a perturbation of reactions

TABLE IV

Partial Reactions of Electron Transport Assayed in Chloroplasts Isolated from Chilled and Unchilled Tomato Leaves

Electron transport reaction	Treatment	Rate of electron transport[f]	Rate of electron transport required to support light- and CO_2-saturated photosynthesis[f]
$H_2O \rightarrow MV$[a]	Control	135 ± 10 (*n* = 14)	170 ± 20 (*n* = 10)
	Chilled[e]	140 ± 15 (*n* = 18)	115 ± 10 (*n* = 7)
DHQ → MV[b]	Control	140 ± 10 (*n* = 8)	170 ± 20 (*n* = 10)
	Chilled[e]	160 ± 40 (*n* = 7)	115 ± 10 (*n* = 7)
DAD → MV[c]	Control	380 ± 35 (*n* = 16)	170 ± 20 (*n* = 10)
	Chilled[e]	420 ± 30 (*n* = 19)	115 ± 10 (*n* = 7)
$H_2O \rightarrow PD^d_{ox}$	Control	235 ± 15 (*n* = 22)	170 ± 20 (*n* = 10)
	Chilled[e]	220 ± 40 (*n* = 8)	115 ± 10 (*n* = 7)

[a] Electron transport was measured as oxygen uptake from the aerobic oxidation of methyl viologen (100 μ*M*). The 2-ml reaction contained 50 m*M* HEPES–KOH (pH 7.5), 100 m*M* sorbitol, 5 m*M* P_i, 2.5 m*M* $MgCl_2$, 5 m*M* KCl, 0.5 m*M* ADP, 300 units superoxide dismutase, 10 m*M* NH_4Cl, 0.03 μ*M* valinomycin, and chloroplasts containing to 32 μ*M* chlorophyll in the 2-ml reacton. The reaction temperature was 20°C.

[b] Conditions are the same as in *a* except that 5 μ*M* nigericin replaced the NH_4Cl and valinomycin, and 5 μ*M* DCMU and 0.5 m*M* DHQ were added.

[c] Conditions are the same as in *b* except 0.5 m*M* diaminodurene (DAD) and 1 m*M* ascorbic acid replaced DHQ.

[d] Electron transport was measured spectrophotometrically as the loss of absorbance at 420 nm to the reduction of ferricyanide. Conditions were similar to *a* with the omission of NH_4Cl, valinomycin, and superoxide dismutase but with the addition of 1 μ*M* dibromothymoquinone, 1.5 m*M* ferricyanide, and 0.5 m*M* PD. The chloroplast content was reduced to 16 μ*M* chlorophyll.

[e] The plants which were chilled were enclosed in a cardboard box in a darkened incubator at 1°C for 16 hr.

[f] Units: mmoles electrons mole chlorophyll^{-1} sec^{-1}.

close to the photochemistry (see Section II,A,2,b), our observation of impaired system II activity was in line with expectations (Table V). Chloroplast thylakoid membranes isolated from leaves at high water potential (i.e., less negative than about −7 bars) exhibited rates of PSII electron transfer which were sufficient to account for 90% of the maximum rate of CO_2 fixation observed in attached leaves of well-watered plants. Photosystem II activity was assayed as the photosystem II-dependent reduction of diiminodurene (see Fig. 7). Photosystem II activity declined with increasing water deficit, with the rate showing a nearly sigmoidal relationship to leaf water potential. For these plants, which were grown under well-watered conditions, leaf water potentials of about −20 bars resulted in about 50% reduction of PSII activity but greater than 90% inhibition

TABLE V

Photosystem II-Dependent Electron Transfer Activity in Chloroplast Thylakoid Membranes Isolated from Sunflower Leaves of Progressively Lower Water Potential

Leaf water potential (bars)	Photosystem II-dependent[a] electron transfer activity
−4.1	210
−7.8	210
−8.7	205
−11.9	185
−14.3	160
−15.3	145
−17.3	130
−17.8	120
−18.7	115
−22.4	110

[a] Electron flow through photosystem II was measured spectrophotometrically by following the photoreduction of ferricyanide (420 nm) mediated by the lipophilic oxidant, diiminodurene (DAD_{ox}) (Ouitrakul and Izawa, 1973). The 2-ml reaction mixture (20 m*M* MOPS–KOH (pH 6.8), 0.1 *M* sorbitol, 10 m*M* KCl, 5 m*M* $MgCl_2$, 1.5 m*M* ferricyanide, 0.5 m*M* DAD_{ox}, 16 μM chlorophyll) was illuminated with saturating red light from a tungsten lamp. Units: mmoles electrons mole chlorophyll^{-1} sec^{-1}.

of CO_2 reduction by the attached leaf, again suggestive of multiple stress-induced lesions affecting chloroplast function.

We tested whether low osmotic potentials imposed on the thylakoid membranes after their isolation could simulate the effects of low leaf water potentials. To accomplish this we added sorbitol to the assay media used for measuring photosystem II activity. Table VI shows that there was only a slight inhibition of photosystem II activity even at osmotic potentials as low as −18.5 bars. Returning the membranes to high osmotic potentials caused an immediate recovery of the control activity (Potter and Boyer, 1973). This is in contrast to thylakoid membranes from leaves having low water potentials, which showed much greater inhibition of activity (Table VI) and did not return to the control activity when assayed without sorbitol (Potter and Boyer, 1973). Thus, low osmotic potentials alone could not simulate the effect of low water potentials, and the decrease in the free energy of water that took place as osmotic potentials decreased is unlikely to have caused the decrease in photosystem II activity (Boyer and Potter, 1973; Potter and Boyer, 1973).

TABLE VI

Effect of Various Osmotic Potentials on Photosystem II Activity of Chloroplast Thylakoid Membranes Isolated from Sunflower Leaves

Leaf water potential (bar)	Assay[a] osmotic potential (bar)	Photosystem II activity (% of control)
−3.2	−2.4	100
	−10.5	100
	−14.5	92
	−18.5	85

[a] Osmotic potentials were produced by adding various concentrations of sorbitol to the assay medium. Assays were conducted at saturating quantum flux densities. Data calculated from Potter and Boyer (1973).

3. *Energy Conservation and Photophosphorylation*

Photophosphorylation is among the most labile chloroplast activities. In light of what we know about the mechanisms of photophosphorylation, its delicate nature is not surprising (for review see Ort and Melandri, 1982). The fragile and highly integrated nature of the process makes it a prime candidate for disurption by unfavorable environmental conditions such as prechilling and water deficiency. If photophosphorylation is indeed impaired in chloroplasts isolated from attached, stress-inhibited leaves, the possible causes of the inhibition should fall into three general categories.

The first possibility is that the ability of the chloroplast to form an energized membrane may be diminished or absent. This condition is most likely to result from an electron transport activity impaired by the stress exposure.

The second possibility is that the lamellar membrane may be unable to maintain an energized state, so that electron transport is said to be "uncoupled" from ATP synthesis. In most cases, uncoupling in higher plant chloroplasts results from an increase in conductance of the lamellar membrane to hydrogen ions. By definition, uncoupling is usually characterized by elevated rates of electron flux with diminished efficiencies of ATP formation (i.e., number of ATP molecules formed per pair of electrons transported). Uncoupling should be accompanied by a rapidly decaying proton accumulation, a stimulation of trypsin-activated, membrane-bound ATP hydrolysis activity, a reduction or a loss of light-induced volume changes of chloroplasts, a reduction in yield of postillumination ATP synthesis, and a decrease in the half time for the dark reduction of cytochrome *f*.

The third general way in which photophosphorylation activity can be impaired pertains to the terminal reactions, that is to the actual formation of the covalent bond in ATP by the membrane-bound ATPase enzyme complex. For example, if the "soluble" subunits (CF_1) of the enzyme complex become detached from the transmembrane proton channel (CF_0), the resulting inhibition of ATP synthesis is superficially similar to uncoupling. In fact, it is a type of uncoupling, but, in this case, the increase in conductivity of the membrane to protons can be reversed by reacting *N,N'*-dicyclohexylcarbodiimide with the proton channel portion of the ATPase enzyme complex (Uribe, 1972). To function properly the enzyme complex must bind substrates (ADP and P_i), release products (ATP), and conduct protons. Furthermore, these processes must occur in a rigid sequence that apparently involves various conformational rearrangements of the enzyme. In most cases, malfunctioning of the ATPase enzyme complex results in considerable inhibition of electron flux because the major process which dissipates the energized state of the membrane is impaired. This special inhibition of electron flux can be differentiated from those discussed earlier because it can be relieved by adding uncoupling agents (Good *et al.*, 1966).

Inhibition of photophyosphorylation appears to be an important aspect of the overall *in vivo* inhibition of photosynthesis brought on by low leaf water potentials. We observed that photophosphorylation activity decayed linearly with decreasing leaf water potential, exhibiting at -25 bars only 40% of the activity found in chloroplasts isolated from well-watered leaves of spinach plants (Younis *et al.*, 1979). There are some indications that photophosphorylation of sunflower is even more sensitive to low water potentials (Keck and Boyer, 1974) than that of spinach. The effect of leaf desiccation on the chloroplast ion concentrations suggests a possible direct mechanism for low water potential-induced losses in photophosphorylation activity. We simulated the effects of low leaf water potentials with ion concentrations that should have increased as water was lost from the tissue. We used magnesium because it is known to be required for photophosphorylation, and magnesium concentrations could have inhibitory effects if they became too high in the chloroplast stroma due to water loss by the leaf tissue. We preincubated the chloroplast thylakoid membranes or coupling factor (CF_1) isolated from the membranes in a medium containing buffer and various magnesium concentrations for 30 min. These were then transferred to the assay medium (Younis *et al.*, 1979) which diluted the preincubation concentrations of magnesium to 1% of the preincubation level. By this procedure, the high magnesium preincubation simulated the anticipated high magnesium concentration in the dehydrated leaf, whereas the assay, after removal of the chloroplast thylakoids or coupling factor from the preincubation medium, simulated the assay conditions after removal of the chloroplast thylakoids or coupling factor from dehydrated leaves. Table VII shows that cyclic photophosphorylation decreased in spinach chloroplasts preincubated in Mg^{2+} concentrations as high as 10 m*M*. The assays for photophosphorylation were carried out in the presence of 3 m*M* Mg^{2+}, which is the approximate concentration of Mg^{2+} in chloroplast stroma at high water potentials (Portis and Heldt, 1976; Portis, 1981), thus assuring uniform conditions for all assays.

TABLE VII

The Effect of Magnesium Preincubation on Phosphorylating Activity of Isolated Spinach Thylakoids and Ca^{2+}-ATPase Activity of Chloroplast Coupling Factor

Mg^{2+} concentration[a] during preincubation (mM)	Photophosphorylating activity[b]	Ca^{2+}-ATPase activity[c]	
		Before heat activation	After heat activation
0	375	5250	5250
5	260	5750	1750
10	185	3500	1250

[a] Preincubation was carried out in the presence of buffer, and a small aliquot of the preparation was transferred to the assay medium after 30 min at room temperature. The assay medium diluted the Mg^{2+} concentration to 1% of the preincubation concentration. Details of methods are given in Younis *et al.* (1979).

[b] Units: mmoles ATP mole $chlorophyll^{-1}$ sec^{-1}.

[c] Units: μmoles ATP mole $chlorophyll^{-1}$ sec^{-1}.

The activity of Ca^{2+}-ATPase of coupling factor (CF_1) was also decreased after preincubation of the polypeptide complex in Mg^{2+} concentrations as high as 10 m*M* (Table VII). The effect was larger if the exposure to Mg^{2+} occurred after heat activation of CF_1, but it was also observed prior to heat activation.

Table VIII shows the relationship of low water potential-induced decreases in photophosphorylation to the estimated concentration of stromal Mg^{2+}. These experiments were carried out with spinach. The expected stromal Mg^{2+} concentrations were calculated from the water content of the tissue prior to chloroplast isolation (Table VIII) and the stromal concentrations of Mg^{2+} [taken to be 3 m*M* based on the measurements of Portis and Heldt (1976)] prior to leaf dehydration. Water loss by spinach leaves at low water potentials was sufficient to cause calculated stromal Mg^{2+} concentrations as high as 9 m*M*. This is similar to the concentrations of Mg^{2+} used in the Mg^{2+} preincubation experiment (Table VII).

The great effects of elevated Mg^{2+} concentrations (to 10 m*M*) on photophosphorylation and coupling factor activity suggest that concentrations of this ion likely to be present in chloroplast stroma of dehydrated leaves could be a factor causing the inhibition of photophosphorylation at low leaf water potentials. Both high Mg^{2+} concentration and low water potential caused persistent changes in chloroplast and coupling factor activities that could be detected in uniform assays, and the effects were similar in magnitude. This is consistent with the alterations in chloroplast activity observed *in vivo* as water potentials and osmotic potentials decreased, because Mg^{2+}, while not contributing signifi-

TABLE VIII

Phosphorylating Activity of Isolated Thylakoids, Approximate Water Content of Leaves, and Calculated Stromal Mg^{2+} Concentrations at Various Leaf Water Potentials in Spinach[a]

Leaf water potential (bar)	Photophosphorylating activity (mmoles ATP mole chlorophyll^{-1} sec^{-1})	Approximate water content (% of turgid)	Calculated stromal concentration of Mg^{2+} (mM)
−2	260	100	3
−15	185	55	6
−25	120	35	9

[a] Chloroplasts were isolated from spinach leaves that had been dehydrated to levels shown. Assays were conducted as in Younis *et al.* (1979). Stromal Mg^{2+} concentrations were calculated from water content of tissue assuming 3m*M* Mg^{2+} in chloroplast stroma of turgid leaves and no changes in compartmentation of the Mg^{2+}.

cantly to the osmotic potential, would have been concentrated in similar fashion. It is also consistent with the lack of effect of low osmotic potentials *in vitro,* because high Mg^{2+} concentrations were not present (Table VI). In other words, as dehydration occurred *in vivo,* low osmotic potentials were likely associated with increased ion concentrations, but, *in vitro* using sorbitol, they did not cause large losses in membrane activity.

Potter and Boyer (1973) found that direct solute–chloroplast interactions were a likely explanation of losses in chloroplast activity in their experiments at low leaf water potentials. Kaiser *et al.* (1981) also concluded that the possible effects of internal ions were consistent with the chloroplast response to water loss caused by osmotic shrinkage of leaf slices, isolated protoplasts, and intact chloroplasts. Mayoral *et al.* (1981) and Shimshi *et al.* (1982) found differences in chloroplast response to low leaf water potentials that were correlated with differences in water content of leaves in wheat and some related wild species. These results provide additional evidence for ion concentrations as central features of photosynthetic response to low water potentials. Although the Mg^{2+} simulation is consistent with the alteration in photosynthesis that is observed at low water potentials, other ions may be involved and need to be investigated. Thus, the results reported here could represent a specific Mg^{2+} effect or could reflect a more general response to high ion concentrations, and these issues remain to be resolved.

4. *In Vitro* Investigation of Other Chloroplast Membrane Activities and Membrane Properties That May Be Altered during Stress Exposure

There are a large number of other parameters which can be measured at the level of the isolated organelle or organelle subpreparation. The relevance of each individual parameter as a factor in the stress-related inhibition of photosynthesis will depend to varying degrees on the particular stress and the species of plant under investigation.

For instance, in the case of chill-induced inhibition of photosynthesis, a role for thylakoid membrane lipid phase transitions has been suggested. There are numerous suggestions in the literature (e.g., Lyons, 1973; Murata and Fork, 1975; Shneyour *et al.*, 1973; Smillie, 1979) that the inhibition of photosytnesis in chill-sensitive organisms may be accounted for by a transition in the physical phase of the membrane from liquid crystalline to solid state. Many of the thermochemical reactions involved in photosynthesis occur between reactants located on or within a membrane and are mediated by membrane-bound enzymes. Intuitively, it seems sensible that the rates of these reactions will respond to temperature as though controlled by the threshold collision energy of the reaction and that the molecular interaction will depend critically on membrane fluidity. So "freezing" membrane lipids should cause an abrupt change in reaction rates. However, the inhibition of photosynthesis by chilling in the dark persists for long periods after the plant is returned to optimal temperatures (e.g., Drake and Raschke, 1974; Martin *et al.*, 1981). This eliminates the fluidity of the membrane as the direct cause of the impairment of photosynthesis, although it does leave open the possibility that changes in membrane fluidity distort metabolism and eventually lead to persistent membrane malfunctions. We have used sensitive differential scanning calorimetry to search for the putative lipid-derived endotherms in chloroplast thylakoid membranes of the chill-sensitive plants cantaloupe, kidney bean, domestic tomato, and soybean (Low *et al.*, 1983). For comparison, calorimetric scans of chloroplast lamellar membranes from the chill-insensitive plants spinach, pea, and wild tomato were conducted. A large reversible endotherm extending from 10 to 35°C was observed in chloroplast membranes from tomatoes of both chilling-sensitive (*Lycopersicon esculentum* Mill. cv. Floramerica) and chill-insensitive (*Lycopersicon hirsutum* LA 1361) species. A much smaller endotherm, approximately one-fifteenth the area of that seen in tomato and extending over a similar temperature range, was detected in chloroplast membranes from spinach and peas. No significant endotherm was observed in this temperature region in chloroplast membranes from the chill-sensitive cantaloupe, kidney bean, and soybean. We were able to conclude from this calorimetric study that no correlation exists between chill sensitivity of photosynthesis and presence or absence of a significant lipid-related endotherm at temperatures above 5°C in chloroplast thylakoids.

We do not at present know if stress-induced malfunctions of one or more of the Calvin cycle enzymes might be a causal element in the inhibitions caused by either prechilling or previous water stress. In neither case is there any change in the apparent K_m for CO_2 of the chloroplast (Martin *et al.*, 1981; Matthews and Boyer, 1983). Possible explanations for this observation are the following: (1)

The amount of activated ribulosebisphosphate (RuP_2) carboxylase is diminished by the stress treatment. The activation level of ribulosebisphosphate carboxylase has been implicated in the regulation of photosynthetic CO_2 assimilation (Jensen and Bahr, 1977). Jensen's group has recently demonstrated (Sicher and Jensen, 1979; Perchorowicz *et al.*, 1981) that the amount of activated RuP_2 carboxylase correlates with the light intensity dependence of CO_2 reduction for intensities greater than 225 μeinsteins m^{-2} sec^{-1}. It should be emphasized that this is a correlative relationship and not necessarily a causative relationship, since the level of activation of RuP_2 carboxylase could as well be a response to lowered photosynthesis as a cause of it. Nevertheless, a finding of high and unaltered levels of activated RuP_2 carboxylase in these plants after stress would strongly imply that the functioning of this enzyme is not an element of chilling inhibition. (2) The concentration of available RuP_2 may be reduced by the stress, partially depriving RuP_2 carboxylase from one of its substrates. In the simplest case, this would lower photosynthesis without changing the apparent K_m for CO_2. It is essential to realize that virtually any perturbation in electron transport, in photophosphorylation, or in the enzymatic steps of CO_2 reduction could lead to altered RuP_2 levels. Thus, a lower RuP_2 level after stress exposure would do little to localize the primary site of stress injury in the chloroplast but would provide a first-level explanation for the observed reduction in photosynthesis. More detailed information on individual reactions of the CO_2 reduction cycle would likely be required. A comparison of pool sizes of Calvin cycle intermediates between chloroplasts isolated from control and stressed leaves may reveal malfunctioning enzymes. This line of experimentation requires the isolation of active, intact chloroplasts from both untreated and stressed plants. Intermediates can be separated from one another and their pool sizes determined by HPLC. The expectation is that the pool size of a substrate of an affected enzyme would increase and that the pool size of the product would decrease. In addition, pulse–chase experiments would show label leaving the substrate pool of an impaired enzyme more slowly than in the control case. As before, it must be considered whether lower activity detected in a particular region of the Calvin cycle is a result of or a cause of chill-impaired photosynthesis. This is of particular concern in those cases of enzymes known to be light activated (Jensen and Bahr, 1977). Thus, it may be useful to assay the *in vitro* performance of a suspected enzyme to help determine whether altered enzyme activity is a cause or an effect.

Different situations will dictate a different set of *in vitro* experiments that have the best chance of uncovering the underlying mechanism of stress-induced inhibition of photosynthesis. Regardless of the nature of stress, the species of the plant, or the particular collection of measurements made, it is essential to verify the meaning of the *in vitro* data with corroborating measurements on intact leaves.

C. Predictions of Whole-Plant Behavior from *in Vitro* Measurements of Photosynthesis

A unifying theme of our chapter has been the need to show a close relationship between whole-plant and *in vitro* experimental data as a criterion for establishing physiological significance. The direction of information flow need not always originate with the whole plant. In many situations the greater advantage is gained through predictions of whole-plant behavior based on *in vitro* measurements of stress-modified photosynthesis.

One example for which this approach is proving to be useful involves the investigation of chilling effects on energy conservation and photophosphorylation in tomato. Many of the plant species in which there may be the greatest interest in studying stress-induced effects on photosynthesis are crop species, and many of these are, to varying degrees, intractable for making subcellular preparations. In the case of tomato, secondary plant products which accumulate in the leaves cause difficulties in the isolation of active organelles. Even in the presence of a variety of protective agents and with plants cultured to minimize the level of these troublesome compounds, chloroplast thylakoid membranes isolated from tomato leaves are not well coupled to ATP formation. Typically, for spinach we observe 1.2 ATP molecules formed per pair of electrons passing through the entire chain (i.e., ATP : $2e^-$ = 1.2), and the maximum achievable is thought to be 1.33 (Ort and Melandri, 1982). For tomato chloroplast thylakoid membranes we observe an ATP : $2e^-$ ratio not greater than 0.8 and found that there was no difference in this ratio for chloroplasts isolated from control or from chill-inhibited plants. While the observation suggests that the inhibition of photophosphorylation is not an element of chill-induced inhibition of photosynthesis, the data are equivocal because of the poor phosphorylation in control and stressed alike. A large stress-induced effect on energy coupling could be hidden in the low ATP : $2e^-$ ratios, and an effect of this size on ATP synthesis would certainly give rise to the level of chill-induced inhibition of CO_2 reduction observed in the whole plant.

Some information about the behavior of photophosphorylation in the attached leaf can come from measurement of the flash-induced electrochromic absorption band shift measured as an absorption change at 518–540 nm. The membrane depolarizing proton efflux through the coupling factor complex associated with phosphorylation results in an acceleration of the decay of the electric field, indicating an absorption change. Further acceleration of proton efflux by uncoupling agents results in even more rapid membrane depolarization, but proton efflux slowed by inhibition of the phosphorylation of ADP on the coupling factor causes relaxation of the field to be slowed (Witt, 1971). We maintain a fairly conservative interpretation of the electrochromic band shift and are reluctant to

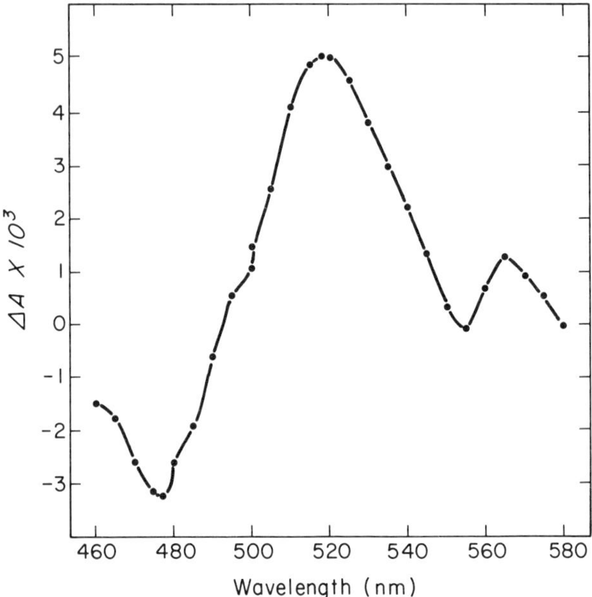

Fig. 8. Wavelength dependence of flash-induced (10 msec) electrochromic absorption change (ΔA) measured on an attached tomato leaf.

estimate actual rates of ATP formation or magnitude of the flash-induced transmembrane electric field from these measurements. However, some qualitative and relative information about photophosphorylation capacity in the attached leaf can be inferred from these measurements. From analysis of the decay of this absorption change, we can decide whether chilling impairs ATP formation in attached leaves and, if so, whether the membranes are uncoupled or if the ATPase reactions have been interfered with. The spectrum of the flash-induced electrochromic absorption band shift for attached tomato leaves is shown Fig. 8 and is in line with published spectra from isolated chloroplast thylakoids of spinach (Witt, 1971). Three compounds which inhibit photophosphorylation *in vitro* were introduced into submerged leaves. The effects of these treatments also confirmed expectations based on the *in vitro* measurements of ATP synthesis and the electrochromic band shift in spinach thylakoid membranes. The decay of the flash-induced electrochromic absorption change in leaves treated with the uncoupler octylamine was accelerated. Leaves treated with triphenyltin, which inhibits proton efflux through CF_0 (Gould, 1976), or with deoxyphlorizin, which inhibits ATP formation by competing with phosphate (Winget *et al.*, 1969), both resulted in a decrease in the decay rate and an increase in the amplitude of the flash-induced absorption change measured at 518–540 nm. These data demonstrate that measurement of the electrochromic absorption band shift in the leaves can be used to investigate effects of chilling on energy conservation which may not be evident after chloroplast isolation. The kinetic traces of the flash-induced absorption changes shown in Fig. 9 and the accompanying computer subtraction (Fig. 9, lowest trace) reveal no chill-induced difference in this sensitive, albeit qualitative, monitor of the energy conservation properties of *in situ* membranes. These data further substantiate the *in vitro* data and indicate that altered energy conservation probably is not an element contributing to chill-induced inhibition of CO_2 reduction. There are also obvious applications of this whole-leaf spectroscopic technique in understanding the effects of other stresses on the energy conservation process, and these experiments remain to be done.

Numerous other noninvasive physical techniques can be applied to analysis of stress-modified whole-leaf photosynthesis. The most frequently used and misused technique is the measurement of whole-leaf chlorophyll fluorescence from PSII. It is much beyond the scope of this chapter to deal with the vicissitudes of such measurements, and we can only say here that the data must be interpreted cautiously since they can be very misleading if dealt with simplistically. Of the recent developments in physiometric techniques to monitor whole-leaf photosynthetic parameters, the technique of delayed light emission imaging (Ellenson and Amundson, 1982; Ellenson and Raba, 1983) may be the most exciting. The unique aspect of this technique is that it is sensitive to the nonuniform response of photosynthesis in leaves to changes in environmental conditions. Of particular interest are spatial nonuniformities and time-dependent variations. This is an important area of research which has received exceedingly little attention, and

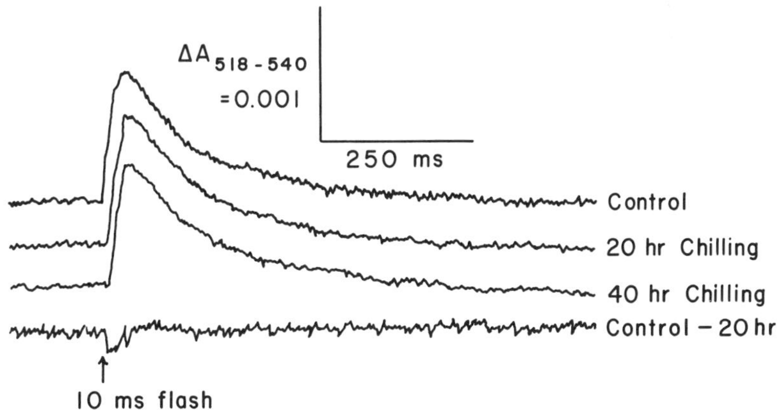

Fig. 9. The lack of effect of dark chilling (1°C) on the flash-induced electrochromic absorption change (ΔA) in attached tomato leaves. The lowest trace in the figure shows the computer subtraction of absorption change measured in a leaf before and after a 20-hr chilling treatment.

the consequences which such nonuniformities might have on space and time-averaged measurements have been heretofore largely ignored.

III. CONCLUSIONS

These studies of photosynthesis after cold nights and during droughts illustrate that the chloroplasts play a primary role and the stomata a secondary role in the stress response of photosynthesis in sunflower and tomato. Others have reached similar conclusions with regard to other kinds of environmental stress. Photosynthesis in garden bean was inhibited by losses in chloroplast activity resulting from the photoinhibitory effects of light on cold days, and these were superimposed on the inherently inhibitory effects of chilling temperatures (Powles *et al.*, 1983). High temperatures caused losses in chloroplast activity in *Atriplex* as thylakoid membranes lost photosystem II activity and the ability to convert excitation energy to stable chemical forms (Pearcy *et al.*, 1977; Badger *et al.*, 1982).

An important observation arising from these studies is that chloroplast activities, while often rate limiting for photosynthesis in extreme environments, display a different response if the plant encounters the conditions for long times or if the stress is imposed gradually. This acclimation is always in a direction that protects chloroplast activity. For example, in sunflower plants exposed to moderately low water potentials for 2 weeks and then subjected to drought, photosynthesis was less affected than in plants that had not experienced low water potentials previously (Matthews and Boyer, 1983). The change in sensitivity was reflected in a smaller response of the quantum yield of CO_2 reduction to low water potentials in acclimated plants than in controls. In addition, chloroplasts isolated from leaves of acclimated plants showed less drought-induced inhibition of PSII-dependent electron transport than did the control plants. The effects of low water potentials on light-saturated rates of CO_2 reduction remained different even when the internal CO_2 concentrations were identical for both types of plants; therefore, acclimation of the stomata was not the underlying mechanism of the adaptation.

These results indicate that acclimation of sunflower photosynthesis to low water potentials occurs largely at the chloroplast level and involves PSII as well as other aspects of photosynthesis. Similar acclimation of photosynthesis in response to stress has also been reported by others. In another study concerning low leaf water potentials (Ackerson and Hebert, 1981), photosynthesis in acclimated cotton plants was inhibited less by drought than in plants grown continuously under well-watered conditions. In yet another study, it was reported that high temperatures caused less loss in chloroplast quantum yield if the plants were grown at high temperatures than if they were grown at low temperatures (Pearcy *et al.*, 1977).

The ability to acclimate to environmental extremes implies a genetic capacity for dealing with these conditions. In *Atriplex,* for example, plant types that were collected from communities that were growing in cool climates were less able to acclimate photosynthetically to high temperatures than those from warm climates, suggesting that there may have been genetic selection for characteristics that fit the climate (Pearcy *et al.,* 1977). The interspecific differences in photosynthetic susceptibility to low water potentials clearly involved large genetic differences (Boyer, 1970; Shimshi, *et al.,* 1982), and there are intraspecific differences in grain production at low water potentials as well (Boyer *et al.,* 1980) that could have a genetic basis in altered photosynthetic performance.

Much more information is needed concerning the genetic variability in photosynthesis in various environments. Although photosynthesis is a constitutive component of plant metabolism, many genetic variants are known and some of these may affect photosynthetic performance in environmental extremes. Large differences in photosynthetic carbon metabolism, such as Crassulacean acid metabolism and C_4 photosynthesis, have evolved in response to environmental extremes, particularly drought and heat, and these undoubtedly involve many genes. With further exploration we may find additional aspects of chloroplast behavior that could be genetically manipulated with favorable consequences in adverse environments.

Although photosynthetic response to drought and cold nights primarily reflects chloroplast changes in sunflower and tomato, stomatal effects are probably primary factors in some other species adapted to extreme environments. Stomatal closure can prevent virtually all gas exchange in plants from xeric habitats, a characteristic that is common among plants exhibiting Crassulacean acid metabolism (Nobel, 1976) and some evergreen species such as *Nerium oleander* (Björkman *et al.,* 1980, 1981).

The large inhibition of plant productivity by adverse agricultural environments (Table 1) illustrates the prevalence of these environments and the promise held by even small improvements in plant performance in these conditions. Because photosynthesis is so sensitive to these environments and is basic to agricultural production, it represents a central and often dominant feature of plant response to stress. With the additional finding that chloroplasts lose activity in environmental extremes and that, in some species, these changes can be rate limiting for plant growth, our attention is focusing on how their susceptibility to extremes could be modified. Genetic manipulation is a possibility, but much more needs to be understood about how plants acclimate to the environment and how acclimation is controlled. Are new proteins synthesized? Does photosynthetic acclimation require altertions at the transcriptional or translational level? Are there regulatory molecules involved, and do they vary with the environment?

The fact that acclimation can occur, regardless of the mechanism, indicates that plant response to environment is partly under internal control and is not simply an inevitability imposed by external factors. This simple fact puts plant performance in adverse environments within reach of the methods of genetics and regulatory biology and is a most promising possibility for significant advances in productivity in the future.

Acknowledgments

The authors thank Drs. B. Martin and M. Matthews for some of the unpublished data presented in this chapter. The work from the laboratories of D. R. Ort and J. S. Boyer described in this chapter was supported in part by grants from the Competitive Research Grants Office of United States Department of Agriculture (D. R. O) and from the National Science Foundation (J. S. B.).

REFERENCES

Ackerson, R. C., and Hebert, R. R. (1981). Osmoregulation in cotton in response to water stress. I. Alterations in photosynthesis, leaf conductance, translocation, and ultrastructure. *Plant Physiol.* **67,** 484–488.

Avron, M., and Chance, B. (1967). Relation of phosphorylation to electron transport in isolated chloroplasts. *Brookhaven Symp. Biol.* **19,** 149–160.

Badger, M. R., Björkman, O., and Armond, P. A. (1982). An analysis of photosynthetic response and adaptation to temperature in higher plants: Temperature acclimation in the desert evergreen *Nerium oleander* L. *Plant, Cell Environ.* **5,** 85–99.

Björkman, O., Downton, W. J. S., and Mooney, H. A. (1980). Response and adaptation to water stress in *Nerium oleander. Year Book—Carnegie Inst. Washington* **79,** 150–157.

Björkman, O., Powles, S. B., Fork, D. C., and Öquist, G. (1981). Interaction between high irradiance and water stress on photosynthetic reactions in *Nerium oleander. Year Book—Carnegie Inst. Washington* **80,** 57–59.

Boyer, J. S. (1970). Differing sensitivity of photosynthesis to low leaf water potentials in corn and soybean. *Plant Physiol.* **46,** 236–239.

Boyer, J. S. (1982). Plant productivity and the environment. *Science* **218,** 443–448.

Boyer, J. S., and Potter, J. R. (1973). Chloroplast response to low leaf water potentials. I. Role of turgor. *Plant Physiol.* **51,** 989–992.

Boyer, J. S., Johnson, R. R., and Saupe, S. G. (1980). Afternoon water deficits and grain yields in old and new soybean cultivars. *Agron. J.* **72,** 981–986.

Crofts, A. R., and Wraight, C. A. (1983). Photosynthesis—The electrochemical domain. *Biochim. Biophys. Acta* (in press).

Diner, B. A., and Wollman, F. A. (1980). Isolation of highly active photosystem II particles from a mutant *Chlamydomonas reinhardtii. Eur. J. Biochem.* **110,** 521–526.

Drake, B., and Raschke, K. (1974). Prechilling of *Xanthium strumarium* L. reduces net photosynthesis and, independently, stomatal conductance, while sensitizing the stomata to CO_2. *Plant Physiol.* **53,** 808–812.

Ehleringer, J., and Björkman, O. (1976). Carbon dioxide and temperature dependence of the quantum yield for CO_2 uptake in C_3 and C_4 plants. *Year Book—Carnegie Inst. Washington* **75,** 418–421.

Ellenson, J. L., and Amundson, R. G. (1982). Delayed light imaging for the early detection of plant stress. *Science* **215,** 1104–1106.

Ellenson, J. L., and Raba, R. M. (1983). Gas exchange and phytoluminography of single red kidney bean leaves during periods of induced stomatal oscillations. *Plant Physiol.* **72,** 90–95.

Gaastra, P. (1959). Photosynthesis of crop plants as influenced by light, carbon dioxide, temperature and stomatal diffusion resistance. *Meded. Landbouwhogesch. Wageningen* **59,** 1–68.

Giaquinta, R. T., Dilley, R. A., Crane, F. L., and Barr, R. (1974). Photophosphorylation not coupled to DCMU-insensitive photosystem II oxygen evolution. *Biochem. Biophys. Res. Commun.* **59,** 985–991.

Good, N. E., and Bell, D. H. (1980). Photosynthesis, plant productivity and crop yield. *In* "The Biology of Crop Productivity" (P. S. Carlson, ed.), pp. 3–51. Academic Press, New York.

Good, N. E., Izawa, S., and Hind, G. (1966). Uncoupling and energy transfer inhibition in photophosphorylation. *In* "Current Topics in Bioenergetics" (D. R. Sanadi, ed.), Vol. I, pp. 75–112. Academic Press, New York.

Gould, J. M. (1976). Inhibition of triphenyltin chloride of a tightly-bound membrane component involved in photophosphorylation. *Eur. J. Biochem.* **62,** 567–575.

Haehnel, W., Pröpper, A., and Krause, H. (1980). Evidence for complexes of plastocyanin as the intermediate electron donor of P-700. *Biochim. Biophys. Acta* **593,** 384–399.

Izawa, S., and Pan, R. L. (1978). Photosystem I electron transport and phosphorylation supported by electron donation to the plastoquinone region. *Biochem. Biophys. Res. Commun.* **83,** 1171–1177.

Izawa, S., Gould, J. M., Ort, D. R., Felker, P., and Good, N. E. (1973). Electron transport and photophosphorylation in chloroplasts as a function of the electron acceptor. III. A dibromothymoquinone-insensitive phosphorylation reaction associated with photosystem II. *Biochim. Biophys. Acta* **305,** 119–128.

Izhar, S., and Wallace, D. H. (1967). Effect of night temperature on photosynthesis of *Phaseolus vulgaris* L. *Crop Sci.* **7,** 546–547.

Jensen, R. G., and Bahr, J. T. (1977). Ribulose 1,5-bisphosphate carboxylase-oxygenase. *Ann. Rev. Plant Physiol.* **28,** 379–400.

Joliot, P., and Kok, B. (1975). Oxygen evolution in photosynthesis. *In* "Bioenergetics of Photosynthesis" (Govindjee, ed.), pp. 388–412. Academic Press, New York.

Joliot, P., Bennoun, P., and Joliot, A. (1973). New evidence supporting energy transfer between photosynthetic units. *Biochim. Biophys. Acta* **305,** 317–328.

Kaiser, W. M., Kaiser, G., Prachuab, P. K., Wildman, S. G., and Heber, U. (1981). Photosynthesis under osmotic stress. *Planta* **153,** 416–422.

Kaniuga, Z., and Michalski, W. (1978). Photosynthetic apparatus in chilling-sensitive plants. II. Changes in free fatty acid composition and photoperoxidation in chloroplasts following cold storage and illumination of leaves in relation to Hill reaction activity. *Planta* **140,** 129–136.

Kaniuga, Z., Sochanowicz, B., Zabek, J., and Krzystyniak, K. (1978). Photosynthetic apparatus in chilling-sensitive plants. I. Reactivation of Hill reaction activity inhibited in the cold and dark storage of detached leaves and intact plants. *Planta* **140,** 121–128.

Keck, R. W., and Boyer, J. S. (1974). Chloroplast response to low leaf water potentials. III. Differing inhibition of electron transport and photophosphorylation. *Plant Physiol.* **53,** 474–479.

Low, P. S., Ort, D. R., Cramer, W. A., Whitmarsh, J., and Martin, B. (1984). Search for an endotherm in chloroplast lamellar membranes associated with chilling-inhibition of photosynthesis. *Arch. Biochem. Biophys.* **231,** 336–344.

Lyons, J. M. (1973). Chilling injury in plants. *Annu. Rev. Plant Physiol.* **24,** 445–466.

McPherson, H. G., and Boyer, J. S. (1977). Regulation of grain yield by photosynthesis in maize subjected to a water deficiency. *Agron. J.* **69,** 714–718.

Malkin, R., and Posner, H. B. (1978). On the site of function of the Rieske iron-sulfur center in the chloroplast electron transport chain. *Biochim. Biophys. Acta* **501,** 552–554.

Margulies, M. M. (1972). Effect of cold-storage of bean leaves on photosynthetic reactions of isolated chloroplasts. Inability to donate electrons to photosystem II and relation to manganese content. *Biochim. Biophys. Acta* **267,** 96–103.

Martin, B., and Ort, D. R. (1982). Insensitivity of water-oxidation and photosystem II activity in tomato to chilling temperatures. *Plant Physiol.* **70,** 689–694.

Martin, B., Ort, D. R., and Boyer, J. S. (1981). Impairment of photosynthesis by chilling-temperatures in tomato. *Plant Physiol.* **68,** 329–334.

Matthews, M. A., and Boyer, J. S. (1984). Acclimation of photosynthesis to low leaf water potentials. *Plant Physiol.* **74,** 161–166.

Mayoral, M. L., Atsmon, D., Gromet-Elhanan, Z., and Shimshi, D. (1981). Effect of water stress on enzyme activities in wheat and related wild species: Carboxylase activity, electron transport, and phosphorylation in isolated chloroplasts. *Aust. J. Plant Physiol.* **8,** 385–394.

Moss, D. N., and Rawlins, S. L. (1963). Concentration of carbon dioxide inside leaves. *Nature (London)* **197,** 1320–1321.

Murata, N., and Fork, D. C. (1975). Temperature dependence of chlorophyll a fluorescence in relation to the physical phase of membrane lipids in algae and higher plants. *Plant Physiol.* **56,** 791–796.

Nobel, P. S. (1976). Water relations and photosynthesis of a desert CAM plant, *Agave deserti*. *Plant Physiol.* **58,** 576–582.

Ort, D. R. (1975). Quantitative relationship between photosystem I electron transport and ATP formation. *Arch. Biochem. Biophys.* **166,** 629–638.

Ort, D. R., and Izawa, S. (1973). Studies on the energy coupling sites of photophosphorylation. II. Treatment of chloroplasts with NH_2OH plus ethylenediaminetetraacetate to inhibit water oxidation while maintaining energy-coupling efficiencies. *Plant Physiol.* **52,** 595–600.

Ort, D. R., and Melandri, B. A. (1982). Mechanisms of ATP synthesis. *In* ''Photosynthesis Energy Conversion by Plants and Bacteria'' (Govindjee, ed.), Vol. I, pp. 537–587. Academic Press, New York.

Ouitrakul, R., and Izawa, S. (1973). Electron transport and photophosphorylation in chloroplasts as a function of the electron acceptor. II. Acceptor-specific inhibition by KCN. *Biochim. Biophys. Acta* **305,** 105–118.

Pearcy, R. W., Berry, J. A., and Fork, D. C. (1977). Effects of growth temperature on the thermal stability of the photosynthetic apparatus of *Atriplex lentiformis* (Torr.) Wats. *Plant Physiol.* **59,** 873–878.

Perchorowicz, J. T., Raynes, D. A., and Jensen, R. G. (1981). Light limitation of photosynthesis and activation of ribulose bisphosphate carboxylase in wheat seedlings. *Proc. Natl. Acad. Sci. U.S.A.* **78,** 2985–2989.

Pfister, K., and Arntzen, C. J. (1979). The mode of action of photosystem II-specific inhibitors in herbicide-resistant weed biotypes. *Z. Naturforsch. C: Biosci.* **34,** 996–1009.

Portis, A. R., Jr. (1981). Evidence of a low stromal Mg^{2+} concentration in intact chloroplasts in the dark. *Plant Physiol.* **67,** 985–989.

Portis, A. R., Jr., and Heldt, H. W. (1976). Light-dependent changes of the Mg^{2+} concentration in the stroma in relation to the Mg^{2+} dependency of CO_2 fixation in intact chloroplasts. *Biochim. Biophys. Acta* **449,** 434–446.

Potter, J. R., and Boyer, J. S. (1973). Chloroplast response to low leaf water potentials. II. Role of osmotic potential. *Plant Physiol.* **51,** 993–997.

Powles, S. B., Berry, J. A., and Björkman, O. (1983). Interaction between light and chilling temperature on the inhibition of photosynthesis in chilling-sensitive plants. *Plant, Cell Environ.* **6,** 117–123.

Saha, S., Ouitrakul, R., Izawa, S., and Good, N. E. (1971). Electron transport and photophosphorylation in chloroplasts as a function of the electron acceptor. *J. Biol. Chem.* **246,** 3204–3209.

Satoh, K., and Butler, W. L. (1978). Low temperature spectral properties of subchloroplast fractions purified from spinach. *Plant Physiol.* **61,** 373–379.

Sharkey, T. D., Imai, K., Farquhar, G. D., and Cowan, I. R. (1982). A direct confirmation of the standard method of estimating intercellular partial pressure of CO_2. *Plant Physiol.* **69,** 657–659.

Shimshi, D., Mayoral, M. L., and Atsmon, D. (1982). Responses to water stress in wheat and related wild species. *Crop Sci.* **22,** 123–128.

Shneyour, A., Raison, J. K., and Smillie, R. M. (1973). The effect of temperature on the rate of photosynthetic electron transfer in chloroplasts of chilling-sensitive and chilling-resistant plants. *Biochim. Biophys. Acta* **292,** 152–161.

Sicher, R. C., and Jensen, R. G. (1979). Photosynthesis and ribulose-1,5-bisphosphate levels in intact chloroplasts. *Plant Physiol.* **64,** 880–883.

Smillie, R. M. (1979). The useful chloroplast: A new approach for investigating chilling stress in plants. *In* "Low Temperature Stress in Crop Plants" (J. M. Lyons, D. Graham, and J. K. Raison, eds.), pp. 187–202. Academic Press, New York.

Trebst, A., Harth, E., and Draber, W. (1970). On a new inhibitor of photosynthetic electron-transport in isolated chloroplasts. *Z. Naturforsch. B: Anorg. Chem., Org. Chem., Biochem., Biophys., Biol.* 25, 1157–1159.

Trebst, A., Reimer, S., and Dallacker, F. (1976). Properties of photoreduction by photosystem II. *Plant Sci. Lett.* **6,** 21–24.

Uribe, E. (1972). The interaction of *N,N'*-dicyclohexylcarbodiimide with the energy conservation system of the spinach chloroplast. *Biochemistry* **11,** 4228–4235.

Whitmarsh, J., and Cramer, W. A. (1979). Photooxidation of the high-potential iron-sulfur center in chloroplasts. *Proc. Natl. Acad. Sci. U.S.A.* **76,** 4417–4420.

Winget, G. D., Izawa, S., and Good, N. E. (1969). The inhibition of photophosphorylation by phlorizin and closely related compounds. *Biochemistry* **8,** 2067–2074.

Witt, H. T. (1971). Coupling of quanta, electrons, fields, ions and phosphorylation in the functional membrane of photosynthesis. Results by pulse spectroscopic methods. *Q. Rev. Biophys.* **4,** 365–477.

Yamashita, T., and Butler, W. L. (1968). Inhibition of the Hill reaction by tris and restoration by electron donation to photosystem II. *Plant Physiol.* **44,** 435–438.

Younis, H. M., Boyer, J. S., and Govindjee (1979). Conformation and activity of chloroplast coupling factor exposed to low chemical potential of water in cells. *Biochim. Biophys. Acta* **548,** 328–340.

Zelitch, I. (1982). The close relationship between net photosynthesis and crop yield. *BioScience* **32,** 796–802.

15

Responses to Environmental Heat Stress in the Plant Embryo

JOSEPH P. MASCARENHAS AND MITCHELL ALTSCHULER

I. INTRODUCTION

In the developing embryos of legumes such as soybean, classes of genes are turned on which are not expressed at any other time during the life of the plant. The midmaturation soybean seed contains a prominent class of superabundant mRNA's, present in about 15,000 to 25,000 copies per mRNA per cell. These mRNA's code for storage proteins and other embryo-specific proteins (Goldberg *et al.*, 1981a,b; Meinke *et al.*, 1981). The protein content of most legume seeds is between 25 and 35% of the weight of the seed. In certain strains the protein content can be as high as 55% of the dry weight of the seed (Derbyshire *et al.*, 1976). There is accordingly a massive amount of developmentally specific protein synthesis during embryo development.

Isolated embryos labeled *in vitro* exhibit a modified response to heat shock as compared to that shown by seedlings or cultured cells (Altschuler and Mascarenhas, 1982a). Storage protein synthesis actually increases during a heat

Changes in Eukaryotic Gene Expression
in Response to Environmental Stress

ISBN 0-12-066290-6

shock of 40°C for 1 hr. This is unlike the situation for most normal proteins in nonembryonic cells and tissues in which normal protein synthesis is greatly reduced during heat shock. In addition to the increase in storage proteins, heat shock proteins are also synthesized during heat shock in embryos (Altschuler and Mascarenhas, 1982a). The developing seed thus seems to react to temperature stress differently from seedlings in that the translation of storage protein and many other mRNA's appears to be insensitive to inhibition by heat shock; and, on the contrary, their synthesis is stimulated.

In this chapter we characterize the changes in transcription and translation of specific mRNA's in the developing embryo during heat shock at various temperatures.

II. STORAGE PROTEIN SYNTHESIS CONTINUES AT HIGHER RATES AT HEAT SHOCK TEMPERATURES IN THE DEVELOPING SOYBEAN EMBRYO

Greenhouse-grown soybean plants (cultivar Tracy) were transferred to a growth chamber maintained at a constant temperature of 28°C and placed under a 12-hr light/12-hr dark regime for 3 days prior to removing seeds for experimentation. Seeds between 100 and 150 mg fresh weight were isolated under sterile conditions, and the seed coats were removed. For the labeling of proteins, the two cotyledons were separated and each was placed on a 5-μl drop of $\frac{1}{10}$-strength Hoagland's solution containing 12.5 μCi of [^{35}S]methionine (Amersham/Searle, 1300 Ci/mmol) on a piece of Parafilm in a Petri dish. An additional 5-μl (12.5 μCi) drop of the label was placed on top of each cotyledon. A filter paper disk soaked in sterile water was placed under the Parafilm in the Petri dish. Embryos were heat shocked for 30 min at temperatures of 28–52°C and then labeled for 2.5 hr with [^{35}S]methionine at the same temperature. A 2.5-hr incubation and labeling was chosen to enable processing of the 11 S and the 7 S storage proteins from their precursors (Beachy *et al.*, 1981; Sengupta *et al.*, 1981). At the end of the incubation, the embryos were processed for sodium dodecyl sulfate(SDS)–polyacrylamide gel electrophoresis and autoradiography as described previously (Altschuler and Mascarenhas, 1982a). The data are presented in Fig. 1. Heat shock proteins (hsp's) are synthesized maximally at 37 and 40°C, and smaller amounts of hsp's are synthesized up to 46°C. Very little if any hsp synthesis occurs at 49°C (Fig. 1). The synthesis of non-hsp's is substantially increased after heat shocks of 37 and 40°C, non-hsp's continue to be synthesized in fairly large amounts up to 46°C, and some protein synthesis occurs even at 49°C. Figure 1, lane 8 (labeled S), which shows the hsp's in seedlings, enables one to

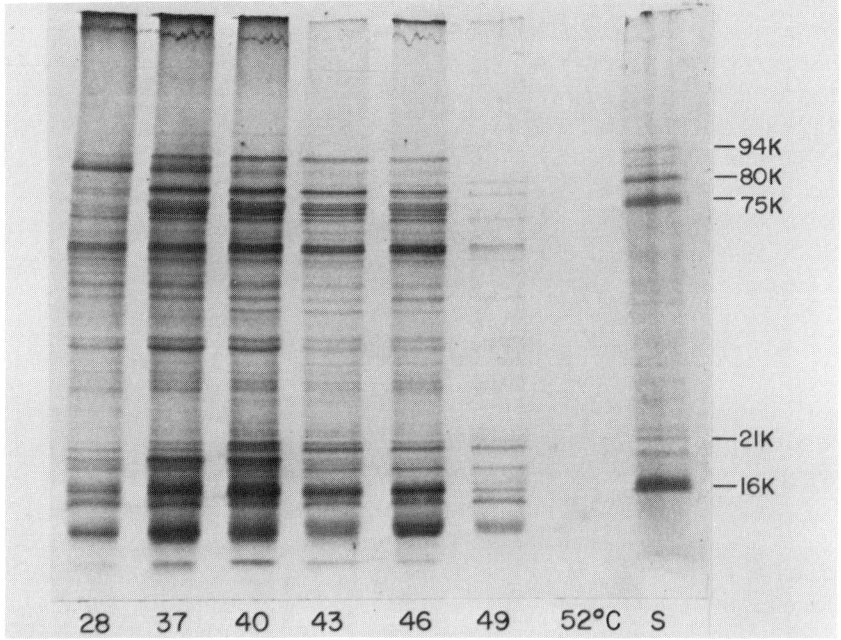

Fig. 1. Protein synthesis in developing soybean embryos during heat shock at various temperatures. Autoradiogram of an SDS–polyacrylamide (12–20% gradient) gel of proteins from embryos preincubated at the specified temperature for 30 min and then labeled with [^{35}S]methionine for 2.5 hr at the same temperature. Extracts of embryos containing equal weights of tissue were loaded in the lanes. Lane S, soybean seedling 1 hr at 40°C to show hsp's.

see more clearly which bands correspond to the hsp's and which to the normal embryo proteins.

The response of developing embryos to heat shock with respect to protein synthesis is very much different from that exhibited by soybean cells in culture (Barnett *et al.*, 1980), by soybean seedlings or parts of seedlings (Altschuler and Mascarenhas, 1982a; Key *et al.*, 1981), or by vegetative tissues of other plants studied (Baszczynski *et al.*, 1982; Cooper and Ho, 1983). In cultured cells and vegetative tissues there is a large reduction in the synthesis of normal non-hsp's during heat shock to 37–43°C, and no protein synthesis whether normal or heat shock is seen by 45°C. In tobacco mesophyll protoplasts transferred from 25 to 40°C, Meyer and Chartier (1983) have identified two groups of proteins on the basis of their regulation by heat shock. The synthesis of the majority of 25°C proteins is greatly reduced, but there are a small number of proteins that are unaffected by the heat shock. This finding is probably generally true for other plant systems, as we have also observed similar proteins in soybean and *Trades-*

cantia vegetative tissue (J. P. Mascarenhas and M. Altschuler, unpublished observations). The developing seed reacts differently than seedlings to heat shock in several respects. First, most normally synthesized proteins in the seed continue to be synthesized at heat shock temperatures. Second, the rate of their synthesis is increased rather than being decreased or remaining constant. Third, the protein synthetic machinery in embryos is stable and functional at temperatures as high as 49°C without any temperature preconditioning such as occurs in seedlings with a gradual temperature increase rather than a sudden heat shock (Altschuler and Mascarenhas, 1982a,b; Key *et al.*, 1982).

To study specifically the effect of heat shock on storage protein synthesis, the 11 S (glycinin) and the 7 S (conglycinin) storage proteins were immunoprecipitated with antiserum to these proteins, followed by SDS–gel electrophoresis and autoradiography. Preparation of the 11 S antiserum and immunoprecipitation of *in vivo*-labeled proteins were as described previously (Altschuler and Mascarenhas, 1982a). The 7 S antiserum was a gift of Dr. Roger N. Beachy. Both antisera were not entirely pure and showed some cross-reactivity. There is an increased synthesis of 11 S proteins at 37 and 40°C relative to that at 28°C, and the precursor to the 11 S proteins (58–60K) continues to be synthesized in large quantities up to 46°C, decreasing thereafter (Fig. 2). The processing of the precursor, however, to the mature acidic (36–38K) and basic (18–20K) subunits seems to be interfered with at temperatures of 43°C and above. Whether this abnormal processing occurs in the intact plant or is an artifact of the *in vitro* growth conditions remains to be determined.

The temperature dependence of 7 S protein synthesis (α- and α^1-polypeptides at bands a,b,c,d, Fig. 3) follows the same pattern as that for the 11 S proteins. The 7 S proteins continue to be synthesized up to 49°C. At heat shock temperatures of 40°C and above, a new band appears (band e, Fig. 3) that immunoprecipitates with the 7 S antiserum. The identity of this band is unknown at present, but it could possibly be a result of faulty processing or proteolytic cleavage of the 7 S precursor molecules at higher temperatures. Whether this also occurs in the intact plant is unknown at present.

The synthesis of the storage proteins (both 11 S and 7 S) is, thus, extremely resistant to temperature shock. The mRNA's for these proteins continue to be translated extensively at 46°C, and their translation occurs, although at reduced levels, even at 49°C. A similar situation seems to exist with respect to most of the other normal temperature proteins synthesized in the embryo (Fig. 1).

The insensitivity of storage protein synthesis to heat shock is not a result of temperature tolerance induced by previous exposure to intermediate or high temperature such as has been found for seedlings (Altschuler and Mascarenhas, 1982a), because the plants were incubated at 28°C for 3 days prior to the removal of seeds for the experiments. It also seems unlikely that the developing embryo synthesis hsp's at normal temperatures as has been described for both *Xenopus*

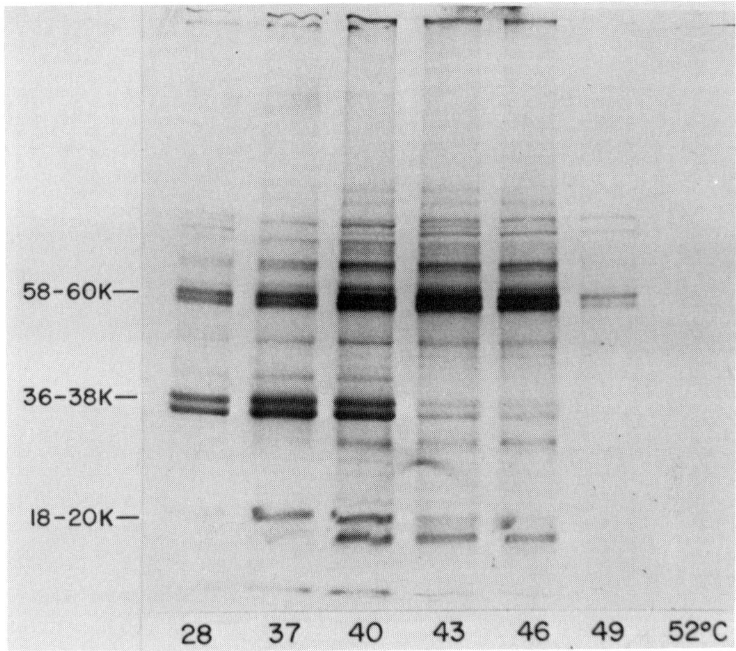

Fig. 2. Analysis of proteins synthesized by developing soybean embryos heat shocked to various temperatures after immunoprecipitation with antiserum to the 11 S proteins. Embryos were preincubated at the various temperatures for 30 min and then labeled with [^{35}S]methionine for 2.5 hr at the same temperature. The 11 S proteins in the extracts from equal weights of tissue were immunoprecipitated and analyzed by SDS–polyacrylamide gel electrophoresis and autoradiography. The 58–60K bands are the precursors of the acidic (36–38K) and the basic (18–20K) subunits of the 11 S protein.

oocytes (Bienz, 1982) and adult ovaries of *Drosophila* (Zimmerman *et al.*, 1983), since the messenger RNA's for at least two heat shock proteins are not present in embryos incubated at 28°C (Fig. 6; J. P. Mascarenhas and M. Altschuler, unpublished observations).

The storage proteins are synthesized on mRNA's that are present in very large numbers of copies per cell (Goldberg *et al.*, 1981a,b), unlike the situation with the majority of mRNA's in the seedling. The difference between hsp synthesis and storage protein synthesis on sudden heat shock to higher temperatures might be caused by the large excess of storage protein mRNA's relative to hsp mRNA's. This could not, however, be the explanation for the relative insensitivity of other normal temperature protein synthesis in the embryo. It is possible that some component or components of the protein synthetic machinery differ between developing embryos and vegetative plant tissue, so that heat shock does not inhibit the translation of normal mRNA's in embryos.

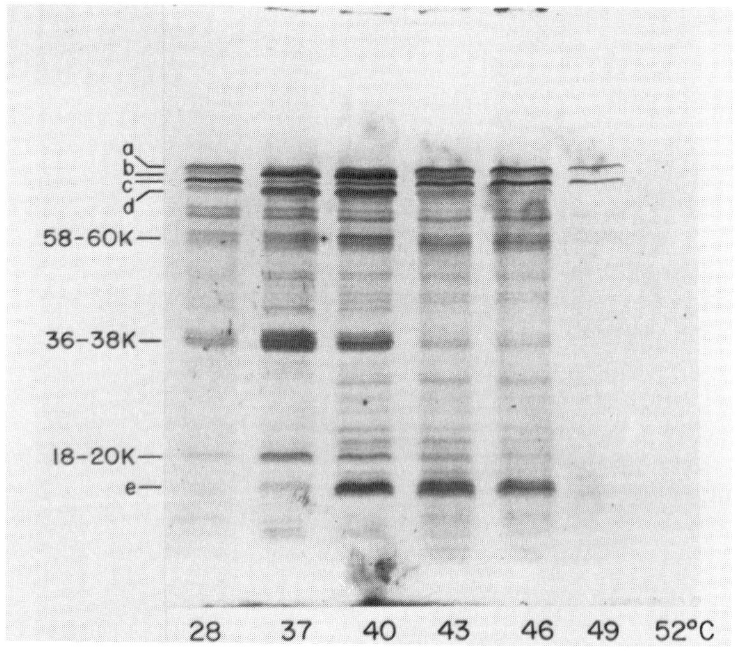

Fig. 3. Analysis of proteins synthesized by developing soybean embryos heat shocked to various temperatures after precipitation with antiserum to the 7 S proteins. (See legend to Fig. 2 for other experimental details.) Bands a,b,c, and d are the α/α' subunits of the 7 S proteins; see text (Section II) for a description of band e.

In rabbit reticulocyte lysates incubated at 40–42°C, an inhibition of the initiation of protein synthesis is observed which appears to correlate with the phosphorylation of initiation factor eIF-2 (Ernst *et al.*, 1982). Other studies indicate that phosphorylation of eIF-2 inhibits initiation by preventing the normal catalytic recycling of eIF-2 during initiation (Ernst *et al.*, 1982; Siekerka *et al.*, 1982). These observations are compatible with the evidence presented by Hickey and Weber (1982a,b); in HeLa cells mRNA's are translated preferentially at 42°C because of their greater ability relative to normal mRNA's to initiate translation under the conditions of reduced initiation found during heat shock. In HeLa cells the heat shock mRNA's seem to be translated in the presence of a large amount of mRNA coding for normal temperature polypeptides under conditions in which the overall rate of protein synthesis is severely inhibited to a level 20–30% of that found at 37°C (Hickey and Weber, 1982b).

The storage protein mRNA's could, thus, be similar to those of the hsp's, that is, relatively insensitive to the conditions of reduced initiation under heat shock. Alternatively, the protein synthetic machinery in the embryo might not be altered by heat shock to cause conditions of reduced initiation of protein synthesis. A situation similar to that found in soybean embryos has been described in *Xenopus,* in which somatic cells continue to translate normal cellular mRNA's at the same rate under heat shock conditions, whereas in the oocyte heat shock results in a decrease in the rate of normal mRNA translation (Bienz, 1982).

III. SYNTHESIS OF SPECIFIC MESSENGER RNA's DURING HEAT SHOCK IN DEVELOPING SOYBEAN EMBRYOS

The synthesis of storage proteins and other proteins is increased during heat shock in developing embryos. Is the increased synthesis of proteins caused by an increase in the rate of translation, or is it in part caused by an increase in the rate of synthesis of mRNA's during heat shock?

Changes in specific transcripts were studied by Northern blot or by dot blot analysis of total RNA from developing embryos heat shocked and probed with cloned DNA's to the 11 S and 7 S storage proteins, to actin, and to one of the small heat shock proteins (16K).

RNA was isolated from 100- to 150-mg midmaturation seeds, after heat shock to various temperatures, by freezing and grinding the tissue in liquid nitrogen followed by homogenization in 50 m*M* Tris (pH 9), 0.2 *M* NaCl, 1 m*M* MgAc, and 1.0% SDS and then extracted with phenol–chloroform–isoamyl alcohol as described in Frankis and Mascarenhas (1980). Total RNA from embryos shocked at different temperatures was analyzed by electrophoresis in denaturing formaldehyde–agarose gels (Lehrach *et al.,* 1977; Goldberg, 1980). The RNA was transferred to nitrocellulose papers and hybridized (Thomas, 1980) to plasmids labeled with [^{32}P]dCTP by nick translation (Maniatis *et al.*, 1975) containing the relevant inserts. Alternatively, the RNA was directly spotted prior to hybridization onto nitrocellulose paper (Thomas, 1980; White and Bancroft, 1982) using a Hybri-Dot apparatus (Bethesda Research Laboratories, Inc.).

The following cloned probes were used:

1. A-16. This probe hybridizes to the mRNA's of the α/α' subunits of the 7 S soybean storage protein (Goldberg *et al.*, 1981a).
2. A-28. This probe hybridizes to the mRNA (2.0 kb in size) for the precursor of the 11 S storage proteins (Goldberg *et al.*, 1981a). Both A-16 and A-28 were kindly provided by Dr. Robert B. Goldberg.
3. pSAc3. This plasmid contains a 3-kb *Hind*III fragment of soybean genomic DNA and was kindly given to us by Dr. R. B. Meagher. The actin coding sequence on this fragment spans approximately 1.4 kb and has been completely sequenced (Shah *et al.*, 1982). It hybridizes to an mRNA 1.75 kb in size.

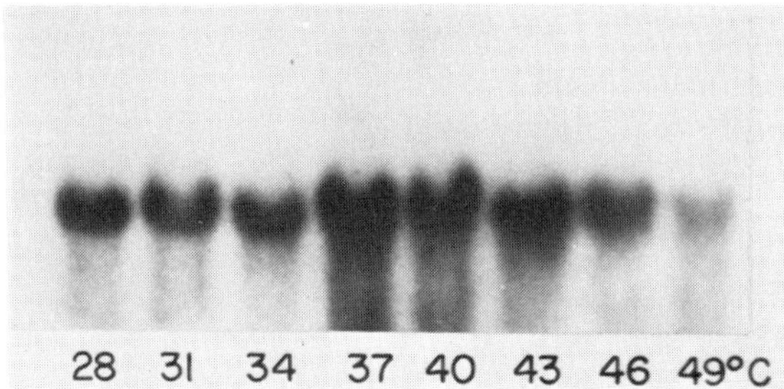

Fig. 4. Northern blot analysis of 5 μg of total RNA from developing seeds heat shocked for 1 hr to various tempertures and probed with the 11 S storage protein plasmid.

4. p0075. The soybean heat shock cDNA plasmid p0075 was generously provided by Dr. Joe L. Key. This plasmid hybrid selects mRNA's of approximately 0.9 kb which are translated into proteins of approximately 15–16K (Schöffl and Key, 1982).

Northern blots of total RNA (5 μg) from developing seeds after a 1-hr heat shock to various temperatures from 28 to 49°C were hybridized to the plasmid containing sequences complementary to the mRNA's for the 11 S storage proteins. The resulting autoradiograph is shown in Fig. 4. There is an increase in the quantity of 11 S transcripts in the developing embryos after a 1-hr heat shock at 37, 40, or 43°C as compared to the level at 28°C. This would seem to indicate either that the mRNA's for the 11 S proteins are synthesized more rapidly during heat shock at these temperatures than at 28°C, or that there is a decrease in the rate of turnover of the mRNA's. At 46°C the quantity of 11 S mRNA in the young embryos is approximately the same as at 28°C, but after heat shock to 49°C there appears to be a reduction in the mRNA quantity relative to that at 28°C. When the levels of 7 S mRNA are studied by dot blot analysis (Fig. 5), a situation very similar to that found with the 11 S mRNA is seen. Heat shock for 1 hr to 37–43°C results in an increase in the total amount of 7 S transcript in the embryos relative to that at 28°C.

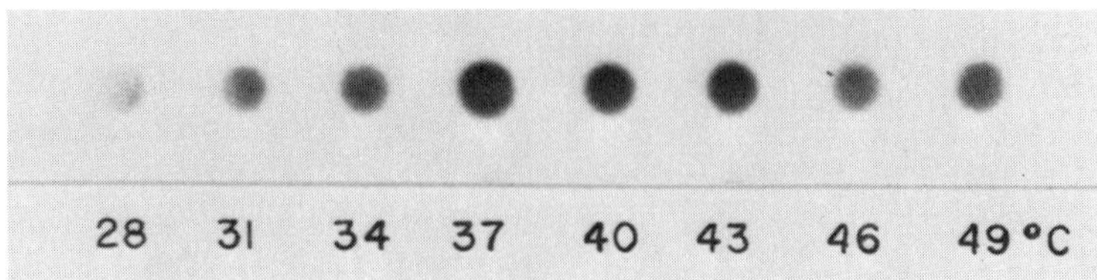

Fig. 5. A dot blot analysis of 2 μg of total RNA from developing seeds heat shocked for 1 hr to various temperatures and probed with the 7 S storage protein plasmid.

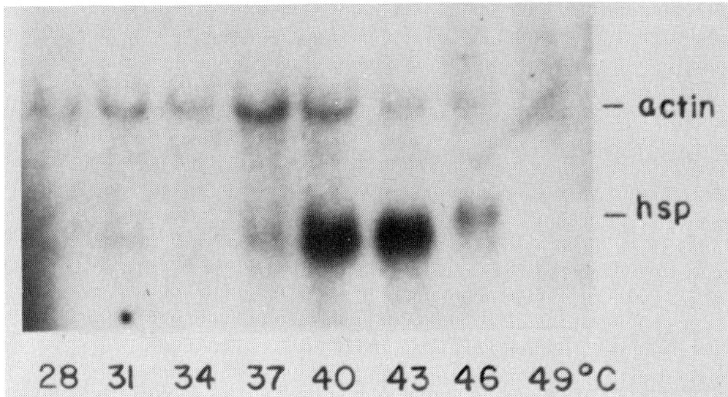

Fig. 6. Northern blot analysis of 5 μg of total RNA from developing seeds heat shocked for 1 hr at various temperatures and probed with the actin and the 15–16K heat shock protein plasmids.

Total RNA (5 μg) from young embryos, after electrophoresis in agarose and transfer to nitrocellulose paper, was hybridized simultaneously to actin and heat shock protein probes (Fig. 6). There appeared to be a slight increase in the quantity of actin transcript after a 1-hr heat shock to 37 and 40°C but not at 43°C, relative to the levels at 28°C.

The mRNA for the 15–16K heat shock proteins is not detectable at 28°C (Fig. 6). A very faint hybridizing band is seen after incubation at 31 and 34°C for 1 hr. Maximum synthesis of heat shock mRNA occurs after a 1-hr heat shock to 40 and 43°C. After 1 hr at 46°C the quantity of heat shock mRNA is greatly decreased. Moreover, the band that hybridizes to the probe is shifted to a larger-sized RNA at 46°C as compared to the RNA hybridizing at lower temperatures. This shift to a higher-molecular-weight RNA at 46°C has been consistently observed. The reasons for this shift are unknown, but it could be caused by a defect in the processing of the transcript. Since neither this gene nor any other plant heat shock gene has yet been sequenced, it is not known whether any processing of the initial transcripts of any of the heat shock genes occurs in plants. It would be of interest to compare the results obtained with the actin and the heat shock mRNA's in developing seeds with what happens in cultured cells or seedling tissue using the same probes. This work is in progress. At the present time, there is no information besides that which has been presented here concerning the synthesis of normal mRNA's in plants during heat shock.

The increase in storage protein content of the developing embryo during heat shock to temperatures as high as 43°C is unexpected and very different from that seen in *Drosophila*. In *Drosophila* there is a suppression of the synthesis of most normally synthesized mRNA's (see review in Ashburner and Bonner, 1979). Within 10 min after heat shock, actin gene transcription drops to about 5% of the 25°C value and remains at this level throughout a 60-min period at the heat shock temperature (Findly and Pederson, 1981).

IV. CONCLUSIONS

The developing embryo reacts to heat shock differently in several respects than do cultured cells or vegetative tissue. The storage proteins and other normally synthesized proteins in the embryo continue to be synthesized at heat shock temperatures. The rates of their synthesis are increased at temperatures up to 43°C rather than decreased, as in vegetative tissue. The protein synthesis machinery in embryos is stable and functional at temperatures as high as 49°C. During heat shock to 49°C storage proteins are translated, although in lesser amounts than at lower temperatures. Heat shock proteins are synthesized at heat shock temperatures and even after heat shock to 46°C, whereas in cells and seedlings no protein synthesis occurs at these temperatures.

In addition, the amount of mRNA's for storage proteins increases after a 1-hr heat shock at 37, 40, or 43°C as compared to the level at 28°C. The mRNA for one of the heat shock proteins (15–16K) is not present in embryos at 28°C, but it is maximally induced after heat shock to 40 or 43°C, and some transcripts are seen at 46°C.

The developing soybean embryo thus seems to be relatively insensitive to the inhibition of both transcription and translation by heat shock, and protein synthesis occurs in the embryo at higher heat shock temperatures than in tissue cultures or seedlings. This fact makes ecological sense, since the seed is the one phase in the life cycle of the plant that is most important for the propagation and the continuity of the species. It is necessary for the seed to develop fully and to synthesize the various storage products including proteins that are required during germination to enable a healthy plant to be formed. Most plants grow and enter their reproductive phase during periods in the year when the temperatures are high. Leaf temperatures of several plants have been measured. In full sunlight it is not uncommon for leaf temperatures to be 10–15°C above air temperatures and to exceed 40°C (Gates, 1968; Levitt, 1980). If protein synthesis in the seed was very sensitive to heat shock, then one would expect the quantity of storage and other proteins accumulating in the maturing seed to be seriously affected. This could seriously affect the viability of the seed and its ability to germinate and produce a plant that was vigorous enough to survive the critical germination and seedling stages. The relative insensitivity of transcription and translation to heat shock in the developing seed would enable the synthesis of proteins to continue even during periods of heat stress, thus ensuring that the seeds contain sufficient reserves for germination.

Acknowledgments

This work was supported by grant 83-CRCR-1-1256 from the Competitive Grants Research Office of the United States Department of Agriculture.

REFERENCES

Altschuler, M., and Mascarenhas, J. P. (1982a). Heat shock proteins and effects of heat shock in plants. *Plant Mol. Biol.* **1,** 103–115.

Altschuler, M., and Mascarenhas, J. P. (1982b). The synthesis of heat shock and normal proteins at high temperatures in plants and their possible roles in survival under heat stress. *In* "Heat Shock, from Bacteria to Man" (M. J. Schlesinger, M. Ashburner, and A. Tissières, eds.), pp. 321–327. Cold Spring Harbor Lab., Cold Spring Harbor, New York.

Ashburner, M., and Bonner, J. J. (1979). The induction of gene activity in *Drosophila* by heat shock. *Cell* **17,** 241–254.

Barnett, T., Altschuler, M., McDaniel, C. N., and Mascarenhas, J. P. (1980). Heat shock induced proteins in plant cells. *Dev. Genet. (Amsterdam)* **1,** 331–340.

Baszczynski, C. L., Walden, D. B., and Atkinson, B. G. (1982). Regulation of gene expression in corn. *Can. J. Biochem.* **60,** 569–579.

Beachy, R. N., Jarvis, N. P., and Barton, K. A. (1981). Biosynthesis of subunits of the soybean 7S storage protein. *J. Mol. Appl. Genet.* **1,** 19–27.

Bienz, M. (1982). The heat-shock response in *Xenopus* oocytes and somatic cells: Differences in phenomena and control. *In* "Heat Shock, from Bacteria to Man" (M. J. Schlesinger, M. Ashburner, and A. Tissières, eds.), pp. 177–181. Cold Spring Harbor Lab., Cold Spring Harbor, New York.

Cooper, P., and Ho, T. D. (1983). Heat shock proteins in maize. *Plant Physiol.* **71,** 215–222.

Derbyshire, E., Wright, D. J., and Boulter, D. (1976). Legumin and vicilin, storage proteins of legume seeds. *Phytochemistry* **15,** 3–24.

Ernst, V., Baum, E. Z., and Reddy, P. (1982). Heat shock protein phosphorylation and the control of translation in rabbit reticulocytes, reticulocyte lysates, and HeLa cells. *In* "Heat Shock, from Bacteria to Man" (M. J. Schlesinger, M. Ashburner, and A. Tissières, eds.), pp. 215–225. Cold Spring Harbor Lab., Cold Spring Harbor, New York.

Findly, R. C., and Pederson, T. (1981). Regulated transcription of the genes for actin and heat-shock proteins in cultured *Drosophila* cells. *J. Cell Biol.* **88,** 323–328.

Frankis, R., and Mascarenhas, J. P. (1980). Messenger RNA in the ungerminated pollen grain: A direct demonstration of its presence. *Ann. Bot. (London)* **45,** 595–599.

Gates, D. M. (1968). Transpiration and leaf temperature. *Annu. Rev. Plant Physiol.* **19,** 211–238.

Goldberg, D. A. (1980). Isolation and partial characterization of the *Drosophila* alcohol dehydrogenase gene. *Proc. Natl. Acad. Sci. U.S.A.* **77,** 5794–5798.

Goldberg, R. B., Hoschek, G., Ditta, G. S., and Breidenbach, R. W. (1981a). Developmental regulation of cloned superabundant embryo mRNAs in soybean. *Dev. Biol.* **83,** 218–231.

Goldberg, R. B., Hoschek, G., Tam, S. H., Ditta, G. S., and Breidenbach, R. W. (1981b). Abundance, diversity and regulation of mRNA sequence sets in soybean embryogenesis. *Dev. Biol.* **83,** 201–217.

Hickey, E. D., and Weber, L. A. (1982a). Modulation of heat shock polypeptide synthesis in HeLa cells during hyperthermia and recovery. *Biochemistry* **21,** 1513–1521.

Hickey, E. D., and Weber, L. A. (1982b). Preferential translation of heat shock mRNAs in HeLa cells. *In* "Heat Shock, from Bacteria to Man" (M. J. Schlesinger, M. Ashburner, and A. Tissières, eds.), pp. 199–206. Cold Spring Harbor Lab., Cold Spring Harbor, New York.

Key, J. L., Lin, C.-Y., and Chen, Y. M. (1981). Heat shock proteins of higher plants. *Proc. Natl. Acad. Sci. U.S.A.* **78,** 3526–3530.

Key, J. L., Lin, C.-Y., Ceglarz, E., and Schöffl, F. (1982). The heat-shock response in plants: Physiological considerations. *In* "Heat Shock, from Bacteria to Man" (M. J. Schlesinger, M. Ashburner, and A. Tissières, eds.), pp. 329–336. Cold Spring Harbor Lab., Cold Spring Harbor, New York.

Lehrach, H., Diamond, D., Wozney, J. M., and Boedtker, H. (1977). RNA molecular weight determination by gel electrophoresis under denaturing conditions, a critical reexamination. *Biochemistry* **16,** 4743–4751.

Levitt, J. (1980). High-temperature or heat stress. *In* "Responses of Plants to Environmental Stresses," Chilling, Freezing and High Temperature Stresses, Vol. I, pp. 347–393. Academic Press, New York.

Maniatis, T., Jeffrey, A., and Kleid, D. G. (1975). Nucleotide sequence of the rightward operator of phage λ. *Proc. Natl. Acad. Sci. U.S.A.* **72,** 1184–1188.

Meinke, D. W., Chen, J., and Beachy, R. N. (1981). Expression of storage-protein genes during soybean seed development. *Planta* **153,** 130–139.

Meyer, Y., and Chartier, Y. (1983). Long-lived and short-lived heat-shock proteins in tobacco mesophyll protoplasts. *Plant Physiol.* **72,** 26–32.

Schöffl, F., and Key, J. L. (1982). An analysis of mRNAs for a group of heat shock proteins of soybean using cloned cDNAs. *J. Mol. Appl. Genet.* **1,** 301–314.

Sengupta, G., DeLuca, V., Bailey, D., and Verma, D. P. S. (1981). Biosynthesis of soybean storage proteins: Post translational processing of 7S and 11S components. *Plant Mol. Biol.* **1,** 19–34.

Shah, D. M., Hightower, R. C., and Meagher, R. B. (1982). Complete nucleotide sequence of a soybean actin gene. *Proc. Natl. Acad. Sci. U.S.A.* **79,** 1022–1026.

Siekerka, J., Mauser, L., and Ochoa, S. (1982). Mechanism of polypeptide initiation in eukaryotes and its control by the phosphorylation of the α-subunit of initiation factor 2. *Proc. Natl. Acad. Sci. U.S.A.* **79,** 2537–2540.

Thomas, P. S. (1980). Hybridization of denatured RNA and small DNA fragments transferred to nitrocellulose. *Proc. Natl. Acad. Sci. U.S.A.* **77,** 5201–5205.

White, B. A., and Bancroft, F. C. (1982). Cytoplasmic dot hybridization. Simple analysis of mRNA levels in multiple small cell or tissue samples. *J. Biol. Chem.* **257,** 8569–8572.

Zimmerman, J. L., Petri, W., and Meselson, M. (1983). Accumulation of a specific subset of *D. melanogaster* heat shock mRNAs in normal development without heat shock. *Cell* **32,** 1161–1170.

16

Physiological and Molecular Analyses of the Heat Shock Response in Plants

JOE L. KEY, JANICE KIMPEL, ELIZABETH VIERLING, CHU-YUNG LIN, RONALD T. NAGAO, EVA CZARNECKA, AND FRIEDRICH SCHÖFFL

I. INTRODUCTION

Organisms ranging from bacteria to man respond to high temperature by synthesizing a new set of proteins, the heat shock proteins (hsp's), and repressing the synthesis of most normal proteins (Schlesinger *et al.*, 1982b). The universality and the evolutionary conservation of the heat shock (hs) response has led to the view that this response plays an important, fundamental role in organism survival of high temperature and, possibly, of stress in general (Ashburner and Bonner, 1979). *Drosophila* has served as the model system for hs studies (Ritossa, 1962; Ashburner and Bonner, 1979) and will be covered at least in part

Changes in Eukaryotic Gene Expression
in Response to Environmental Stress

ISBN 0-12-066290-6

by chapters in this volume. However, while the *Drosophila* system has provided a basis for studies in many systems, there are significant differences between the hs response in *Drosophila* and in plants. A major difference is the relative abundance of low-molecular-weight and high-molecular-weight proteins synthesized during heat shock. In *Drosophila* the major hsp's are 83, 70, 68, 27, 26, 23, and 22 k with hsp70 synthesis clearly dominating the response. Although plants produce heat shock proteins with M_r values of about 92K, 84K, 70K, and 68K, the majority of hsp synthesis is devoted to the production of a complex group of 20–30 low-molecular-weight hsp's (15K–27K). It appears that this group of proteins may be unique to plants.

In this chapter we provide a descriptive overview of the hs response in plants, emphasizing specifically the response in soybeans, covering both physiological and molecular analyses. A number of papers from this and other laboratories have been published recently on hs in plants (Dawson and Grantham, 1981; Barnett *et al.*, 1980; Key *et al.*, 1981, 1982, 1983a,b; Altschuler and Mascarenhas, 1982; Scharf and Nover, 1982; Baszczynski *et al.*, 1982; Schöffl and Key, 1982; Kelley and Freeling, 1982; Cooper and Ho, 1983; Meyer and Chartier, 1983; Nover *et al.*, 1983).

II. RESULTS

A. Heat Shock Protein Synthesis

When the growth temperature of soybean seedlings is shifted abruptly from 28 or 30°C to one of several higher temperatures ranging up to 45°C, hsp synthesis appears at temperatures above 32.5 to 35°C (Fig. 1). At 40°C hsp's represent the majority of newly synthesized proteins, as visualized on SDS gels. Above 40°C there is a dramatic decrease in amino acid incorporation into protein; however, the residual incorporation at these elevated temperatures is primarily into hsp's (Key *et al.*, 1982). From these analyses it is clear that there are distinct molecular weight groups of hsp's: the high-molecular-weight proteins of approximately 92K, 84K, 70K, and 68K, and the low-molecular-weight proteins of about 27K, 23K, 21K, 18K, and 15K. The two-dimensional O'Farrell gel analysis presented in Fig. 2 provides further insight into the complexity of soybean hsp's. While some of these electrophoretic components may represent *in vivo* posttranslational protein modification, our hybrid selection/translation and Southern hybridization analysis suggest that most of the low-molecular-weight (15K–27K) hsp's are distinct but, in some cases, related protein sequences. We have not yet analyzed the 68K, 70K, 84K, and 92K groups of hsp's of soybean in detail.

Many plant species synthesize a complex pattern of hsp's in response to a temperature shift (Key *et al.*, 1983a). The patterns for pea and millet are shown

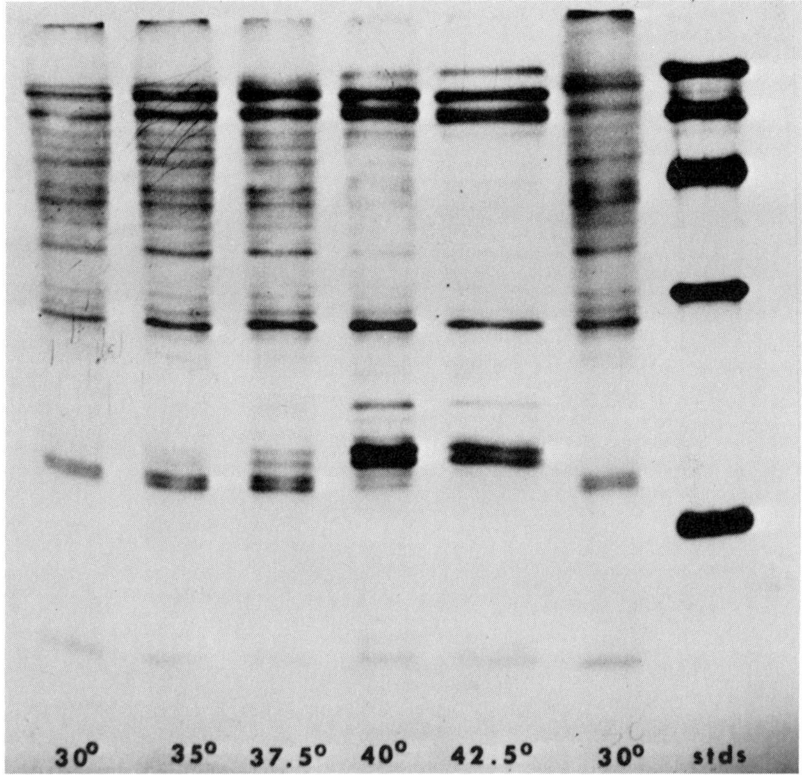

Fig. 1. The temperature of induction of heat shock protein synthesis. Etiolated, detached hypocotyls were incubated at the indicated temperature for 3 hr and 300 μCi of [^{3}H]leucine was added during the final 2 hr. Molecular weight standards are 92.5K, 69K, 46K, 30K, and 12.3K. (From Key *et al.*, 1981.)

in Fig. 3. While the high-molecular-weight hsp's appear electrophoretically similar in all species, the pattern of low-molecular-weight hsp's shows considerable heterogeneity. We have noted that there is a temperature at which amino acid incorporation drops precipitously. This "optimum" or "breakpoint" temperature for hsp synthesis varies considerably between species, from about 37°C for pea to about 46°C for millet (Key *et al.*, 1983a). We do not know if this difference has any physiological significance relative to the climatic adaptations of these species.

The induction of hsp synthesis is a rapid process (Fig. 4). During the initial 30 min following a shift from 30 to 41°C, production of hsp's becomes a substantial proportion of the total protein synthesis; over the next 3–4 hr at 41°C, the overall pattern of protein synthesis continues to reflect primarily hsp's. There are more

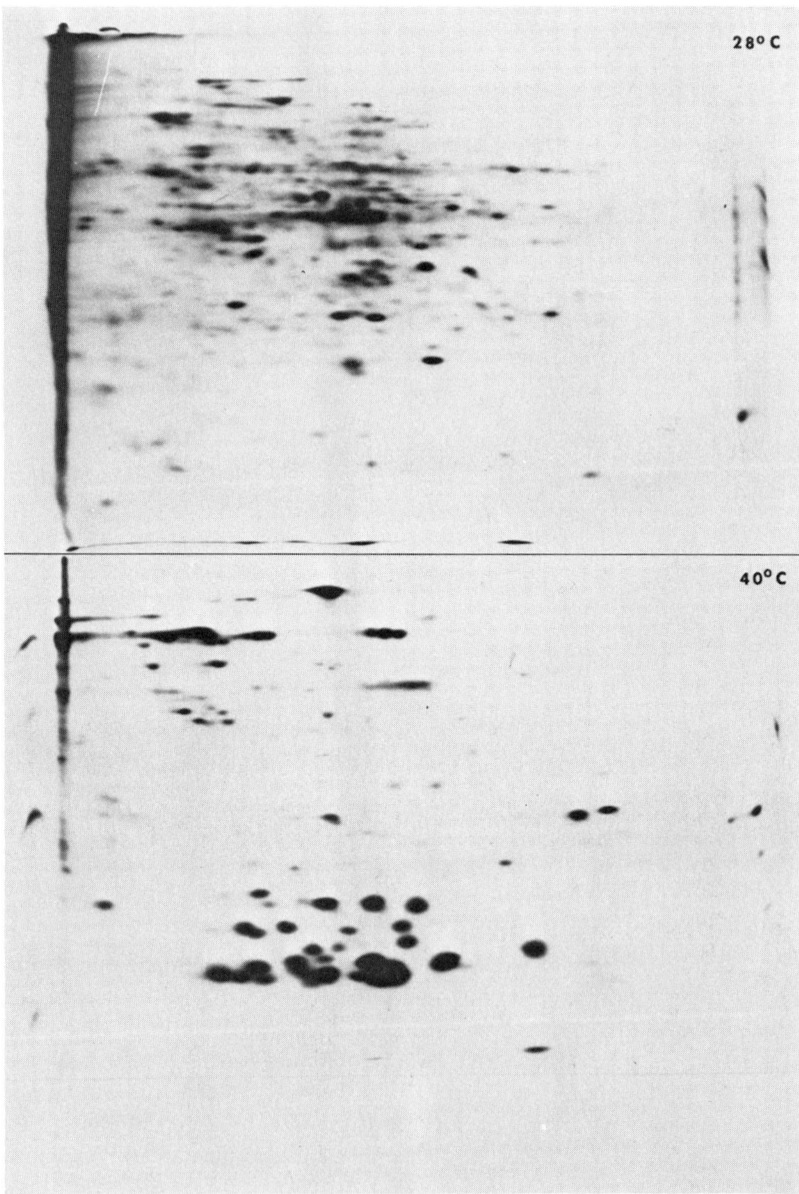

Fig. 2. Two-dimensional O'Farrell analysis of control and heat-shocked soybean proteins. Soybean hypocotyls were incubated at the indicated temperature for 2 hr in the presence of 300 μCi of [^{3}H]leucine. Proteins were extracted, electrophoresed, and fluorographed as described in Schöffl and Key (1982).

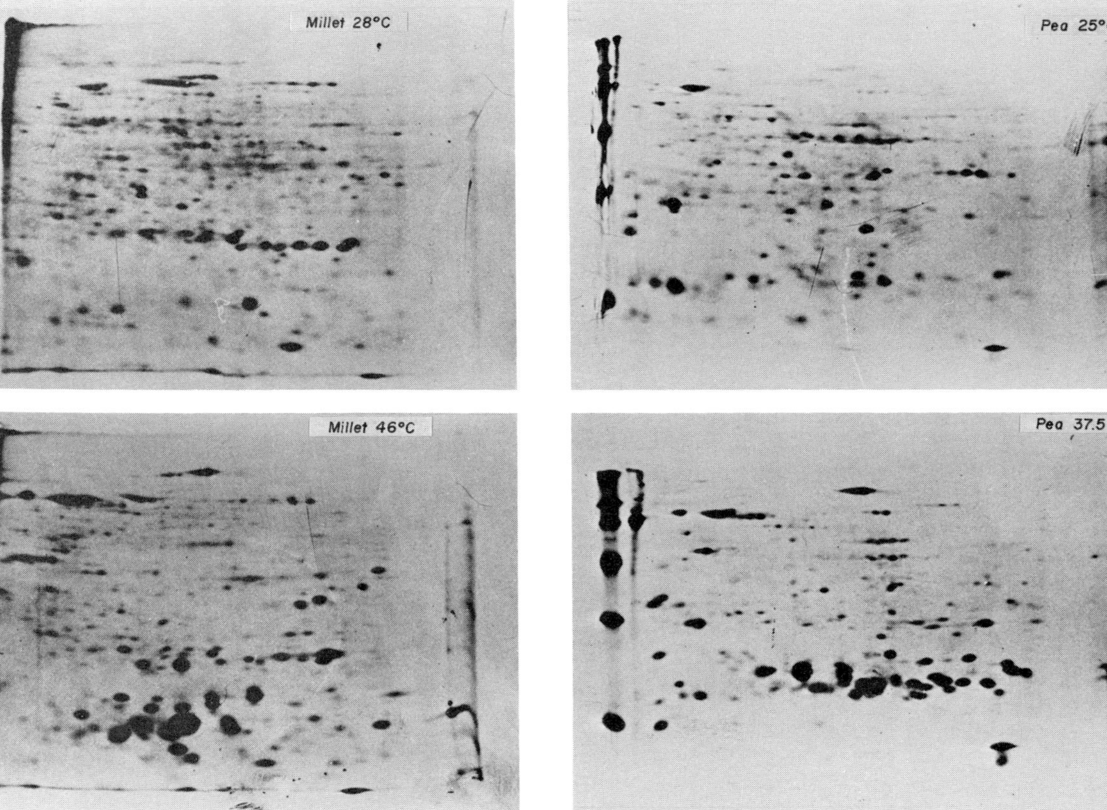

Fig. 3. Two-dimensional O'Farrell analysis of control and heat-shocked pea and millet proteins. (From Key *et al.*, 1983a.)

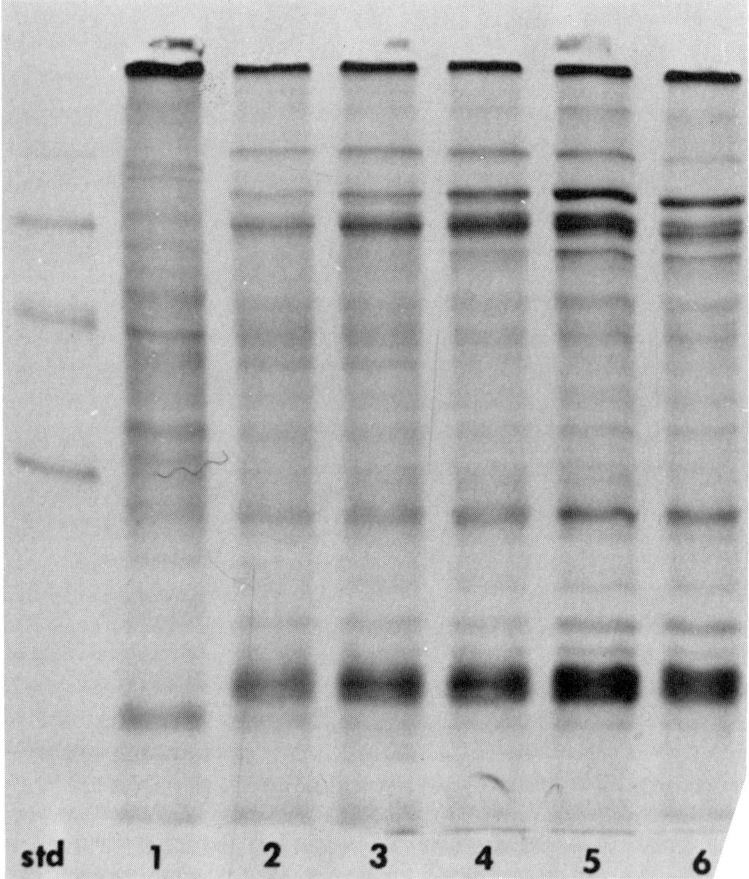

Fig. 4. Temporal changes in heat shock protein synthesis at 41°C. Proteins were labeled *in vivo* with [^{3}H]leucine during the indicated time period at 41°C. Molecular weight standards: 92.5K, 69K, 46K, 30K, and 12.3K. Lane 1, labeled 4 hr at 30°C; lane 2, labeled 0–30 min at 41°C; lane 3, labeled 30–60 min at 41°C; lane 4, labeled 1–2 hr at 41°C; lane 5, labeled 2–3 hr at 41°C; lane 6, labeled 3–4 hr at 41°C.

subtle, quantitative changes in the pattern during this time, clearly indicating the differential expression of individual hsp's. For example, by the fourth hour at 41°C, synthesis of the 92K hsp has largely ceased. The response during a continuous hs is transient, since after 8 hr at 41°C the pattern of protein synthesis is essentially indistinguishable from that at 30°C (C.-Y. Lin, unpublished observations; see also Altschuler and Mascarenhas, 1982). Since actual rates of protein synthesis were not measured in these and the earlier experiments, it is not possible to determine whether the increased proportion of normal proteins results from an increase in the rate of normal protein synthesis after long-term continuous hs, a marked decrease in hsp synthesis, or a combination of the two. Analysis of hs mRNA's during continuous hs indicates, however, that a dramatic decrease in the amount of hs mRNA's present in the tissues likely accounts for a substantial part of this altered pattern of protein synthesis (J. Kimpel, unpublished observations).

When soybean seedlings are returned to the normal growth temperature after an hs treatment (e.g., 40°C for 2 hr), there is a rapid return to the normal pattern of protein synthesis (Fig. 5). Within the first hour after the shift back to 30°C, there is an increasing proportion of normal protein synthesis, and after 3 to 4 hr at 30°C, hsp synthesis is not detected, as assayed by one-dimensional SDS–gel electrophoresis. Thus, the hs response is both rapidly induced on a shift to hs conditions and rapidly repressed on the shift back to normal growing conditions.

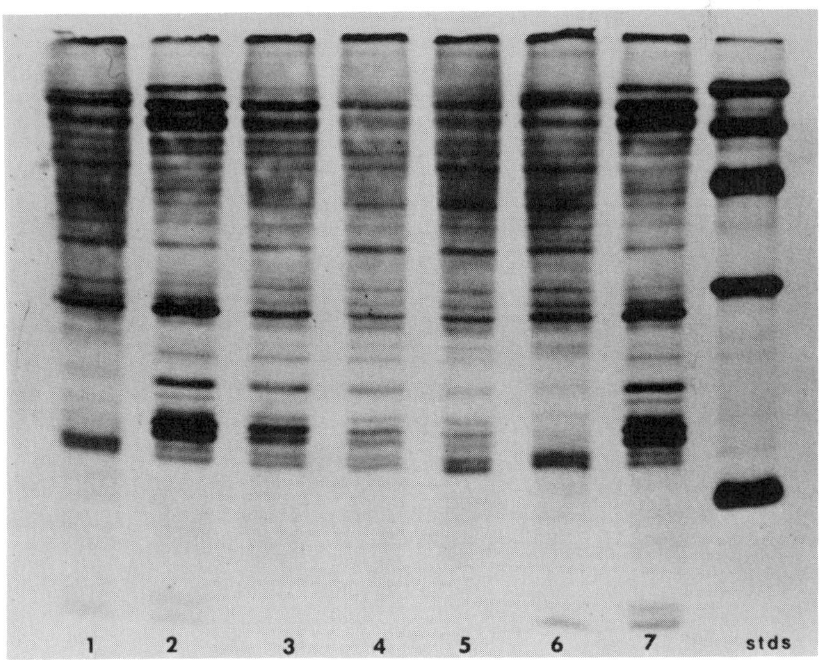

Fig. 5. Protein synthesis during recovery from heat shock. Plants were initially heat shocked at 40°C for 2 hr and then returned to 28°C to monitor recovery. Hypocotyls were labeled with [^{3}H]leucine at the indicated time, and then their proteins were extracted and fractionated on an SDS gel. Lane 1, labeled 4 hr at 28°C; lane 2, labeled 4 hr at 40°C; lane 3, labeled during first hour at 28°C after 4 hr at 40°C; lane 4, labeled during second hour at 28°C; lane 5, labeled during third hour at 28°C; lane 6, labeled during fourth hour at 28°C; lane 7, labeled 4 hr at 40°C. Molecular weight standards: 92.5K, 69K, 46K, 30K, 12.3K. (From Key *et al.*, 1981.)

Although seedlings quickly return to normal protein synthesis following a shift back to 30°C, many of the hsp's synthesized during the hs treatment are still present because they are quite stable proteins (Fig. 6). Over 80% of the total proteins labeled with [^{3}H]leucine during an initial 2-hr hs are still present after a 21-hr chase with nonradioactive leucine (1 m*M*), and a high percentage of these proteins are hsp's (Figs. 1 and 2). We have not yet performed a detailed two-dimensional electrophoretic analysis under chase conditions, so there may be

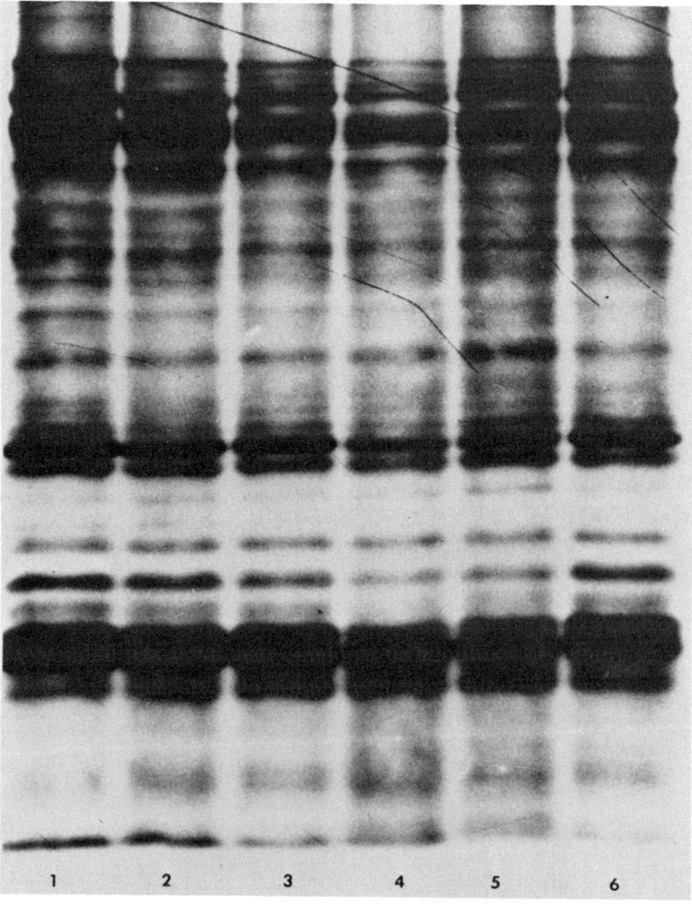

Fig. 6. Stability of heat shock proteins. Hypocotyls were labeled for 2 hr at 40°C and then moved to a chase solution (1 m*M* unlabeled leucine) at either 28 or 40°C. Lane 1, 2 hr at 40°C labeled proteins; lane 2, 2 hr at 40°C, chased at 28°C for 4 hr; lane 3, 2 hr at 40°C, chased at 28°C for 8 hr; lane 4, 2 hr at 40°C, chased at 28°C for 21 hr; lane 5, 2 hr at 40°C, chased at 40°C for 8 hr; lane 6, 2 hr at 40°C, chased at 40°C for 21 hr.

some turnover of select, individual hsp's during the chase. Nonetheless, a large fraction of the hsp's made during a 2-hr hs are stable at either normal or hs temperatures for at least 21 hr.

An abrupt rise in temperature of 10°C or more is only a laboratory condition and plants under normal growing conditions experience only a gradual rise in temperature. In the laboratory, when soybean seedlings are shifted gradually, that is, 2.5–3°C/hr, from 30 to 47.5°C (Fig. 7), hsp synthesis at a given temperature coincides generally with that observed when there is an abrupt shift in temperature (Fig. 1). Low levels of hsp synthesis occur at 35 and 37.5°C with the

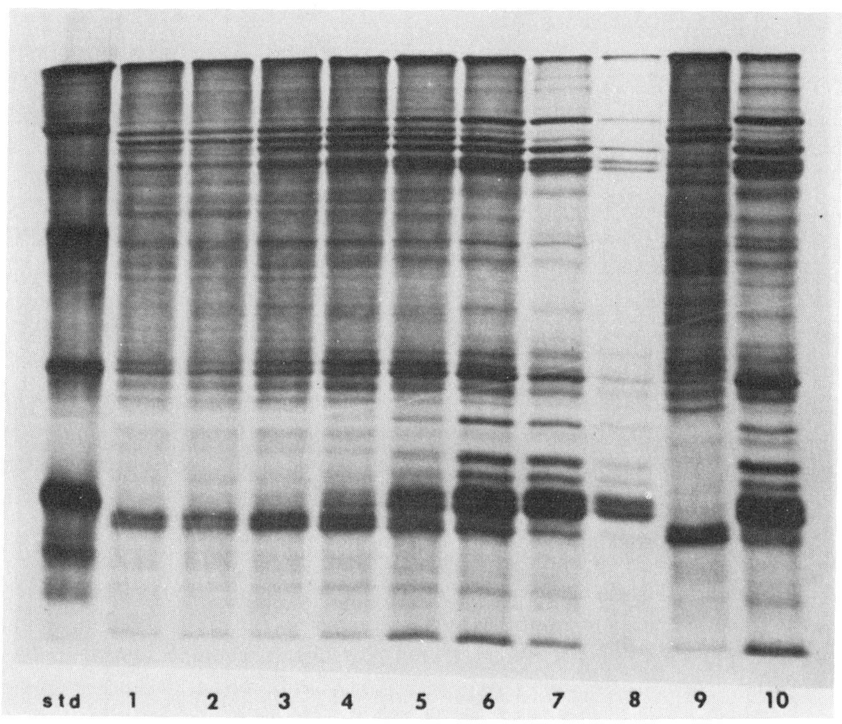

Fig. 7. Induction of heat shock proteins during a gradual rise in temperature. Hypocotyls were incubated for 1 hr at 28°C and then moved each hour to a higher temperature. [^{3}H]Leucine (200 μCi) was added 10 min after hypocotyls were moved to the last bath, and hypocotyls were harvested 50 min later. Lane 1, 1 hr at 28°C; lane 2, 1 hr at 28°C and 1 hr at 31°C; lane 3, 1 hr at 28°C, 1 hr at 31°C, and 1 hr at 34°C; lane 4, 1 hr at 28°C, 1 hr at 31°C, 1 hr at 34°C, and 1 hr at 37°C; lane 5, 1 hr at 28°C, 1 hr at 31°C, 1 hr at 34°C, 1 hr at 37°C, and 1 hr at 40°C; lane 6, 1 hr at 28°C, 1 hr at 31°C, 1 hr at 34°C, 1 hr at 37°C, 1 hr at 40°C, and 1 hr at 42.5°C; lane 7, 1 hr at 28°C, 1 hr at 31°C, 1 hr at 34°C, 1 hr at 37°C, 1 hr at 40°C, 1 hr at 42.5°C, and 1 hr at 45°C; lane 8, 1 hr at 28°C, 1 hr at 31°C, 1 hr at 34°C, 1 hr at 37°C, 1 hr at 40°C, 1 hr at 42.5°C, 1 hr at 45°C, and 1 hr at 47.5°C; lane 9, 8 hr at 28°C (labeled during eighth hour); lane 10, 2 hr at 40°C (labeled during second hour).

proportion of total protein synthesis devoted to hsp's increasing dramatically at temperatures of 40°C and above. However, the gradual increase in temperature apparently enables the seedlings to maintain a much higher level of hsp synthesis at 42.5, 45, and 47.5°C than if the seedlings had been abruptly shifted to these temperatures.

B. Analysis of Heat Shock mRNA's during Heat Shock and Recovery

Three approaches have been used to analyze changes in mRNA populations during hs: (1) *in vitro* translation of poly(A) RNA in the wheat germ system followed by two-dimensional gel analysis of the products; (2) cDNA/poly(A) RNA kinetic hybridization analyses; and (3) Northern blot analyses using cloned cDNA's made to hs poly(A) RNA's (Schöffl and Key, 1982). The translation analyses showed the marked accumulation, during 2 hr at 40°C, of translatable poly(A) RNA's whose products were not detectable in similar analyses of control poly(A) RNA (Schöffl and Key, 1982; F. Schöffl and J. L. Key, unpublished observations). The cDNA/poly(A) RNA hybridization analyses showed the accumulation of a very abundant component of poly(A) RNA in hs tissue, representing 20–25 different sequences of poly(A) RNA of about 800 nucleotides each, which is the mean size of these hs poly(A) RNA's. These accumulate to about 20,000 copies per cell and represent over 20% of the total poly(A) RNA mass within 2 hr (Schöffl and Key, 1982). In contrast, the sequences of the most abundant fraction in control soybean tissue are present at about 2000 copies per cell, on the average. Additional cDNA hybridization analyses and saturation hybridization of single-copy tracer DNA to excess poly(A) RNA indicate that there is very little loss of total complexity during a 2-hr hs at 40°C, while there is loss in complexity at higher temperatures (e.g., 42.5°C or greater; F. Schöffl and J. L. Key, unpublished observations).

Northern blot hybridizations with hs cDNA's confirm that the rapid labeling of hsp's during the initial 30 min of hs (Fig. 4) results in part from a rapid induction and accumulation of hs mRNA's following the shift to hs temperatures (Fig. 8). The hs mRNA's are detected during the initial 3–10 min of hs depending on the temperature of hs (Schöffl and Key, 1982).

When the tissue is shifted back to the normal growing temperature after a 2-hr hs, there is a rapid depletion of the hs mRNA's with a half-time of about 1 hr (Schöffl and Key, 1982; Key *et al.*, 1982). The relative contributions of changes in rates of synthesis and turnover to this rapid depletion of hs mRNA are unknown. This depletion of hs mRNA during the 4-hr incubation at 30°C after a 2-hr hs at 40°C coincides with the loss of hsp synthesis over this time period (Fig. 5). The level of hs mRNA accumulation at hs temperatures ranging from 30 to 45°C (Fig. 9) corresponds well with the relative level of hsp synthesis at these

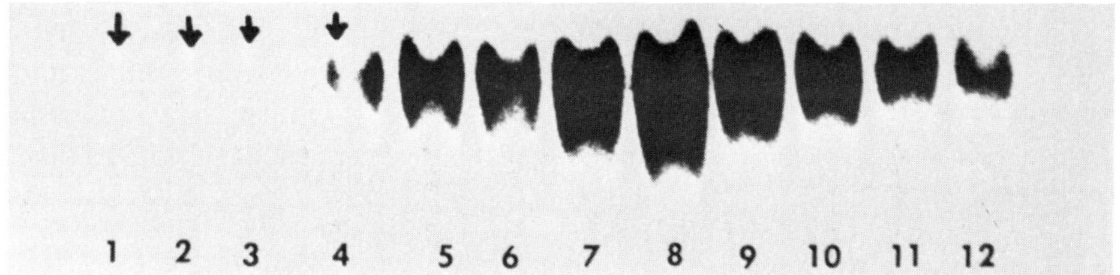

Fig. 8. The time course of induction of heat shock mRNA's during a 40°C heat shock. Poly(A) RNA was isolated from 40°C heat-shocked hypocotyls after the indicated time of treatment. A 1-μg poly(A) RNA sample was electrophoresed through 2% agarose denaturing gels, blotted onto nitrocellulose, and probed with ^{32}P-labeled nick-translated pCE53, an hs cDNA plasmid (Czarnecka *et al.*, 1984). Lane 1, 2 hr at 28°C; lane 2, 3.5 min; lane 3, 7.5 min; lane 4, 15 min; lane 5, 30 min; lane 6, 1 hr; lane 7, 2 hr; lane 8, 4 hr; lane 9, 4 hr at 40°C followed by 1 hr at 28°C; lane 10, 4 hr at 40°C followed by 2 hr at 28°C; lane 11, 4 hr at 40°C followed by 3 hr at 28°C; lane 12, 4 hr at 40°C followed by 4 hr at 28°C.

temperatures, based on patterns of protein synthesis (Fig. 1) and relative levels of amino acid incorporation at those temperatures (Key *et al.*, 1982, 1983a).

The accumulation of hs mRNA increases dramatically from about 37.5°C up to about 42.5°C; above this temperature there is little accumulation of hs mRNA's in an abrupt hs treatment. A gradual increase in the incubation temperature (2.5–3°C/hr), however, results in a gradual increase in the level of hs mRNA up to about 40°C (Fig. 10). Under these experimental conditions, high levels of hs mRNA's persist at 45 and 47.5°C, in contrast to the lack of accumulation of hs mRNA's at those temperatures in a single temperature shift experiment (Fig. 9). This gradual hs treatment certainly more nearly approximates a potential hs state in the normal environment than the dramatic temperature change used in most hs experiments.

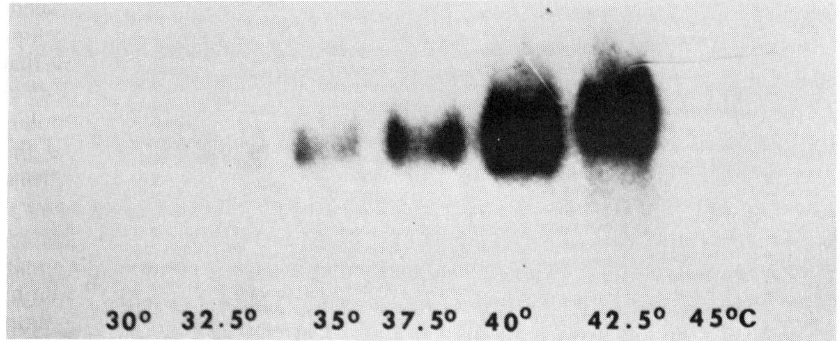

Fig. 9. Temperature of induction of hs mRNA. Poly(A) RNA isolated from hypocotyls incubated for 2 hr at the indicated temperatures was probed with pCE53 (Key *et al.*, 1983b).

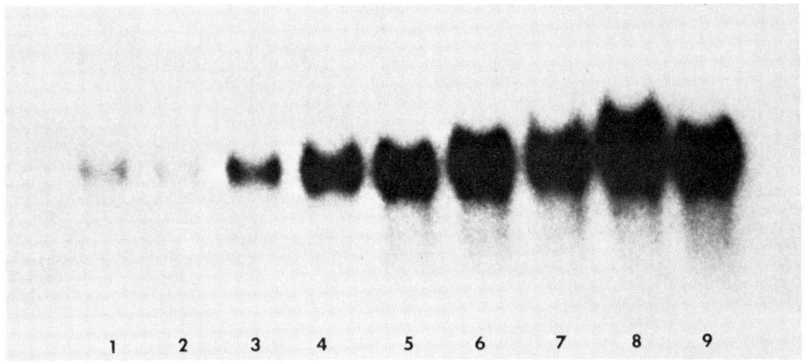

Fig. 10. Induction of hs mRNA's during a gradual rise in temperature. Poly(A) RNA, isolated from hypocotyls, incubated for 1 hr at 28°C and then moved each hour to successively higher temperatures. Northern blots were hybridized with pCE53. Lanes 1–8 correspond exactly to lanes 1–8 in Fig. 7; lane 9, 2 hr at 40°C.

C. Development of Thermal Tolerance in Response to Heat Shock

More recently, in studies focusing on the possible function of hsp's, the concept of thermotolerance has emerged. In this context, thermotolerance is the ability of an organism to survive an otherwise lethal temperature by first being exposed to an hs at a nonlethal temperature. Growth (length) of soybean seedlings 72 hr after treatment has been used as a parameter to investigate thermotolerance. Seedlings given a 2-hr hs at 40°C are protected during a subsequent 2-hr 45°C treatment, the latter being a treatment that is normally lethal to seedling growth (Key *et al.*, 1983b; Lin *et al.*, 1984). Such thermotolerance develops in a time- and temperature-dependent manner, consistent with the hypothesis that accumulation of hsp's is an essential component of the protection process (Key *et al.*, 1983b). Other conditions of hs also lead to the development of thermotolerance (Table I; Fig. 11). As an example, an hs treatment of 10 min at the "nonpermissive" temperature of 45°C followed by 2 hr at 28°C provides thermal protection for seedling growth from a subsequent (otherwise lethal) treatment at 45°C for 2 hr. A similar 10-min hs treatment at 40°C, however, is not sufficient to activate the hs system for hs mRNA and hsp accumulation to levels which provide substantial thermal protection (Lin *et al.*, 1984). Even a 2- or 4-min hs treatment at 45°C is sufficient to activate the synthesis and the accumulation of hs mRNA's (J. Kimpel, unpublished observations), while, as described above, continuous hs for 2 hr at 45°C results in the accumulation of only trace levels of hs mRNA's and only extremely low levels of incorporation of ^{3}H-labeled amino acids into hsp's.

TABLE I

Protection of Seedling Growth by a Brief 45°C Treatment Followed by Incubation at 28°C and a Subsequent 2-hr Heat Shock at 45°C[a]

	Length (cm)			Percentage of seedlings (72 hr)		
	Time of incubation (hr)			Range (cm)		
Heat shock condition	24	48	72	<5	5–10	>10
Normal 28°C	5.2 ± 0.7	10.9 ± 1.5	19.1 ± 2.5	0	3	97
45°C (10 min)	4.9 ± 0.4	9.7 ± 1.4	16.1 ± 2.0	0	0	100
45°C (2 hr)	2.2 ± 0.3	2.2 ± 0.3	3.2 ± 1.2	94	6	0
45°C (10 min) → 28°C (30 min) → 45°C (2 hr)	2.2 ± 0.3	2.9 ± 0.7	4.3 ± 2.1	70	30	0
45°C (10 min) → 28°C (1 hr) → 45°C (2 hr)	2.9 ± 0.5	4.1 ± 1.3	6.8 ± 3.0	47	40	13
45°C (10 min) → 28°C (2 hr) → 45°C (2 hr)	3.2 ± 0.3	6.1 ± 0.7	11.2 ± 1.5	0	17	83
45°C (10 min) → 28°C (3 hr) → 45°C (2 hr)	3.3 ± 0.4	5.9 ± 0.7	11.0 ± 1.4	0	13	87

[a] Thirty germinating seedlings with the embryonic axes protruding about 1 cm from the seed coats were subjected to the temperature regimes shown above. After each treatment, seedlings were planted in moist Chem-pak rolls and grown in a dark incubator at 28°C, and the lengths measured at the indicated times. Data from Lin *et al.*, 1984.

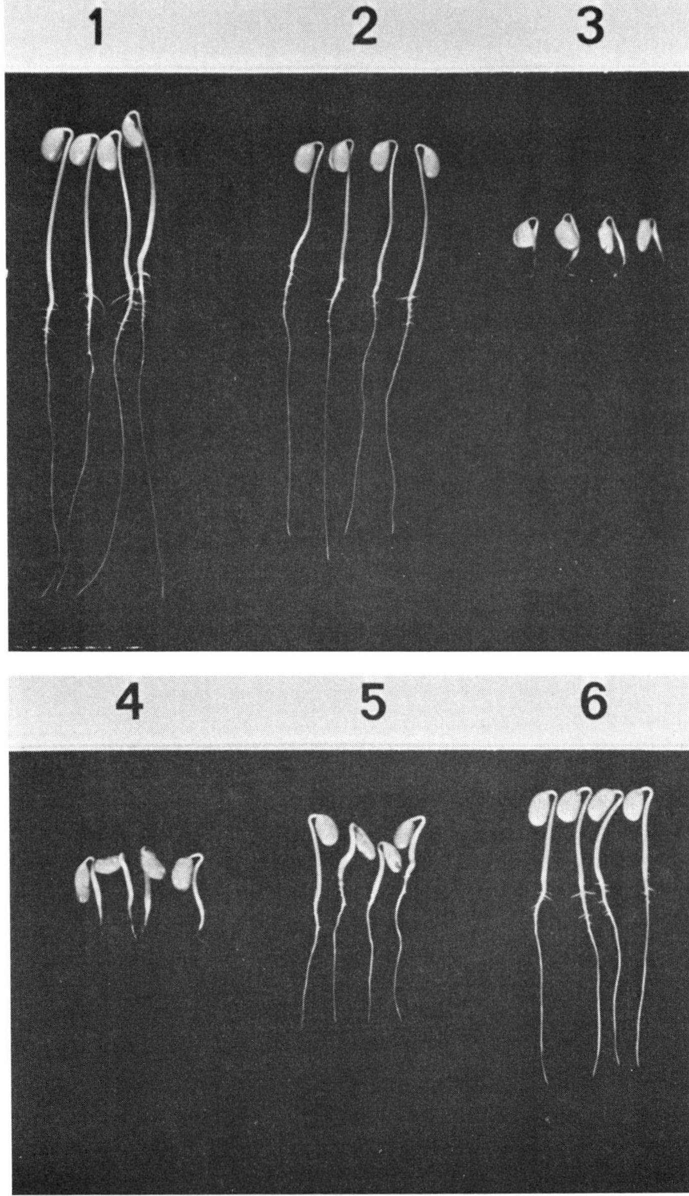

Fig. 11. Thermal protection of soybean seedling growth to an otherwise lethal 45°C treatment. Seedling groups numbered 1–6 correspond to heat shock conditions 1–6, respectively, of Table I.

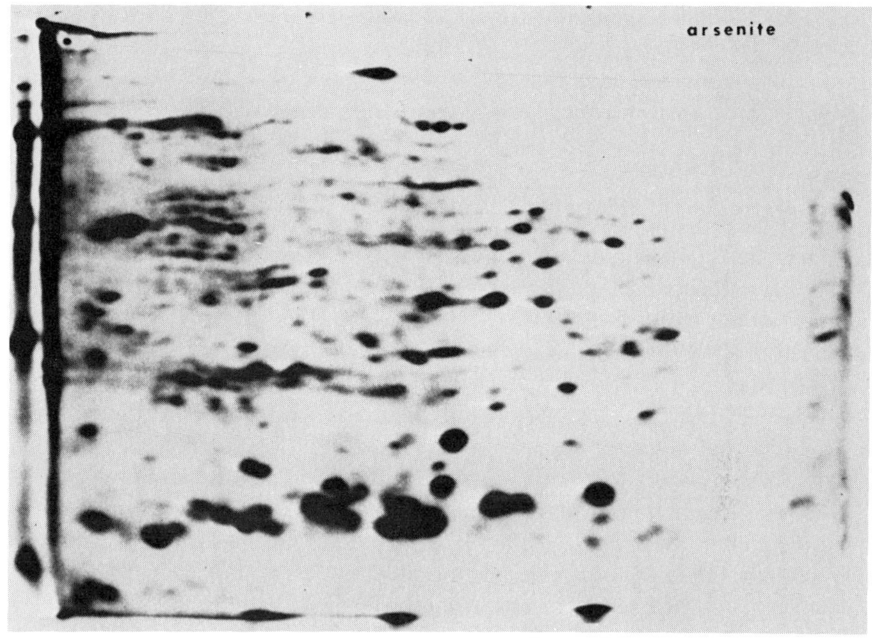

Fig. 12. Two-dimensional O'Farrell analysis of proteins produced during arsenite treatment. Hypocotyls were incubated for 3 hr in the presence of 50 μ*M* arsenite and [^{3}H]leucine. Proteins were extracted and electrophoresed as described by Key *et al.*, 1981.

One other "stress" treatment, among many which we have studied to date (Czarnecka *et al.*, 1984), also provides substantial thermoprotection for soybean seedlings. Seedlings treated with 50–75 μ*M* arsenite for a few hours develop tolerance to subsequent heat treatment at otherwise lethal temperatures, (for example, 45°C; Lin *et al.*, 1984). These arsenite-treated seedlings produce a pattern of proteins very similar to that from an hs treatment. As seen in Fig. 12, most of the major hsp's are synthesized in response to arsenite, although some only accumulate at reduced levels. Additionally, a few other proteins accumulate label above the control levels, and normal protein synthesis seems not to be as depressed in arsenite-treated seedlings as in hs-treated tissue. Northern blot analyses of poly(A) RNA's isolated from arsenite-treated tissue show that arsenite induces the hs mRNA's to levels comparable to a 40°C, 2-hr hs (Czarnecka *et al.*, 1984; E. Czarnecka, unpublished observations). As with the protein analyses, there are some quantitative differences between the level of expression of hs mRNA's in response to arsenite as contrasted to hs. Other stresses induce varying but generally low levels of some hs mRNA's (e.g., cadmium, anaerobiosis, amino acid analogs; see Czarnecka *et al.*, 1984), but these treatments have not been used in an analysis of thermotolerance.

D. Localization of Heat Shock Proteins

From the studies described in Section II,B, it is reasonable to propose that hsp's play a major role in thermotolerance. To further investigate the molecular mechanism of their function, we (Key *et al.*, 1982; Lin *et al.*, 1984) and others (Nover *et al.*, 1983) have examined the intracellular localization of hsp's. Purified nuclei, mitochondria, and ribosomes have a rather common set of low-molecular-weight hsp's associated with them under hs conditions; [Some hsp's (e.g., 27K and 84K) are "soluble" proteins under all conditions.] During a 30°C chase, hsp's dissociate from the organelle fractions in a kinetically similar fashion to the resumption of normal patterns of protein synthesis on return to the normal growth temperature (Lin *et al.*, 1984 and Fig. 5). This dissociation or loss of hsp's from the organelle fractions does not occur if the hs temperature is continued. After chase treatment is complete (e.g., 4 hr), a second hs results in a very rapid (within 15 min) reassociation of the hsp's with the organelle fractions. While it is clear that the hsp's are differentially localized during heat shock versus non-heat-shock conditions, the nature of this association with the various organelle fractions during hs is not understood. At least a part of this "selective localization" may result from the aggregation of hsp's into "granules" during hs (see Nover *et al.*, 1983), which might fractionate with the various organelles in gradients of sucrose or Percoll.

E. Other Work in Progress

To date, our experiments have focused primarily on the low-molecular-weight heat shock proteins. There are clearly a large number of related but different hs gene families in plants reflecting different hs "domains" (Schöffl and Key, 1982, 1983), and there are a large number of genes in some of these families, based on hybridization selection/translation analyses and Southern blot hybridization analyses (Schöffl and Key, 1982, 1983; F. Schöffl and J. L. Key, unpublished observations). We are currently sequencing both cDNA and genomic clones for several of the low-molecular-weight hsp's (R. T. Nagao, E. Czarnecka, F. Schöffl, and J. L. Key, unpublished observations), but it is too early to relate the hybridization selection and Southern blot analyses to specific gene structure relationships. A considerable amount of additional work is required in order to fully understand the expression and the organization of the genes encoding these proteins. Analyses of the high-molecular-weight hsp's and their genes is also an area of current research activity.

An understanding of organelle protein synthesis during plant hs is especially needed in view of the work on chloroplast function at elevated temperatures (Berry and Björkman, 1980; Feierabend, 1982). Toward this end, we have initiated work on the influence of hs on synthesis of chloroplast proteins encoded by both nuclear and chloroplast genes (E. Vierling and B. Rutti, unpublished observations). We have also demonstrated that the hsp's of etiolated soybean seedlings, to which we have cDNA clones, are not encoded by the mitochondrial genome and that the mitochondrial proteins may continue to be synthesized during hs (J. L. Key, J. Kimpel, E. Vierling, C.-Y. Lin, R. T. Nagao, E. Czarnecka, and F. Schöffl, unpublished observations) as has been observed in *Drosophila* cells (Spradling *et al.*, 1977).

In view of the complexity of the organelle fractionation/localization of hsp's and the interpretation of data from those experiments, monoclonal antibodies to some hsp's are being prepared. These will be used both in cellular localization studies of the hsp's and in determination of the concentrations of some of the hsp's under a wide range of physiological conditions.

III. DISCUSSION AND SUMMARY

Plants respond to several conditions of high-temperature treatment, collectively referred to here as heat shock or hs, by synthesizing new sets of mRNA's and proteins. The most abundant hs mRNA's accumulate in amounts of up to 20,000 copies or more per cell following hs, and they are easily detected on Northern blots within 3 to 10 min of hs, depending on the temperature of hs. The hsp's reflect a high proportion of total protein synthesis even during the initial 30 min of hs and quantitatively dominate protein synthesis after about 1 hr of hs at or slightly above the "breakpoint" temperature (40–41°C) in soybean seedlings (Key *et al.*, 1982). The major hs conditions which elicit these responses and which lead to the development of thermotolerance are (1) a shift from the normal growing temperature of 28–30°C directly to a permissive hs temperature of 40–41°C for about 2 hr, (2) a brief shift of up to 10 min directly from 30 to 45°C followed by transfer back to 28–30°C for 2 hr, and (3) a gradual shift of 2.5–3°C/hr from the normal growing temperature up to as high as 47.5°C. The levels and kinetics of hs mRNA and hsp accumulation vary somewhat among these treatments. Finally, field conditions, in which the afternoon temperature may reach 40°C, elicit an apparently normal hs response relative to the laboratory-generated responses presented here (J. Kimpel and M. Mansfield, unpublished observations). Additionally, a treatment with 50–75 μM arsenite for a few hours also elicits a rather normal "hs" response leading to the accumulation of hs mRNA's and hsp's and the development of thermotolerance.

Thus, very different treatments elicit cellular events which turn on the transcription and the translation of hs mRNA's result in varying levels of reduced translation of normal mRNA's, and finally lead to the development of thermotolerance. While the fundamental mechanisms operative in the regulation of these responses are not understood (see Schlesinger *et al.*, 1982a,b), the dramatic differences in the responses to a 10-min (or less) hs at 45 and at 40°C (as contrasted to a 2-hr 40°C hs) by plants should provide a basis for analyzing cellular events associated with these fundamental regulatory mechanisms.

Analyses of the hs response in both *Escherichia coli* and *Drosophila* suggest that hsp production is regulated in part by the hsp's themselves. In *E. coli* the hs response and the level of accumulation of hsp's are altered by mutations in the *dnaK* gene (Tilly *et al.*, 1983). The dnaK protein, which regulates DNA and RNA synthesis by an undetermined mechanism, is among those proteins transiently overproduced during hs of *E. coli.* Bacteria carrying mutations in this gene fail to turn off this response and, therefore, continue to overproduce hsp's and underproduce normal proteins at 43°C. Also, *E. coli* cells that overproduce the dnaK protein at all temperatures make markedly reduced levels of hsp's at hs temperatures. Accordingly, Tilly *et al.* (1983) conclude that "the dnaK protein is an inhibitor of the heat shock response in *E. coli.*" The heat shock response itself is under the positive control of a protein produced from the *htp* (*hin*) locus (Neidhardt and Van Bogelen, 1981; Yamomori and Yura, 1982). At least in *E. coli,* and apparently only in *E. coli,* the specific functional identity of some hsp's is known (see Tilly *et al.*, 1983).

It has been known for some time that hsp70 of *Drosophila* is highly conserved across a broad spectrum of organisms (see Schlesinger *et al.*, 1982a). There is about 50% homology between the dnaK protein of *E. coli* (referred to by Tilly *et al.*, 1983) and the *Drosophila* hsp70 (Craig *et al.*, 1982). Interestingly, DiDomenico *et al.* (1982) have presented evidence consistent with the conclusion that hsp70 of *Drosophila* is a negative regulator of the hs response. In these studies hsp synthesis continued beyond normal levels when nonfunctional hsp's were made in the presence of amino acid analogs; and hs mRNA's and hsp's overaccumulated to vast quantities, while normally their synthesis is turned off after some period of hs. After removal of the amino acid analog, a full complement of functional hsp70 must accumulate before its synthesis ceases and before the hsp70 mRNA's and preexisting hs mRNA's are destabilized.

Data from our own work and unpublished observations on hs mRNA levels during continuous hs are certainly consistent with the conclusions reached by DiDomenico *et al.* (1982) and Tilly *et al.* (1983) on "autoregulation" or "self-regulation" in the hs response; those authors used the term "self-regulation" to avoid specific mechanistic connotations. This is a plausible mechanism to account for the decline in hsp synthesis during continuous or repeated cycles of hs and is also consistent with the apparently more rapid decline in hsp synthesis compared to the loss of hs mRNA's (J. Kimpel and C.-Y. Lin, unpublished observations). The observations that the accumulation of hs mRNA's is lower on successive rounds of hs during cycles of 40°C followed by 28°C are also consistent with the proposition that a sufficient level of hsp's "autoregulates" the hs response. This general supposition is also supported by the observations presented earlier on the stability of the hsp's. However, until more detailed analyses are made, including measurements of the rates of hsp and hs mRNA synthesis and turnover, conclusions relating to regulatory mechanisms in the plant system must be viewed as tentative.

While much remains to be done to understand regulation in the hs response, there is a developing consensus that hsp's are involved in the development of thermotolerance or thermal adaptation to otherwise nonpermissive temperatures (Yarwood, 1967; Mitchell *et al.*, 1979; McAlister and Finkelstein, 1980; Li and Werb, 1982; Key *et al.*, 1982; Lin *et al.*, 1984). Again, the mechanism of development of protection to lethality and to anomalies in developmental patterns is unknown. However, selective cellular localization of some of the hsp's seems to be involved in the development of thermotolerance (Arrigo *et al.*, 1980; Velazquez *et al.*, 1980; Loomis and Wheeler, 1982a,b; McAlister and Finkelstein, 1980; Key *et al.*, 1982; Lin *et al.*, 1984). In at least one case there is detailed evidence for the association of hsp70 with the nucleus and within the interband regions of the chromosomes (Velazquez *et al.*, 1980; DiDomenico *et al.*, 1982). In the case of plants, we have presented evidence that hsp's become associated in some way with purified organelle fractions, although the nature of this association is not understood (Key *et al.*, 1982; Lin *et al.*, 1984). The kinetics of association or localization of the hsp's during hs and dissociation during recovery from hs are consistent with the assumption that this association/localization process is important to thermotolerance and to resumption of normal RNA and protein synthesis. Another line of evidence suggestive of the influence of selective organelle localization of some hsp's during hs in the development of thermotolerance relates to results with arsenite-induced "hsp's." During induction by arsenite at 30°C, these proteins remain as soluble supernatant proteins, and during a subsequent hs, they rapidly associate with organelles. However, in view of results of Sanders *et al.* (1982) and Nover *et al.* (1983), the results concerning localization must be interpreted with caution. These workers have observed "aggregate" or "granule-like" structures under hs conditions but not at control temperatures following a pre-heat shock (Nover *et al.*, 1983), which is consistent with our observations that selective localization is strictly hs dependent (Lin *et al.*, 1984). These aggregates may be important under normal physiological conditions of hs, but they might also be artifactually produced during manipulation of tissue under some conditions such as high salt. It is interesting to note, however, that detailed structural analyses of *Drosophila* hsp genes (e.g., Corces *et al.*, 1980; Ingolia *et al.*, 1980; Ingolia and Craig, 1982; Craig *et al.*, 1982; Voellmy *et al.*, 1983) are consistent with these proteins possibly forming "aggregate" structures under some conditions. There is considerable homology between the 22K–27K *Drosophila* hsp's and the vertebrate lens α-crystallins (see Craig *et al.*, 1982; Southgate *et al.*, 1983); the α-crystallins form large aggregates that play important structural roles in eye lens function. Another very interesting structural feature of these proteins is the alternating regions with hydrophobic and hydrophilic properties (e.g., Southgate *et al.*, 1983). These properties could have important implications for hsp structural organization and function as well as the general property or properties resulting in "aggregate" formation. An understanding of the nature of selective localization and its role in the development of thermotolerance will only be extended by specific methods such as immunofluorescent antibody localization and structural analyses of these proteins and their genes. It should be reemphasized here that hsp70 represents a major proportion of hsp synthesis in systems like *Drosophila,* while in plant systems, the highly related group of 20 to 30 (or more) 15K–27K proteins dominates hsp synthesis. Thus, fundamental differences in mechanisms of regulation of hs and development of thermotolerance may exist between plant and animal systems.

Acknowledgments

This research was supported by DOE contract DE-A509-SDER10678 and by Agrigenetics Research Associates.

REFERENCES

Altschuler, M., and Mascarenhas, J. P. (1982). Heat shock proteins and effects of heat shock in plants. *Plant Mol. Biol.* **1,** 103.

Arrigo, A. P., Fakan, S., and Tissières, A. (1980). Localization of the heat shock-induced proteins in *Drosophila melanogaster* tissue culture cells. *Dev. Biol.* **78,** 86.

Ashburner, M., and Bonner, J. J. (1979). The induction of gene activity in *Drosophila* by heat shock. *Cell* **17,** 241.

Barnett, T., Altschuler, M., McDaniel, C. N., Mascarenhas, J. P. (1980). Heat shock induced proteins in plant cells. *Dev. Genet. (Amsterdam)* **1,** 331.

Baszczynski, C. L., Walden, D. B., and Atkinson, B. G. (1982). Regulations of gene expression in corn (*Zea mays* L.) by heat shock. *Can. J. Biochem.* **60,** 569.

Berry, J., and Björkman, O. (1980). Photosynthetic response and adaptation to temperature in higher plants. *Annu. Rev. Plant Physiol* **31,** 491–543.

Cooper, P., and Ho, D. T. H. (1983). Heat shock proteins in maize. *Plant Physiol.* **71,** 215.

Corces, V., Holmgren, R., Freund, R., and Morimoto, R., and Meselson, M. (1980). Four heat shock proteins of *Drosophila melanogaster* coded within a 12-kilobase region in chromosome subdivision 67B. *Proc. Natl. Acad. Sci. U.S.A.* **77,** 5390.

Craig, E. A., Ingolia, T. D., Slater, M., and Manseau, L. J., and Bardwell, J. (1982). *Drosophila,* yeast, and *E. coli* genes related to the *Drosophila* heat-shock genes. *In* "Heat Shock, from Bacteria to Man" (M. J. Schlesinger, M. Ashburner, and A. Tissières, eds.), p. 11. Cold Spring Harbor Lab., Cold Spring Harbor, New York.

Czarnecka, E., Edelman, L., Schöffl, F., and Key, J. L. (1984). Comparative analysis of physical stress responses in soybean seedlings using cloned heat shock cDNAs. *Plant Mol. Biol.* **3,** 45.

Dawson, W. O., and Grantham, G. L. (1981). Inhibition of stable RNA synthesis and production of a novel RNA in heat stressed plants. *Biochem. Biophys. Res. Commun.* **100,** 23.

DiDomenico, B. J., Bugaisky, G. E., and Lindquist, S. (1982). The heat shock response is self-regulated at both the transcriptional and post-transcriptional levels. *Cell* **31,** 593–603.

Feierabend, J. (1982). Inhibition of chloroplast ribosome formation by heat in higher plants. *In* "Methods in Chloroplast Molecular Biology" (M. Edelman, R. B. Hallick, and N.-Ha.Chua, eds.), p. 671. Elsevier, Amsterdam.

Ingolia, T. D., and Craig, E. A. (1982). Four small *Drosophila* heat shock proteins are rleated to each other and to mammalian α-crystallin. *Proc. Natl. Acad. Sci. U.S.A.* **79,** 2360.

Ingolia, T. D., Craig, E. A., and McCarthy, B. J. (1980). Sequence of three copies of the gene for the major *Drosophia* heat shock induced protein and their flanking regions. *Cell* **21,** 669.

Kelley, P. M., and Freeling, M. (1982). A preliminary comparison of maize anaerobic and heat shock proteins. *In* "Heat Shock, from Bacteria to Man" (M. J. Schlesinger, M. Ashburner, and A. Tissières, eds.), p. 315. Cold Spring Harbor Lab., Cold Spring Harbor, New York.

Key, J. L., Lin, C.-Y., and Chen, Y. M. (1981). Heat shock proteins of higher plants. *Proc. Natl. Acad. Sci. U.S.A.* **78,** 3526.

Key, J. L., Lin, C.-Y., Ceglarz, E., and Schöffl, F. (1982). The heat shock response in plants. *In* "Heat Shock, from Bacteria to Man" (M. J. Schlesinger, M. Ashburner, and A. Tissières, eds.), p. 329. Cold Spring Harbor Lab., Cold Spring Harbor, New York.

Key, J. L., Czarnecka, E., Lin, C.-Y., Kimpel, J., Mothershed, C., and Schöffl, F. (1983a). A comparative analysis of the heat shock response in crop plants. *In* "Current Topics in Plant Biochemistry and Physiology Symposium" (D. D. Randall, D. G. Blevins, R. L. Larson, and B. J. Rapp, eds.), p. 107. Univ. of Missouri Press, Columbia.

Key, J. L., Lin, C.-Y., Ceglarz, E., and Schöffl, F. (1983b). The heat shock response in soybean seedlings. *In* "NATO Advanced Studies Workshop on Genome Organization and Expression in Plants" (L. Dure, ed.), p. 25. Plenum, New York.

Li, G. C., and Werb, Z. (1982). Correlation between synthesis of heat shock proteins and development of thermotolerance in Chinese hamster fibroblasts. *Proc. Natl. Acad. Sci. U.S.A.* **79,** 3218.

Lin, C.-Y., Roberts, J. K., and Key, J. L. (1984). Acquisition of thermotolerance in soybean seedlings: Synthesis and accumulation of heat shock proteins and their cellular localization. *Plant Physiology* **74,** 152.

Loomis, W. F., and Wheeler, S. A. (1982a). Chromatin-associated heat shock proteins of *Dictyostelium. Dev. Biol.* **90,** 412.

Loomis, W. F., and Wheeler, S. A. (1982b). The physiological role of heat-shock proteins in *Dictyostelium. In* "Heat Shock, from Bacteria to Man" (M. J. Schlesinger, M. Ashburner, and A. Tissières, eds.), p. 353. Cold Spring Harbor Lab., Cold Spring Harbor, New York.

McAlister, L., and Finkelstein, D. B. (1980). Heat shock proteins and thermal resistance in yeast. *Biochem. Biophys. Res. Commun.* **93,** 819.

Meyer, Y., and Chartier, Y. (1983). Long-lived and short-lived heat-shock proteins in tobacco mesophyll protoplasts. *Plant Physiol.* **72,** 26.

Mitchell, H. K., Moller, G., Peterson, N. S., and Lipps-Sarmiento, L. (1979). Specific protection from phenocopy induction by heat shock. *Dev. Genet. (Amsterdam)* **1,** 181.

Niedhardt, F. C., and Van Bogelen, R. A. (1981). Positive regulatory gene for temperature-controlled proteins in *Escherichia coli. Biochem. Biophys. Res. Commun.* **100,** 894.

Nover, L., Scharf, K.-D., and Neumann, D. (1983). Formation of cytoplasmic heat shock granules in tomato cell cultures and leaves. *Mol. Cell. Biol.* **9,** 1648.

Ritossa, F. (1962). A new puffing pattern induced by heat shock and DNP in *Drosophila. Experientia* **18,** 571.

Sanders, M. M., Feeney-Trimer, D., Olsen, A. S., and Farrell-Towt, J. (1982). Changes in protein phosphorylation and histone H2b disposition in heat shock in *Drosophila*. *In* "Heat Shock, from Bacteria to Man" (M. J. Schlesinger, M. Ashburner, and A. Tissières, eds.), p. 235. Cold Spring Harbor Lab., Cold Spring Harbor, New York.

Scharf, K.-D., and Nover, L. (1982). Heat-shock-induced alternations of ribosomal protein phosphorylation in plant cell cultures. *Cell* **30,** 427.

Schlesinger, M. J., Aliperti, G., and Kelley, P. M. (1982a). The response of cells to heat shock. *Trends in Biochem. Sci. (Pers. Ed.)* **7,** 222.

Schlesinger, M. J., Ashburner, M., and Tissières, A., eds. (1982b). "Heat Shock, from Bacteria to Man." Cold Spring Harbor Lab., Cold Spring Harbor, New York.

Schöffl, F., and Key, J. L. (1982). An analysis of mRNAs for a group of heat shock proteins of soybean using cloned cDNAs. *J. Mol. Appl. Genet.* **1,** 301.

Schöffl, F., and Key, J. L. (1983). Identification of a multigene family for small heat shock proteins in soybean and physical characterization of one individual coding region. *Plant Mol. Biol.* **2,** 269.

Southgate, R., Ayme, A., and Voellmy, R. (1983). Nucleotide sequence analysis of the *Drosophila* small heat shock gene cluster at locus 67B. *J. Mol. Biol.* **165,** 35.

Spradling, A., Pardue, M. L., and Penman, S. (1977). Messenger RNA in heat-shocked *Drosophila* cells. *J. Mol. Biol.* **109,** 559.

Tilly, K., McKittrick, N., Zylicz, M., and Georgopoulos, C. (1983). The dnaK protein modulates the heat-shock response of *Escherichia coli*. *Cell* **34,** 641.

Velazquez, J. M., DiDomenico, B. J., and Linquist, S. (1980). Intracellular localization of heat shock proteins in *Drosophila*. *Cell* **20,** 679.

Voellmy, R., Bromley, P., and Kocher, H. P. (1983). Structural similarities between corresponding heat-shock proteins from different eucaryotic cells. *J. Biol. Chem.* **258,** 3516.

Yamamori, T., and Yura, T. (1982). Genetic control of heat-shock protein synthesis and its bearing on growth and thermal resistance in *Escherichia coli* K-12. *Proc. Natl. Acad. Sci. U.S.A.* **79,** 860.

Yarwood, C. E. (1967). Adaptation of plants and plant pathogens to heat. *In* "Molecular Mechanisms of Temperature Adaptation" (C. L. Prosser, ed.), p. 75. Publ. No. 84, Amer. Assoc. Advance. Sci., Washington, D.C.

17

Maize Genome Response to Thermal Shifts

CHRIS L. BASZCZYNSKI, DAVID B. WALDEN, AND BURR G. ATKINSON

I. INTRODUCTION

Changes in the types of proteins synthesized in response to heat shock have been studied extensively in animal systems (for reviews see Ashburner and Bonner, 1979; Schlesinger *et al.*, 1982) and, more recently, in plant and fungal systems (Barnett *et al.*, 1980; Loomis and Wheeler, 1980; Key *et al.*, 1981;

ISBN 0-12-066290-6

Altschuler and Mascarenhas, 1982; Baszczynski *et al.*, 1982; Cooper and Ho, 1983; Kapoor, 1983; Meyer and Chartier, 1983; Silver *et al.*, 1983). Environmental stresses such as those resulting from rapid changes in nutrition (Webster, 1980), water (Dhinsda, 1976; Botha, 1979; Bewley *et al.*, 1983), anaerobiosis (Sachs and Freeling, 1978; Sachs *et al.*, 1980), and plant growth regulators (Zurfluh and Guilfoyle, 1980, 1982) also elicit dramatic changes in the types of proteins synthesized by higher plants. The work in our laboratories has focused primarily on changes in gene expression induced by rapid elevations in temperature (heat shock). We have utilized various tissues from a well-defined, standard inbred cultivar of maize (Oh43) for these investigations.

In this chapter we provide an overview of the changes in gene expression in maize brought about by brief elevations in the temperature to which the maize seedlings are exposed. However, since maize (and many other plant systems) can grow and develop normally over a wide range of temperatures, we explored the possibility that the term "heat shock" may be limiting in this system; maize grown at one "normal" temperature may respond to a thermal shift differently than when grown at another "normal" temperature. The results from these investigations permit us to introduce the concept that synthesis of some so-called heat shock proteins (hsp's) of maize may represent a general response to thermal shift or stress, while the expression of other maize hsp's may represent a true high-temperature- or heat shock-induced class of polypeptides. Finally, our studies provide some evidence that suggests that the genotype of maize may be important in determining the types of polypeptides synthesized under different temperature conditions.

II. CHARACTERIZATION OF THE HEAT SHOCK RESPONSE IN MAIZE (cv. Oh43) SEEDLINGS

A. Response of Five-Day-Old Maize Plumules to Heat Shock

The response of plumules from intact, 5-day-old maize seedlings (or of plumules excised from similar seedlings) to a rapid increase in incubation tem-

Fig. 1. Fluorographic analysis of the electrophoretically separated polypeptides synthesized *in vivo* by maize (cv. Oh43) plumules at 27°C or following a 1-hr shift to the temperatures indicated. (A) SDS–PAGE (3–15% gradient) separation of the polypeptides synthesized by maize plumules during a 1-hr shift of 5-day-old seedlings from 27°C (control) to each of the indicated temperatures. (B) Two-dimensional IEF–SDS–PAGE separation of the polypeptides synthesized in maize plumules at 27°C and during a 1-hr shift from 27 to 41°C. The relative molecular weight (M_r) of each hsp class is shown on the right sides of (A) and (B) Relative molecular weight values were determined from coelectrophoresed protein standards (Std.). Approximately 25,000 cpm of acid-precipitable [^{14}C]leucine-labeled lysate was added to each well of the gels in (A) and approximately 100,000 cpm to each of the gels in (B).

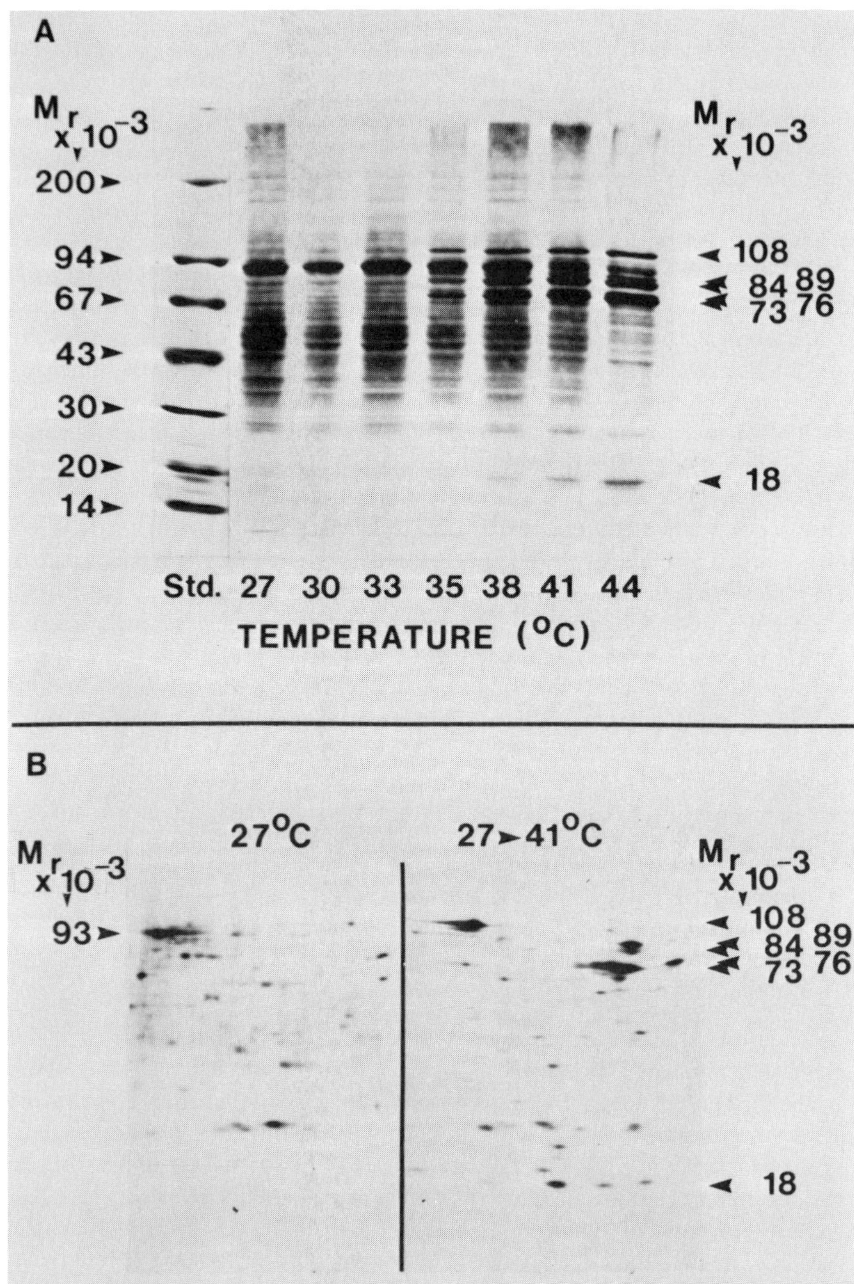
A
M_r x 10^{-3}
200
94
67
43
30
20
14
M_r x 10^{-3}
108
84 89
73 76
18
Std. 27 30 33 35 38 41 44
TEMPERATURE (°C)
B
27°C
27►41°C
M_r x 10^{-3}
93
M_r x 10^{-3}
108
84 89
73 76
18
–
+
–
+

perature (heat shock) involves enhanced and/or new synthesis of a small group of polypeptides (the heat shock proteins; hsp's) and a concomitant depression in the synthesis of proteins previously made at the control temperature. The synthesis of polypeptides with relative molecular weights (M_r's) of 108,000, 89,000, 84,000, 76,000, 73,000, and 18,000 following heat shock (27–41°C; Fig. 1) is dramatic; these hsp's constitute greater than 15% of the total amount of polypeptides synthesized at 41°C. Two-dimensional electrophoretic separation of newly synthesized polypeptides resolves the six molecular weight classes of hsp's into a large array of polypeptides (Fig. 1B); depending on the pH gradient employed, the 18K hsp class resolves into a family of four to eight distinct polypeptides. The heat shock response is clearly temperature dependent since the hsp's are initially detected at or above 35°C (following a shift from 27°C) and intensify with increasing temperatures (Fig. 1A). The response is also dependent on the duration of the heat treatment, since synthesis of some hsp's is detectable after only 15 min at 41°C while synthesis of all six hsp classes requires 60–120 min at 41°C (Baszczynski *et al.*, 1982). These heat shock-induced changes in the patterns of polypeptide synthesis are reversible; when seedlings are returned to 27°C (the control temperature shown in Fig. 1), recovery of a normal (preshift) pattern of synthesis resumes after 6–8 hr. The demonstration (Baszczynski *et al.*, 1982) that the types of polypeptides synthesized by excised or intact plumules are the same, indicates that the synthetic changes observed are attributable to heat shock or temperature shift and not a result of wounding of the tissue, as shown in some animal systems (Currie and White, 1981; Hightower and White, 1981; Dean and Atkinson, 1983). For the purpose of this discussion, it is assumed that the polypeptides are the products of specific gene activity, although these genes have not yet been mapped in maize.

B. Detection of Polypeptides Synthesized *in Vivo* by Control and Heat-Shocked Maize Plumules Using Different Radioactively Labeled Amino Acid Precursors

Our initial studies, aimed at describing the heat shock response in maize plumules, employed [^{14}C]leucine as an amino acid precursor for detecting newly synthesized polypeptides (Baszczynski *et al.*, 1982). Although this report clearly defined a group of hsp's analogous to those reported in other plant and animal systems (as described in Section I), we were intrigued by the possibility that the use of different amino acid precursors might reveal additional differences in the types of polypeptides synthesized. The need to actually compare fluorographic spectra of newly synthesized polypeptides detected by the use of different radioactively labeled amino acids became apparent when *in vitro* translation studies requiring the use of an isotope with a high specific activity (e.g., [^{35}S]methionine) were undertaken. The fluorograms from two-dimensional electrophoretic separations of polypeptides synthesized in the presence of different radioactively labeled amino acids by control and heat-shocked plumules are shown in Fig. 2. These fluorographic results reveal a number of quantitative and qualitative differences among the newly synthesized polypeptides. Although a similar array of differences is detected among non-heat shock proteins in heat-shocked plumules, the newly synthesized hsp's are similar in number and distribution regardless of whether methionine, valine, lysine, or leucine is used as the amino acid precursor (Fig. 2). These results support the contention that the six major molecular weight classes of hsp's reported earlier in maize plumules (Baszczynski *et al.*, 1982, 1983a) probably represent the full complement of hsp's capable of being induced in this tissue under this temperature shift regime.

According to Burr *et al.* (1982), zein lacks the amino acid lysine. The fact that the low-molecular-weight hsp's can be detected easily when radioactively labeled lysine is used as the amino acid probe indicates that the low-molecular-weight hsp's are probably not the zein proteins. The possibility that some of these 18 K hsp's might correspond to the zeins was based on the observations that (1) the low-molecular-weight hsp's of maize have molecular weights (18K–20K) and isoelectric points (7.0–7.5) similar to some zein polypeptides (Burr and Burr, 1982) and (2) other workers have shown that heat shock in soybean leads to enhanced synthesis of storage proteins in this crop (Altschuler and Mascarenhas, 1982). However, further experiments utilizing antibodies to zein proteins and/or cDNA probes to the zein genes will need to be conducted before any further conclusions about the relatedness of these polypeptides to hsp's can be made.

C. Comparison of Polypeptides Synthesized in Different Tissues from Control and Heat-Shocked Maize Seedlings

Reports on the synthesis of hsp's in various animal systems suggest that the quantity and the type of hsp's expressed may be dependent on the type of tissue involved or the developmental and/or the differentiative state of the induced cells (Atkinson, 1981; Heilmann and Infante, 1981; Roccherri *et al.*, 1981; Atkinson *et al.*, 1983; Dean and Atkinson, 1983). To assess whether the hsp's synthesized in plumules of intact maize seedlings are specific for and/or restricted to plumules, the mesocotyls, plumules, radicles, and young leaves from seedlings were studied in a similar manner. A fluorographic comparison of the polypeptides synthesized by these tissues in control (30°C growing temperature) and heat-shocked (1-hr shift from 30°C to 42°C) seedlings is shown in Fig. 3. Although the newly synthesized polypeptides detected on these two-dimensional polyacrylamide gels indicate differences in the proteins synthesized at control

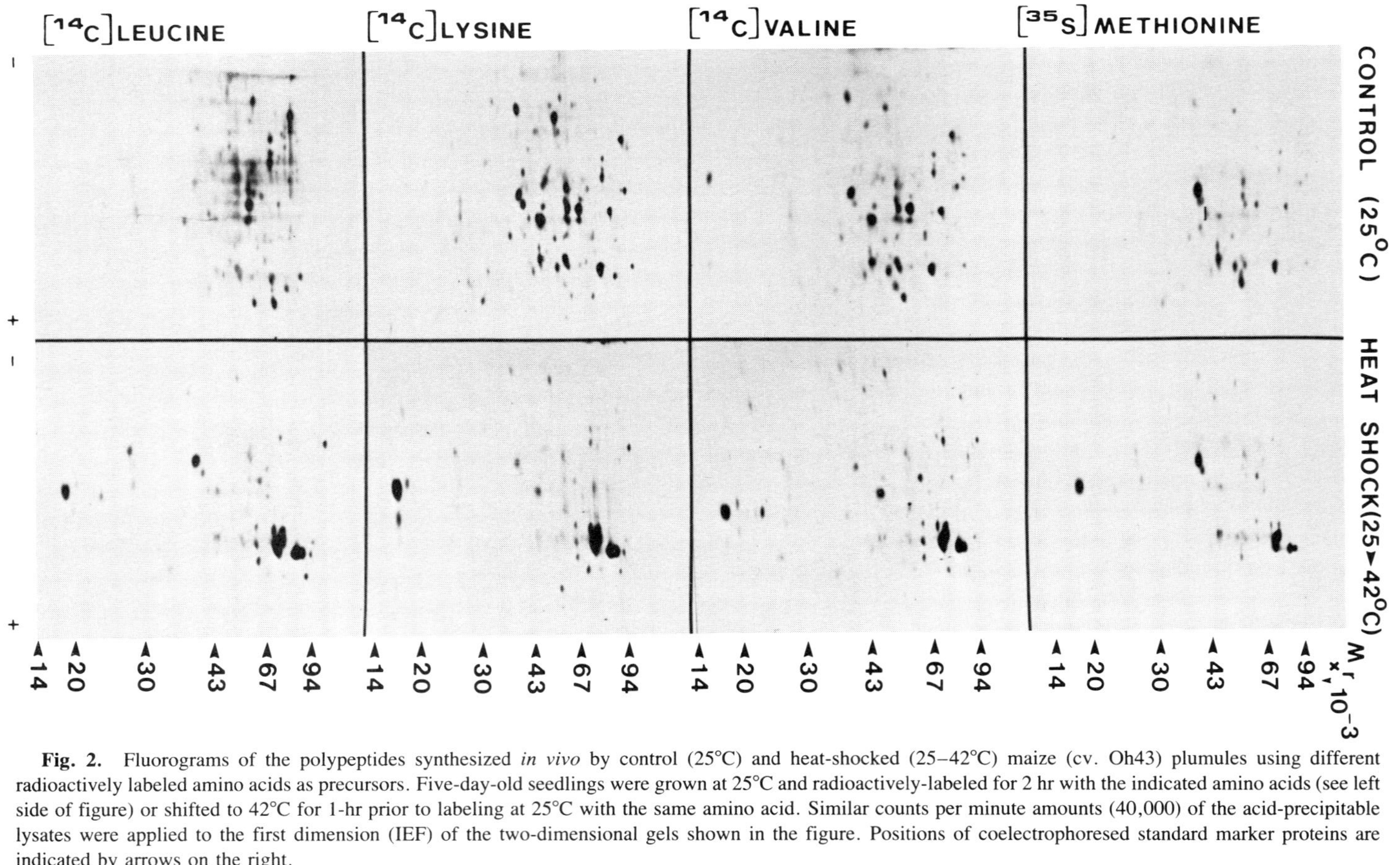

Fig. 2. Fluorograms of the polypeptides synthesized *in vivo* by control (25°C) and heat-shocked (25–42°C) maize (cv. Oh43) plumules using different radioactively labeled amino acids as precursors. Five-day-old seedlings were grown at 25°C and radioactively-labeled for 2 hr with the indicated amino acids (see left side of figure) or shifted to 42°C for 1-hr prior to labeling at 25°C with the same amino acid. Similar counts per minute amounts (40,000) of the acid-precipitable lysates were applied to the first dimension (IEF) of the two-dimensional gels shown in the figure. Positions of coelectrophoresed standard marker proteins are indicated by arrows on the right.

temperatures by each tissue, all of the tissues exhibit a similar protein synthetic response to heat shock. In each tissue of maize examined, a 1-hr temperature shift from 30 to 42°C resulted in new and/or enhanced synthesis of hsp's with M_r and p*I* values identical to those observed in 5-day-old plumules.

The qualitative similarity of the heat shock response in different maize tissues is supported further by immunochemical studies employing monospecific antibodies prepared against hsp's extracted from maize plumules (Baszczynski, 1984). Figure 4 shows the products obtained from immunoprecipitation of total polypeptide lysates (Kelley and Schlesinger, 1982; Atkinson *et al.*, 1983) from control (lanes 1–4) or heat-shocked (lanes 5–8) plumules, mesocotyls, radicles, and leaves with antibodies prepared against the 18K hsp's isolated from maize plumules. Studies with antibodies prepared against other plumule hsp's (not shown) yielded similar results with all four tissues. These findings establish that the hsp's synthesized by maize plumules, mesocotyls, radicles, and young leaves have similar immunochemical properties. Thus, the similarity in the isoelectric points, molecular weights, and immunochemical properties of the hsp's from each of these tissues in maize strongly argues against any tissue specificity in the synthesis of the major maize hsp classes.

D. Synthesis of Heat Shock Proteins in Maize Tissues Is Dependent on Appearance of Heat Shock Protein mRNA in Cytoplasm

The dramatic synthesis of hsp's observed in tissues from heat-shocked maize seedlings was shown to result from the increased availability of mRNA's for the hsp's. Cell-free translations of total RNA and poly(A)$^+$ mRNA from control and heat-shocked maize plumules were carried out in both the rabbit reticulocyte and the wheat germ systems (Baszczynski *et al.*, 1983a). No detectable hsp mRNA's were found in control cells. However, the mRNA's isolated from heat-shocked cells were translated *in vitro* into the major hsp classes observed previously *in vivo*. The hsp-like polypeptides synthesized in cell-free systems exhibited fewer isoelectric point variants than the hsp's of similar molecular weight synthesized *in vivo*. Nevertheless, there appears to be good fidelity in the translation of mRNA's from heat-shocked tissues into polypeptides which are similar in size and charge to the hsp's observed *in vivo* (Baszczynski *et al.*, 1983a).

We have isolated free and membrane-associated polyribosomes from plumules of control and heat-shocked seedlings. Translations of these polyribosomes in a cell-free reticulocyte lysate system reveal that the high-molecular-weight classes of hsp's appear to be translated from mRNA's distributed on both free and membrane-bound polyribosomes, while the low-molecular-weight classes of hsp's (i.e., the 18K hsp's) are translated from mRNA's primarily associated with

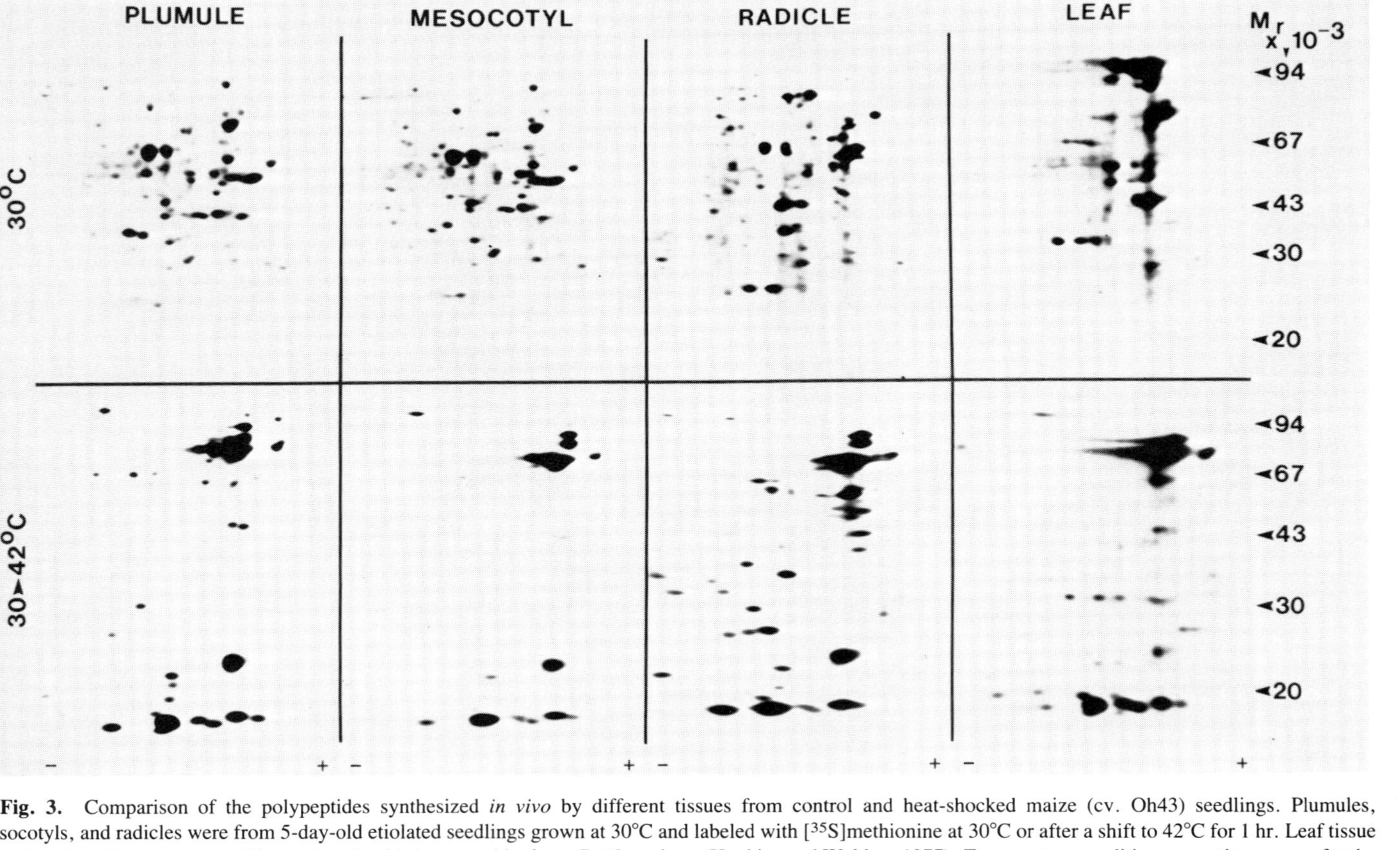

Fig. 3. Comparison of the polypeptides synthesized *in vivo* by different tissues from control and heat-shocked maize (cv. Oh43) seedlings. Plumules, mesocotyls, and radicles were from 5-day-old etiolated seedlings grown at 30°C and labeled with [^{35}S]methionine at 30°C or after a shift to 42°C for 1 hr. Leaf tissue was taken from light-grown seedlings when the third emerged leaf was 7–10 cm long (Hopkins and Walden, 1977). Temperature conditions were the same as for the other tissues. Similar counts per minute amounts (100,000) of acid-precipitable lysate from each tissue were applied to the first dimension of each two-dimensional gel.

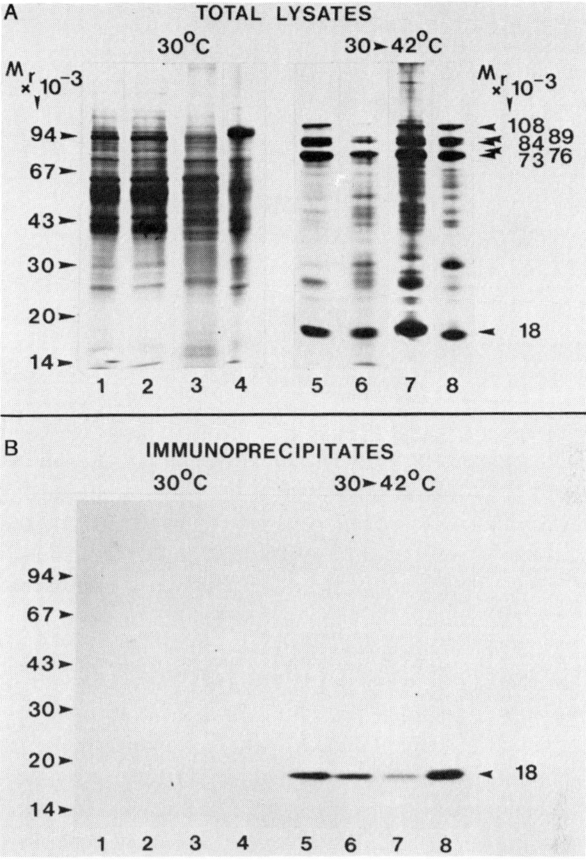

Fig. 4. Immunoprecipitation of the newly synthesized 18K class of hsp's from lysates of various maize (cf. Oh43) tissues with antibodies prepared against maize plumule 18K hsp's. Shown are fluorograms from one-dimensional SDS–PAGE separations of the polypeptides synthesized by various tissues from control (30°C) and heat-shocked (30–42°C) maize seedlings (A) and of polypeptides precipitated from these lysates by antibodies prepared against the 18K hsp's from maize plumules (B). The maize tissues used in this experiment include plumules (lanes 1 and 5), mesocotyls (lanes 2 and 6), radicles (lanes 3 and 7), and young leaves (lanes 4 and 8) prepared as described in the legend to Fig. 3. Immunoprecipitation was carried out according to Kelley and Schlesinger (1982) and Atkinson *et al.* (1983) using approximately 100,000 cpm of acid-precipitable lysate for the immunoprecipitations shown in part (B) of this figure.

membrane-bound polyribosomes (Fig. 5). Although the observed distribution of mRNA's for the high-molecular-weight hsp's may reflect possible contamination of one fraction with the other, the apparent exclusive translation of the 18K hsp's from the membrane-bound polyribosomal fraction suggests that the mRNA's for the 18K hsp's are membrane-associated cytoplasmic constituents. The lack of any detectable hsp mRNA's among the cell-free translational products of free or membrane-bound polyribosomes from control maize tissues supports the contention that the synthesis of hsp's is dependent on new and/or enhanced synthesis and/or processing of hsp mRNA's.

III. INFLUENCE OF GROWING TEMPERATURE AND THERMAL SHIFTS ON GENE EXPRESSION IN MAIZE (cv. Oh43) SEEDLINGS

A. Thermal Shifts Lead to New and/or Enhanced Synthesis of Specific Temperature Shift Polypeptides

The dramatic changes in synthesis of polypeptides and mRNA's in maize following a shift to an elevated incubation temperature are analogous to those observed in other animal and plant systems. Maize (and many other plant systems) grows normally over a wide range of temperatures (from less than 15°C to greater than 35°C). This feature makes it possible to assess the influence of various growing temperatures and temperature shifts on gene expression (Baszczynski *et al.*, 1983b; Walden *et al.*, 1984).

The one-dimensional electrophoretic patterns of newly synthesized polypeptides from plumules of etiolated seedlings grown at 15, 20, 25, or 30°C are shown in the first four lanes of Fig. 6. Although many similarities exist, each growing temperature leads to several unique differences in the polypeptides synthesized. Recognizing that temperature influences the rate of seedling growth, all seedlings used in this study were chosen at a morphologically uniform stage (plumules were approximately 1.5 cm long). A number of qualitative differences are evident among the 35K–45K polypeptides synthesized at these different growing temperatures, and synthesis of a 93K polypeptide apparently is more pronounced in plumules from seedlings grown at 25 and 30°C than from seedlings grown at 15 or 20°C. These results indicate that "control" polypeptide synthetic patterns are a characteristic feature of the temperature at which the organism is cultured. Therefore, the possibility exists that the initial "control" or "preshift" temperature may influence the final array of polypeptides obtained following heat shock or other temperature shifts.

When seedlings grown at these four temperatures are subjected to a 1-hr shift to a common and an established heat shock temperature (42°C), enhanced synthesis of the six molecular weight classes of hsp's (described previously) is observed (see last four lanes of Fig. 6). Additional quantitative and qualitative changes in polypeptide synthesis are also noted. For example, a 31K polypeptide exhibits enhanced synthesis following a shift from 20 to 42°C but is less pro-

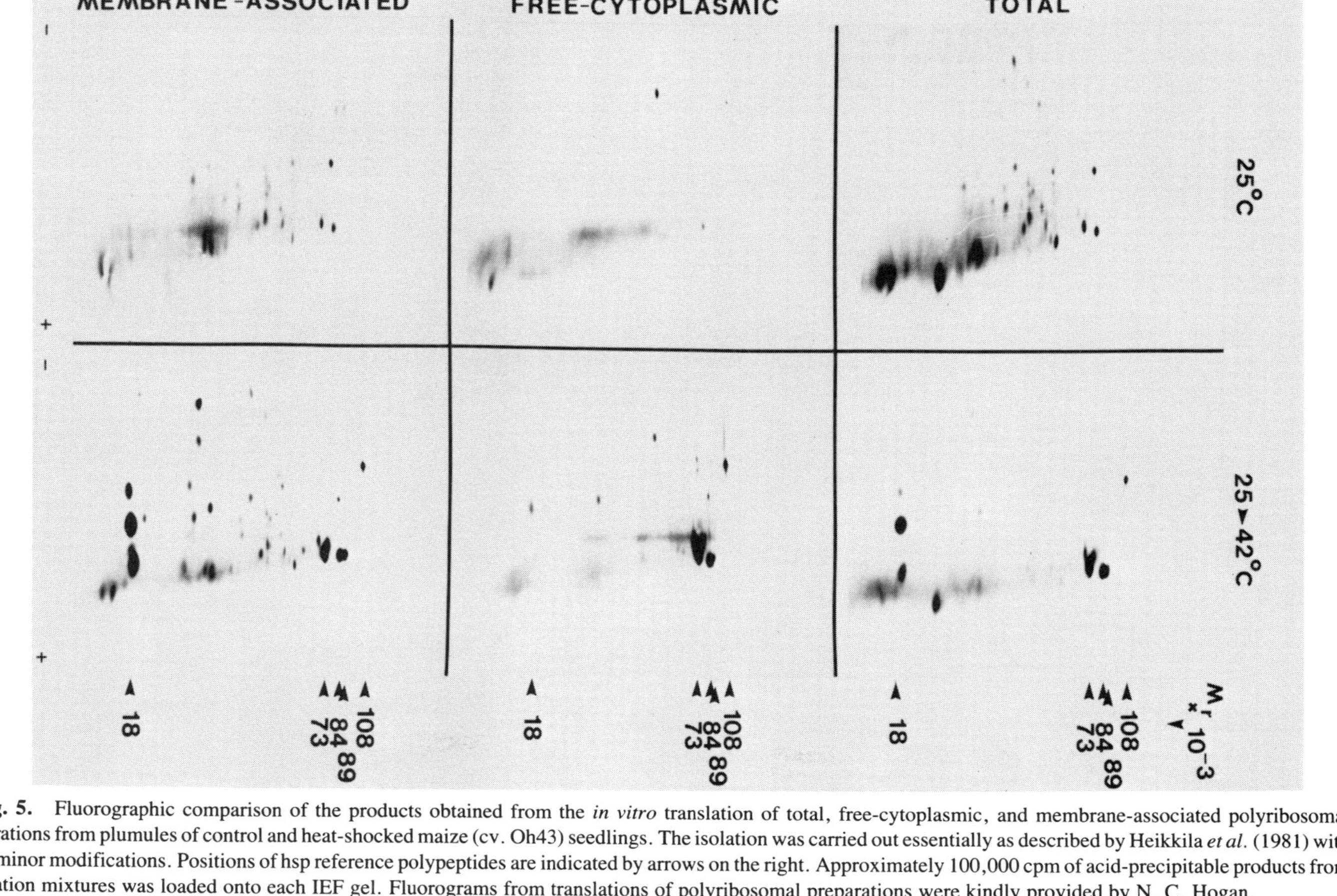

Fig. 5. Fluorographic comparison of the products obtained from the *in vitro* translation of total, free-cytoplasmic, and membrane-associated polyribosomal preparations from plumules of control and heat-shocked maize (cv. Oh43) seedlings. The isolation was carried out essentially as described by Heikkila *et al.* (1981) with some minor modifications. Positions of hsp reference polypeptides are indicated by arrows on the right. Approximately 100,000 cpm of acid-precipitable products from translation mixtures was loaded onto each IEF gel. Fluorograms from translations of polyribosomal preparations were kindly provided by N. C. Hogan.

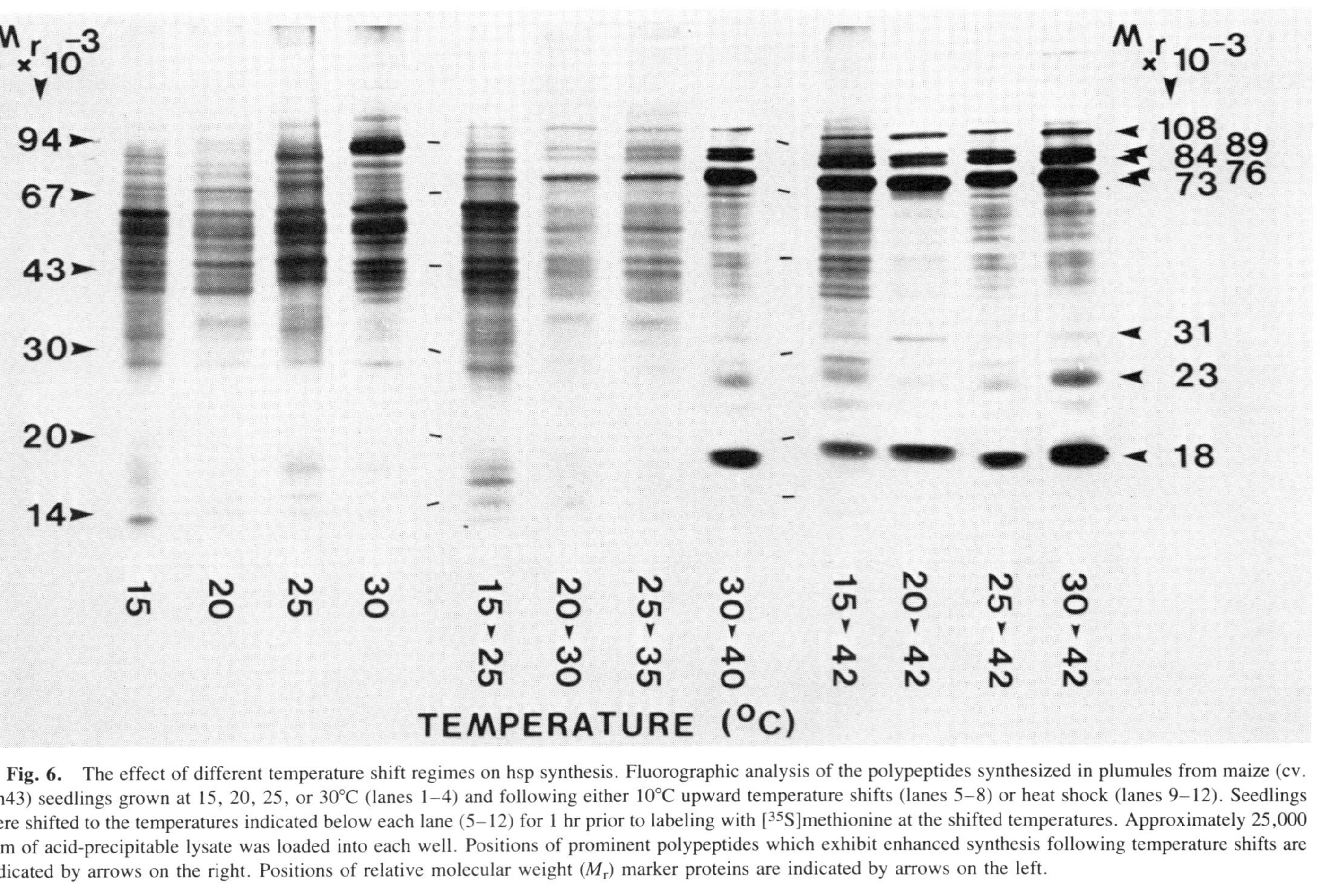

Fig. 6. The effect of different temperature shift regimes on hsp synthesis. Fluorographic analysis of the polypeptides synthesized in plumules from maize (cv. Oh43) seedlings grown at 15, 20, 25, or 30°C (lanes 1–4) and following either 10°C upward temperature shifts (lanes 5–8) or heat shock (lanes 9–12). Seedlings were shifted to the temperatures indicated below each lane (5–12) for 1 hr prior to labeling with [^{35}S]methionine at the shifted temperatures. Approximately 25,000 cpm of acid-precipitable lysate was loaded into each well. Positions of prominent polypeptides which exhibit enhanced synthesis following temperature shifts are indicated by arrows on the right. Positions of relative molecular weight (M_r) marker proteins are indicated by arrows on the left.

nounced following shifts from 15, 25, or 30 to 42°C. In addition, a 23K hsp (not described as a prominent hsp in our previous reports) appears to exhibit enhanced synthesis following a shift to 42 from 15, 20, 25, or 30°C, and its level of synthesis is influenced by the preshift growing temperature. It appears from these results that the final array of hsp's (i.e., those polypeptides which exhibit enhanced synthesis following heat shock) expressed by plumules of heat-shocked seedlings is dependent on the initial culture temperature.

These findings prompted us to address the question of whether various thermal shifts could be employed to demonstrate a specific threshold temperature required for hsp synthesis. Seedlings grown at 15, 20, 25, or 30°C were subjected to a 10°C upward shift for 1 hr. As shown in the middle four lanes of Fig. 6, these shifts (which are within the normal growing range for maize) lead to changes in the patterns of polypeptide synthesis. An important finding is that some of the high-molecular-weight maize hsp's (108K, 89K, 84K, and 73K) exhibit enhanced synthesis following thermal shifts in some of the lower temperature ranges (note especially the 20–30°C and the 25–35°C shifts). On the other hand, the low-molecular-weight 18K hsp class apparently is not synthesized following these low-temperature shifts. The results from the fluorogram of a two-dimensional IEF–SDS–PAGE separation of polypeptides from seedlings shifted from 20 to 30°C and from 30 to 40°C (Fig. 7) further supports the above findings. There is no evidence for synthesis of the 18K hsp's following the 20–30°C shift, while these same polypeptides exhibit marked enhancements of synthesis following the 30–40°C shift (Fig. 7).

It appears from these studies with maize seedlings that the new and/or enhanced synthesis of hsp's is not regulated in a coordinate fashion. This conclusion supports similar findings regarding hsp synthesis in *Drosophila* (Lindquist, 1980) and corroborates our previous observations—dealing with the kinetics of hsp appearance during heat shock and disappearance during recovery—which suggested that the expression of the different hsp molecular weight classes is noncoordinate (Baszczynski *et al.*, 1982). It seems likely that the new and/or enhanced synthesis of some maize hsp's may represent a general response to various temperature shifts, while the expression of other maize hsp's, such as the 18K class, may represent a true high-temperature-induced class of polypeptides.

B. Heat Shock Protein mRNA's Are Present but Not Always Translated in Maize Plumules Following Thermal Shifts Other Than Heat Shock

It has been demonstrated in maize (Baszczynski *et al.*, 1983a) as well as in other systems (Ashburner and Bonner, 1979; Key *et al.*, 1981; Krüger and Benecke, 1981; Kelley *et al.*, 1980; Schöffl and Key, 1982; Kapoor, 1983) that hsp synthesis following heat shock results from the increased availability and the

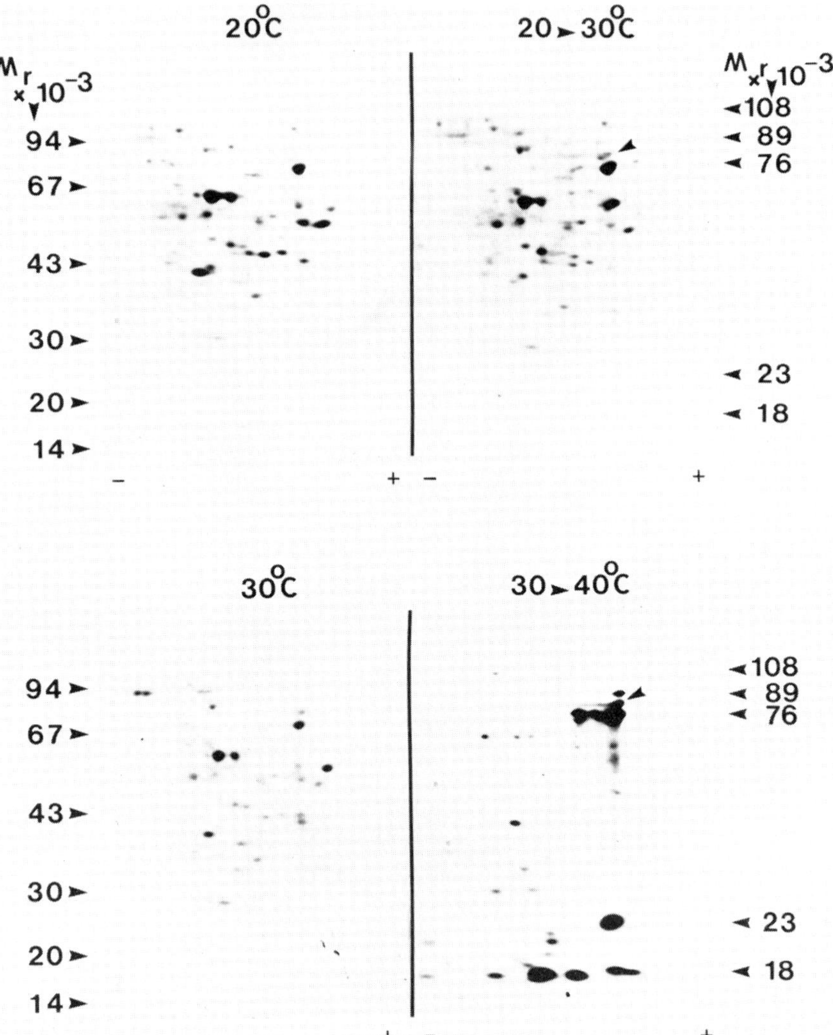

Fig. 7. Temperature-dependent synthesis of low-molecular-weight hsp's. Fluorograms of two-dimensional IEF–SDS–PAGE separations of polypeptides synthesized in plumules of maize (cv. Oh43) seedlings grown at 20 or 30°C and following a 1-hr temperature shift from 20 to 30°C or from 30 to 40°C. Positions of relative molecular weight (M_r) marker proteins are indicated by arrows on the left, while positions of major hsp's are shown by arrows on the right. Approximately 100,000 cpm of acid-precipitable lysate was loaded onto each first-dimensional IEF tube gel.

preferential translation of hsp mRNA's. The apparent absence of any newly synthesized 18K hsp's and the marginal synthesis of the high-molecular-weight hsp's following the 15–25°C or the 20–30°C shifts in maize (Figs. 6 and 7) suggest that (1) hsp mRNA levels may be lower, or in some cases absent, in seedlings subjected to these thermal shift or (2) translation of the hsp mRNA's is not enhanced. *In vitro* translations (using a reticulocyte cell-free system) of the mRNA's isolated from plumules following a shift from 15 to 25°C clearly show that mRNA's for all the previously defined hsp's (including the 18K class) are present but are not translated *in vivo* following this low-temperature shift (Fig. 8). The relative levels of hsp's synthesized *in vitro* are very similar for both the 15- to 25°C-shifted and the 25- to 42°C-shifted seedlings (not shown). Thus, while the level of transcription of mRNA's for hsp's may be relatively similar for a variety of temperature shifts, the translation of these mRNA's into hsp's appears to be noncoordinate and under some form of temperature-dependent translational control.

The emerging feature from these studies is that changes in gene expression may be a normal first response to an environmental stress condition. These changes may be subtle in the case of a mild stress or dramatic in the case of a major stress condition such as heat shock. Detection, however, is clearly dependent on the level at which one monitors responses. We now realize that in maize a variety of temperature shifts, in the range which has been examined (15–45°C), leads to the new and/or enhanced synthesis of hsp mRNA's but may or may not lead to the synthesis of hsp's. The array of gene products and the degree to which they are synthesized are determined by the interaction of several factors which include in the minimum: (1) the initial growing or "preshift" temperature; (2) the temperature increment over which the tissues are shifted; (3) the temperature regime in which the shift is carried out (e.g., 15–25°C versus 30–40°C); (4) the rate at which the temperature changes from the initial to the final temperature; and (5) the duration of the temperature shift.

IV. IMPACT OF GENOTYPE ON POLYPEPTIDE SYNTHESIS IN MAIZE SEEDLINGS

The results from the studies outlined in the preceding sections indicated that the gene products which are expressed in different tissues of maize seedlings are markedly influenced by the growing temperature as well as by a variety of thermal shifts. In order to investigate the biological and the molecular basis of the response of maize to different thermal conditions, the majority of our studies have employed a single, well-defined, standard inbred cultivar of maize (Oh43). The plethora of available genetic stocks makes maize particularly amenable to a critical investigation of the impact of genotype on (1) the types of gene products

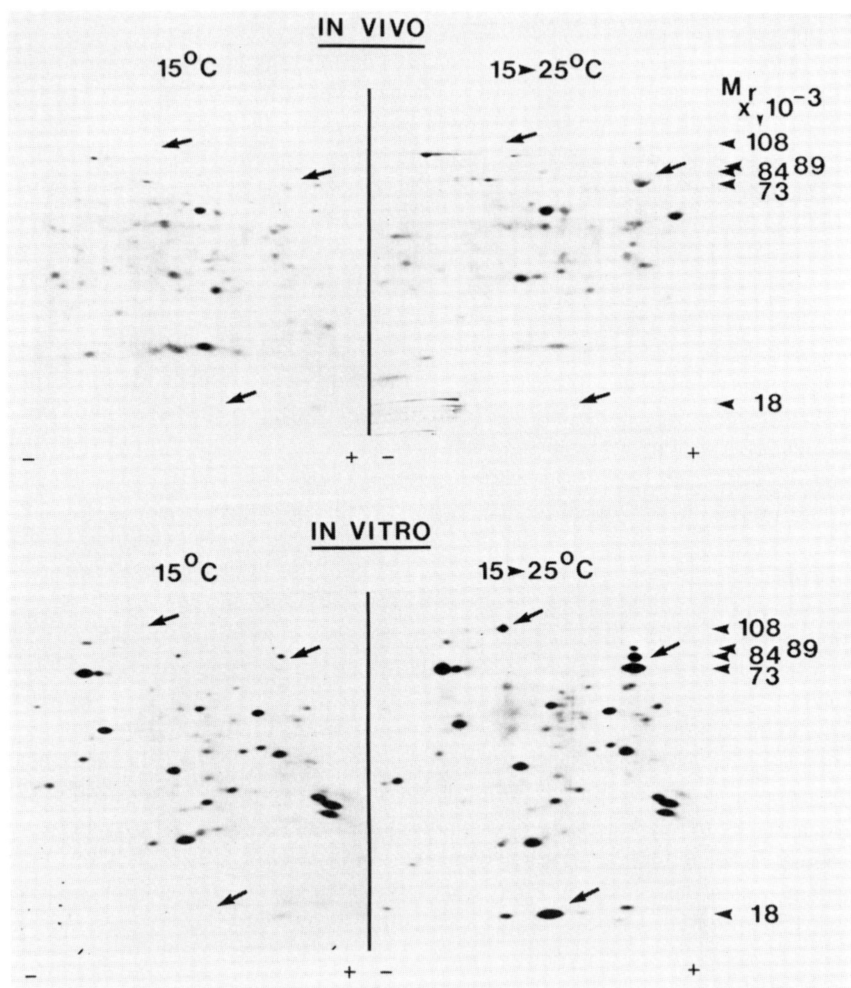

Fig. 8. Temperature-dependent and noncoordinate appearance and translation of maize hsp mRNA's. Fluorograms of two-dimensional IEF–SDS–PAGE separations of the polypeptides synthesized *in vivo* (top) or the products derived from *in vitro* translations of mRNA's (bottom) isolated from plumules of maize (cv. Oh43) seedlings grown at 15°C or following a 1-hr shift from 15 to 25°C are shown. Molecular weights (M_r) of the hsp's are indicated by arrows on the right. The arrows on the photographs point to corresponding spots in each fluorgram and indicate the expected positions of hsp classes. Approximately 100,000 cpm of acid-precipitable lysate was loaded onto each first-dimensional IEF gel.

which are expressed during normal-growth and development and (2) the response to adverse environmental conditions including temperature shifts or stresses. We, therefore, have initiated a study to evaluate the differences in the patterns of newly synthesized polypeptides in 50 inbreds, several unique cytogenetic variants, and several reciprocal F_1's. Thus far, the data from our work and that of others (Walden and Atkinson, 1983; Nebiolo and White, 1983) demonstrate that the genotype constitutes another important parameter in studies of gene regulation.

A first comparison of the polypeptides synthesized in 5-day-old maize plumules from a variety of inbred cultivars at 20 and 30°C and following a heat shock to 45°C from each of these temperatures is presented in Fig. 9. A considerable number of the polypeptides (greater than 60%) synthesized in these plumules appear to be the same in all the genotypes, although some qualitative and several quantitative differences can be detected. The polypeptides which exhibit striking differences between genotypes are indicated by small arrows beside the bands on the figure. It is also evident that the growing temperature (20 or 30°C) has a marked effect on the array of polypeptides which are synthesized in the plumules of these different cultivars.

A shift to 45°C from either 20 or 30°C leads to new and/or enhanced synthesis of the same six major hsp classes described previously. Additional genotype-specific differences (largely quantitative) are also noted following the thermal shifts. These differences, in general, reflect the differences observed in the control or the "preshift" patterns of polypeptide synthesis (note, for example, the prominent 45K polypeptide in cv. Wilbur's flint; Fig. 9, lane 12).

An examination of newly synthesized polypeptides in radicles (Fig. 10) from the same genotypes used in the studies on plumules (Fig. 9) confirms that the source of genotype has an impact on the array of gene products which are expressed. The patterns obtained from radicles exhibit noticeably less variation between genotypes than do the patterns obtained from plumules. This may reflect the relative differences in developmental fates of the two tissue sources; the aerial portion of the plant has evolved the greater ability to respond to fluctuations in temperature than the more temperature-buffered soilborne organs and tissues. A shift to 45°C from either 20 or 30°C leads to new and/or enhanced synthesis of the same hsp classes in radicles as observed in plumules.

Results from two-dimensional gel electrophoretic separations of polypeptides (not shown) from some of these genotypes indicate an even greater array of differences between the various genotypic sources.

The results from these studies suggest that the genotype controls the array of polypeptides synthesized under different temperature conditions. There appears to be little or no difference in the ability of plumules or radicles from all genotypes to synthesize *in vivo* the six major classes of heat shock polypeptides. It remains to be established, however, whether quantitative differences exist in

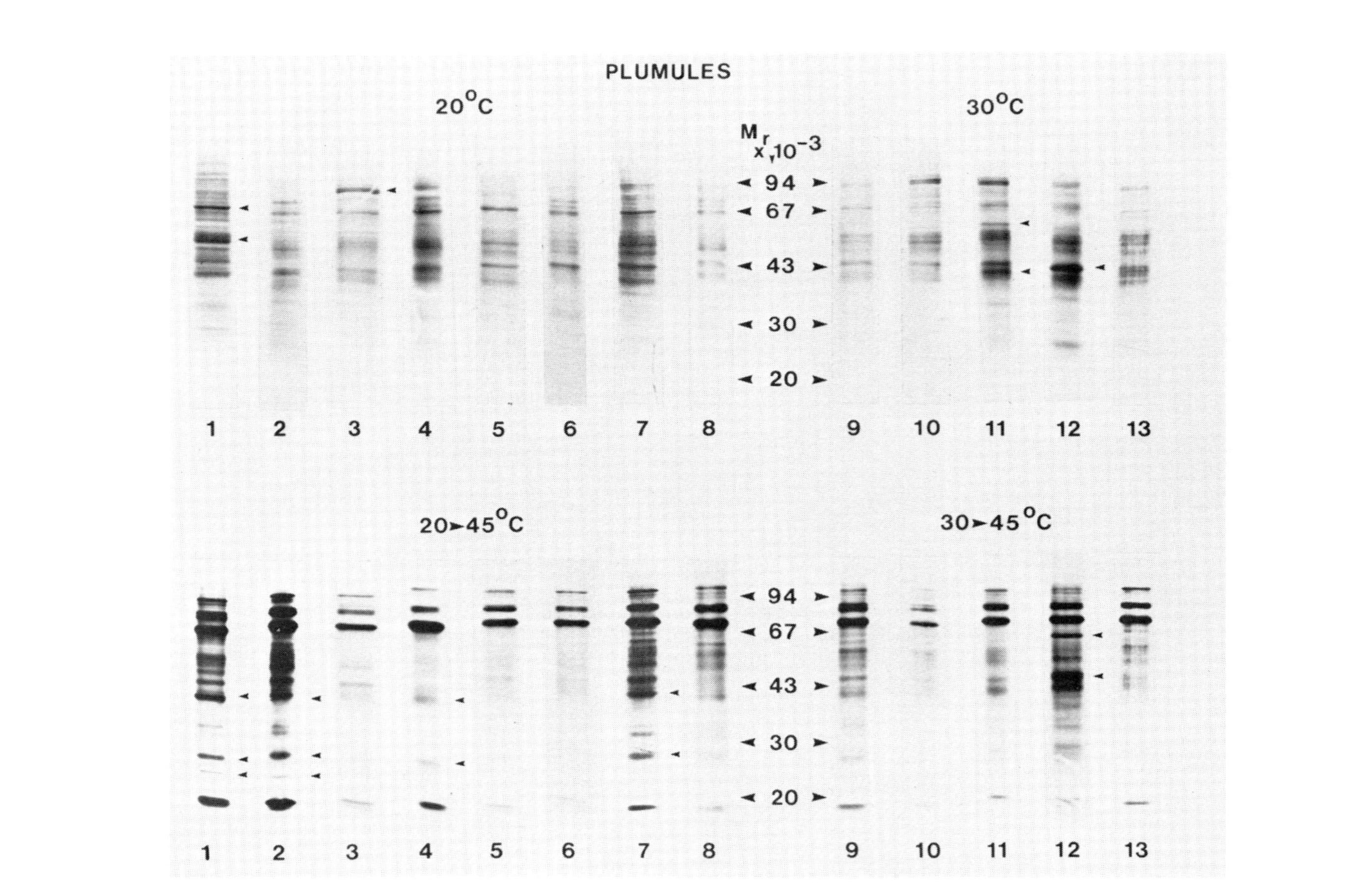
PLUMULES
20°C
30°C
M_r x 10^-3
94
67
43
30
20
1 2 3 4 5 6 7 8 9 10 11 12 13
20➤45°C
30➤45°C
94
67
43
30
20
1 2 3 4 5 6 7 8 9 10 11 12 13

the relative rates of synthesis of the various hsp classes in the different genotypes.

V. SUMMARY

The conditions of maize culture and the availability of domesticated and nondomesticated cultivars provide the opportunity to extend studies of heat shock induction and gene regulation into agronomically important areas. Stress-induced polypeptides will continue to receive attention, as will differences attributable to heterosis and other genetic events that have been engineered so skillfully over the past 50 years by the maize breeders. Of particular merit may be the opportunity to transfer the genetic potential to produce a specific polypeptide, once the peptide is ascertained and its cDNA synthesized, from a "have" to a "have not" cultivar or possibly to another crop plant.

The temperature shift work with maize leads to the conclusion that small increments (we have detected differences with increments of 5°C) or decrements (10°C) and a duration of 10 or more min are sufficient to induce most genotypes to respond with an altered array of polypeptides. It is likely that most organisms produce classes of polypeptides that are temperature-specific (TSP's) as a result of the culture temperature and/or a shift (Walden *et al.*, 1985).

Our data tend to obfuscate the seemingly straightforward idea of heat shock and heat shock-induced gene regulation. In the maize system, heat shock refers to an upshift of more than 5°C above 35°C—but hsp's can be produced at temperatures as low as 25°C. Thus, while heat shock may be considered an extreme shift regime, the polypeptides so elicited are also elicited within the normal, optimal growth range. Noncoordinate regulation of thermally induced gene expression appears to be the rule rather than the exception in maize.

The demonstration (Fig. 8) that some of the hsp mRNA's are present at 25°C but not translated *in vivo* requires further clarification and quantification. If such was the general case, restraint must be exercised in undertaking the declaration of differences (or as importantly, perhaps, the lack of differences) among treatments (e.g., an environmental factor) and genotypes. It may be that the re-

---

Fig. 9. Fluorograms comparing the patterns of polypeptide synthesis in control and heat-shocked plumules from five-day-old seedlings of a series of inbred cultivars grown at 20 or 30°C (top) and following a 1-hr shift to 45°C from each of the growing temperatures (bottom). Positions of standard marker proteins are indicated by arrows in the middle of the figures. Positions of polypeptides exhibiting prominent differences between genotypes are indicated by arrows pointing to bands beside individual lanes. The genotypes included are as follows: Oh43 (lanes 1 and 9), W64a (lanes 2 and 10), B73 (lanes 3 and 11), H95 (lane 4), Gaspé flint (lane 5), Wilbur's flint (lanes 6 and 12), Rhode Island White flint (lanes 7 and 13), and Tana's flint (lane 8).

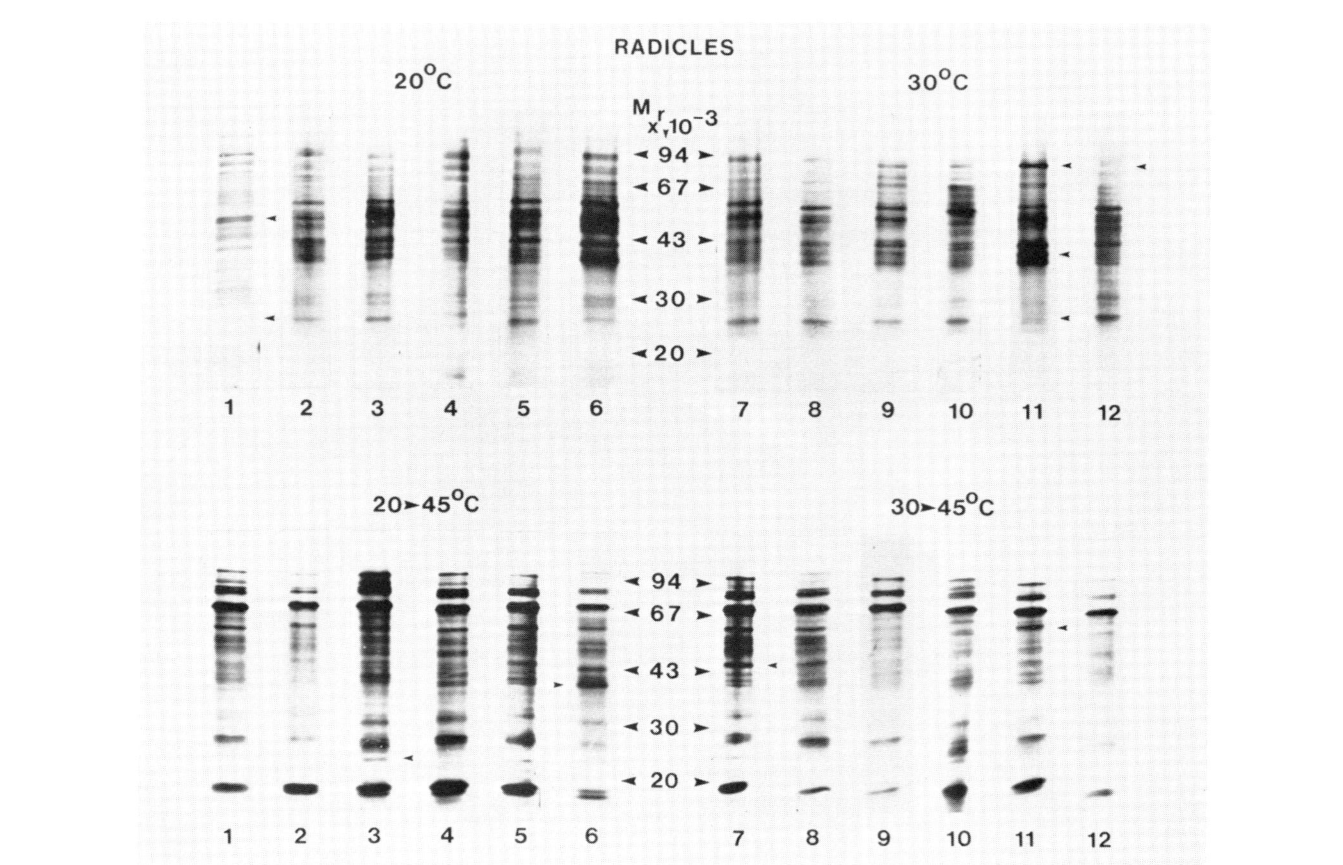

RADICLES
20°C
30°C
Mr x 10-3
94
67
43
30
20
1 2 3 4 5 6 7 8 9 10 11 12
20▸45°C
30▸45°C
94
67
43
30
20
1 2 3 4 5 6 7 8 9 10 11 12

sponses detected by fluorography represent the fixation of specific allele forms by the plant breeders whose selection criterion may have been "adaptability," "combining ability," or other important agronomic features. While such breeder involvement is less likely in the flints (lanes 5–8, 12, and 13, Fig. 9) than in the dents (lanes 1–4, 9–11, Fig. 9), fixation of allele forms cannot be discounted. Our preliminary data with reciprocal hybrids (to be reported later) support this view.

Acknowledgments

This work was supported by NSERC (Canada) operating grants to D.B.W. and B.G.A.

REFERENCES

Altschuler, M., and Mascarenhas, J. P. (1982). Heat shock proteins and effects of heat shock in plants. *Plant Mol. Biol.* **1,** 103–115.

Ashburner, M., and Bonner, J. J. (1979). The induction of gene activity in *Drosophila* by heat shock. *Cell* **17,** 241–254.

Atkinson, B. G. (1981). Synthesis of heat shock proteins by cells undergoing myogenesis. *J. Cell Biol.* **89,** 666–673.

Atkinson, B. G., Cunningham, T., Dean, R. L., and Sommerville, M. (1983). Comparison of the effects of heat shock and metal-ion stress on gene expression in cells undergoing myogenesis. *Can. J. Biochem. Cell Biol.* **61,** 404–413.

Barnett, T., Altschuler, M., McDaniel, C. N., and Mascarenhas, J. P. (1980). Heat shock induced proteins in plant cells. *Dev. Genet. (Amsterdam)* **1,** 331–340.

Baszczynski, C. L. (1984). Thermal shift induction and regulation of gene expression in maize. Ph.D. thesis, Univ. of W. Ontario, London, Ontario, Canada.

Baszczynski, C. L., Walden, D. B., and Atkinson, B. G. (1982). Regulation of gene expression in corn (*Zea mays* L.) by heat shock. *Can. J. Biochem.* **60,** 569–579.

Baszczynski, C. L., Walden, D. B., and Atkinson, B. G. (1983a). Regulation of gene expression in corn (*Zea mays* L.) by heat shock. II. *In vitro* analysis of RNAs from heat-shocked seedlings. *Can. J. Biochem. Cell Biol.* **61,** 395–403.

Baszczynski, C. L., Walden, D. B., and Atkinson, B. G. (1983b). Temperature shifts within the normal growing range of maize lead to novel or enhanced synthesis of heat shock proteins and their mRNAs. *J. Cell Biol.* **97** (Part 2), 153a (abstr. #579).

Fig. 10. Fluorograms comparing the patterns of polypeptide synthesis in control and heat-shocked radicles from 5-day-old seedlings of a series of inbred cultivars grown at 20 or 30°C (top) and following a 1-hr shift to 45°C from each of the growing temperatures (bottom). Positions of standard marker proteins are indicated by arrows in the middle of the figures. Positions of polypeptides exhibiting prominent differences between genotypes are indicated by arrows pointing to bands beside individual lanes. The genotypes included are as follows: Oh43 (lanes 1 and 7), W64a (lanes 2 and 8), Co220 (lane 3), B73 (lane 9), H95 (lanes 4 and 10), Gaspé flint (lane 5), Wilbur's flint (lanes 6 and 11), and Rhode Island White flint (lane 12).

Bewley, J. D., Larsen, K. M., and Papp, J. E. T. (1983). Water-stress-induced changes in the patterns of protein synthesis in maize seedling mesocotyls: A comparison with the effects of heat shock. *J. Exp. Bot.* **34,** 1126–1133.

Botha, F. C. (1979). The effect of drought conditions on water soluble proteins of two maize lines. *Z. Pflanzenphysiol.* **95,** 371–375.

Burr, F. A., and Burr, B. (1982). Three mutations in *Zea mays* affecting zein accumulation: A comparison of zein polypeptides, *in vitro* synthesis and processing, mRNA levels, and genomic organization. *J. Cell Biol.* **94,** 201–206.

Burr, B., Burr, F. A., St. John, T. P., Thomas, M., and Davis, R. W. (1982). Zein storage protein gene family of maize. An assessment of heterogeneity with cloned messenger RNA sequences. *J. Mol. Biol.* **154,** 33–49.

Cooper, P., and Ho, T.-H. D. (1983). Heat shock proteins in maize. *Plant Physiol.* **7,** 215–222.

Currie, R. W., and White, F. P. (1981). Trauma-induced protein in rat tissues: A physiological role for a "Heat Shock" protein? *Science* **214,** 72–73.

Dean, R. L., and Atkinson, B. G. (1983). The acquisition of thermal tolerance in larvae of *Calpodes ethlius* (Lepidoptera) and the *in situ* and *in vitro* synthesis of heat-shock proteins. *Can. J. Biochem. Cell Biol.* **61,** 472–479.

Dhinsda, R. S. (1976). Water stress and protein synthesis. III. Subcellular distribution of inhibition of protein synthesis. *Z. Pflanzenphysiol.* **78,** 82–84.

Heikkila, J. J., Cosgrove, J. W., and Brown, I. R. (1981). Cell-free translation of free- and membrane-bound polysomes and polyadenylated mRNA from rabbit brain following the administration of D-lysergic acid diethylamide *in vivo*. *J. Neurochem.* **36,** 1229–1238.

Heilmann, L. J., and Infante, A. A. (1981). Developmental regulation of competence for induction of heat shock proteins in sea urchin embroys. *Fed. Proc. Fed. Am. Soc. Exp. Biol.* **40,** 1702.

Hightower, L. E., and White, F. P. (1981). Cellular responses to stress: Comparison of a family of 71–73 kilodalton proteins rapidly synthesized in rat tissue slices and canavanine-treated cells in culture. *J. Cell. Physiol.* **108,** 261–275.

Hopkins, W. G., and Walden, D. B. (1977). Temperature sensitivity of virescent mutants of maize. *J. Hered.* **68,** 283–286.

Kapoor, M. (1983). A study of the heat-shock response in *Neurospora crassa*. *Int. J. Biochem.* **15,** 639–649.

Kelley, P. M., and Schlesinger, M. J. (1982). Antibodies to two major chicken heat shock proteins cross-react with similar proteins in widely divergent species. *Mol. Cell. Biol.* **2,** 267–274.

Kelley, P. M., Aliperti, G., and Schlesinger, M. J. (1980). *In vitro* synthesis of heat shock proteins by mRNAs from chicken embryo fibroblasts. *J. Biol. Chem.* **255,** 3230–3233.

Key, J. L., Lin, C.-Y., and Chen, Y. M. (1981). Heat shock proteins of higher plants. *Proc. Natl. Acad. Sci. U.S.A.* **78,** 3526–3530.

Krüger, C., and Benecke, B. J. (1981). *In vitro* translation of *Drosophila* heat shock and non-heat shock mRNAs in heterologous and homologous cell-free systems. *Cell* **23,** 595–603.

Lindquist, S. (1980). Varying patterns of protein synthesis in *Drosophila* during heat shock: Implications for regulation. *Dev. Biol.* **77,** 463–479.

Loomis, W. F., and Wheeler, S. (1980). Heat shock response of *Dictyostelium*. *Dev. Biol.* **79,** 399–408.

Meyer, Y., and Chartier, Y. (1983). Long-lived and short-lived heat-shock proteins in tobacco mesophyll protoplasts. *Plant Physiol.* **72,** 26–32.

Nebiolo, C. M., and White, E. M. (1983). Protein synthesis by maize mitochondria in response to heat shock. *J. Cell Biol.* **97,** (Part 2), 153a (abstr. #577).

Roccheri, M. C., Di Bernardo, M. G., and Guidice, G. (1981). Synthesis of heat shock proteins in developing sea urchins. *Dev. Biol.* **83,** 173–177.

Sachs, M. M., and Freeling, M. (1978). Selective synthesis of alcohol dehydrogenase during anaerobic treatment of maize. *Mol. Gen. Genet.* **161,** 111–115.

Sachs, M. M., Freeling, M., and Okimoto, R. (1980). The anaerobic proteins of maize. *Cell* **20,** 761–767.

Schlesinger, M. J., Aliperti, G., and Kelley, P. M. (1982). The response of cells to heat shock. *Trends Biochem. Sci.* (Pers. Ed.) **7,** 222–225.

Schöffl, F., and Key, J. L. (1982). An analysis of mRNAs for a group of heat shock proteins of soybean using cloned cDNAs. *J. Mol. Appl. Genet.* **1,** 301–314.

Silver, J. C., Andrews, D. R., and Pekkala, D. (1983). Effect of heat shock on synthesis and phosphorylation of nuclear and cytoplasmic proteins in the fungus *Achlya. Can. J. Biochem. Cell Biol.* **61,** 447–455.

Walden, D. B., and Atkinson, B. G. (1983). Genetical aspects of the temperature shift and heat shock response in maize. *J. Cell Biol.* **97** (Part 2), 153a (abstr. #578).

Walden, D. B., Baszczynski, C. L., Boothe, J., and Atkinson, B. G. (1985). Thermal regulation of gene expression in maize. (Submitted for publication.)

Webster, P. L. (1980). "Stress" protein synthesis in pea root meristem cells. *Plant Sci. Lett.* **20,** 141–146.

Zurfluh, L. L., and Guilfoyle, T. J. (1980). Auxin-induced changes in the patterns of protein synthesis in soybean hypocotyl. *Proc. Natl. Acad. Sci. U.S.A.* **77,** 357–361.

Zurfluh, L. L., and Guilfoyle, T. J. (1982). Auxin-induced changes in the population of translatable messenger RNA in elongating maize coleoptile sections. *Planta* **156,** 525–527.

Index

V

W

X

Y

Z

CELL BIOLOGY: A Series of Monographs

EDITORS

G. M. Padilla, G. L. Whitson, and I. L. Cameron (editors). THE CELL CYCLE: Gene-Enzyme Interactions, 1969

A. M. Zimmerman (editor). HIGH PRESSURE EFFECTS ON CELLULAR PROCESSES, 1970

I. L. Cameron and J. D. Thrasher (editors). CELLULAR AND MOLECULAR RENEWAL IN THE MAMMALIAN BODY, 1971

I. L. Cameron, G. M. Padilla, and A. M. Zimmerman (editors). DEVELOPMENTAL ASPECTS OF THE CELL CYCLE, 1971

P. F. Smith. THE BIOLOGY OF MYCOPLASMAS, 1971

Gary L. Whitson (editor). CONCEPTS IN RADIATION CELL BIOLOGY, 1972

Donald L. Hill. THE BIOCHEMISTRY AND PHYSIOLOGY OF *TETRAHYMENA*, 1972

Kwang W. Jeon (editor). THE BIOLOGY OF AMOEBA, 1973

Dean F. Martin and George M. Padilla (editors). MARINE PHARMACOGNOSY: Action of Marine Biotoxins at the Cellular Level, 1973

Joseph A. Erwin (editor). LIPIDS AND BIOMEMBRANES OF EUKARYOTIC MICROORGANISMS, 1973

A. M. Zimmerman, G. M. Padilla, and I. L. Cameron (editors). DRUGS AND THE CELL CYCLE, 1973

Stuart Coward (editor). DEVELOPMENTAL REGULATION: Aspects of Cell Differentiation, 1973

I. L. Cameron and J. R. Jeter, Jr. (editors). ACIDIC PROTEINS OF THE NUCLEUS, 1974

Govindjee (editor). BIOENERGETICS OF PHOTOSYNTHESIS, 1975

James R. Jeter, Jr., Ivan L. Cameron, George M. Padilla, and Arthur M. Zimmerman (editors). CELL CYCLE REGULATION, 1978

Gary L. Whitson (editor). NUCLEAR–CYTOPLASMIC INTERACTIONS IN THE CELL CYCLE, 1980

Danton H. O'Day and Paul A. Horgen (editors). SEXUAL INTERACTIONS IN EUKARYOTIC MICROBES, 1981

Ivan L. Cameron and Thomas B. Pool (editors). THE TRANSFORMED CELL, 1981

Arthur M. Zimmerman and Arthur Forer (editors). MITOSIS/CYTOKINESIS, 1981

Ian R. Brown (editor). MOLECULAR APPROACHES TO NEUROBIOLOGY, 1982

Henry C. Aldrich and John W. Daniel (editors). CELL BIOLOGY OF *PHYSARUM* AND *DIDYMIUM*, Volume I: Organisms, Nucleus, and Cell Cycle, 1982; Volume II: Differentiation, Metabolism, and Methodology, 1982

John A. Heddle (editor). MUTAGENICITY: New Horizons in Genetic Toxicology, 1982

Potu N. Rao, Robert T. Johnson, and Karl Sperling (editors). PREMATURE CHROMOSOME CONDENSATION: Application in Basic, Clinical, and Mutation Research, 1982

George M. Padilla and Kenneth S. McCarty, Sr. (editors). GENETIC EXPRESSION IN THE CELL CYCLE, 1982

David S. McDevitt (editor). CELL BIOLOGY OF THE EYE, 1982

P. Michael Conn (editor). CELLULAR REGULATION OF SECRETION AND RELEASE, 1982

Govindjee (editor). PHOTOSYNTHESIS, Volume I: Energy Conversion by Plants and Bacteria, 1982; Volume II: Development, Carbon Metabolism, and Plant Productivity, 1982

John Morrow. EUKARYOTIC CELL GENETICS, 1983

John F. Hartmann (editor). MECHANISM AND CONTROL OF ANIMAL FERTILIZATION, 1983

Gary S. Stein and Janet L. Stein (editors). RECOMBINANT DNA AND CELL PROLIFERATION, 1984

Prasad S. Sunkara (editor). NOVEL APPROACHES TO CANCER CHEMOTHERAPY, 1984

Burr G. Atkinson and David B. Walden (editors). CHANGES IN EUKARYOTIC GENE EXPRESSION IN RESPONSE TO ENVIRONMENTAL STRESS, 1985